CW00381400

International Environmental Law and Policy Series

International Law and the Conservation of Biological Diversity

International Environmental Law and Policy Series

International Law and the Conservation of Biological Diversity

Edited by

Michael Bowman
Senior Lecturer in Law, University of Nottingham

Catherine Redgwell
Senior Lecturer in Law, University of Nottingham

KLUWER LAW INTERNATIONAL

LONDON – THE HAGUE – BOSTON

Published by Kluwer Law International
Sterling House
66 Wilton Road
London SW1V 1DE
United Kingdom

Kluwer Law International incorporates
the publishing programmes of
Graham & Trotman Ltd,
Kluwer Law & Taxation Publishers
and Martinus Nijhoff Publishers

Sold and distributed in the USA and Canada
by Kluwer Law International
675 Massachusetts Avenue
Cambridge MA 02139
USA

In all other countries, sold and distributed
by Kluwer Law International
P.O. Box 85889
2508 CN The Hague
The Netherlands

ISBN 90-411-0863-7
© Kluwer Law International 1996
First published 1996

**British Library Cataloguing in Publication Data and Library of Congress
Cataloging-in-Publication Data is available**

This publication is protected by international copyright law. All rights reserved. No part of this publication may be reproduced, stored in a retrieval system, or transmitted in any form or by any means, electronic, mechanical, photocopying, recording or otherwise, without the prior permission of the publisher.

Typeset in 11/10 Bembo by EXPO Holdings, Malaysia.
Printed and bound in Great Britain by Hartnolls Ltd, Bodmin, Cornwall.

Contents

Contributors

Professor Patricia Birnie was from 1989–1992 Director of the IMO International Law Institute of Maritime Law in Malta. She was previously a lecturer in public international law at the University of Edinburgh (1973–1983) and then senior lecturer in law at the London School of Economics. Her extensive publications on the law of the sea and international environmental law include: *The Maritime Dimension* (edited with R.P. Barston, 1980); *International Regulation of Whaling* (2 volumes, 1985); and, with Alan Boyle, *International Law and the Environment* (1992).

Michael Bowman is a senior lecturer in law at the University of Nottingham, where he has recently completed a term as Vice-Dean of the Faculty of Law and Social Sciences. He is co-director of the University of Nottingham Treaty Centre, founded in 1983 to conduct research into the law and practice of treaty making, and joint author with Professor D.J. Harris of *Multilateral Treaties: Index and Current Status* and its annual cumulative supplements. He is also a member of the University's Centre for Environmental Law and offers courses in the area of wildlife conservation and environmental protection at both undergraduate and postgraduate level. He has undertaken advisory work on international and environmental law for various non-governmental organisations and foreign governments and has recently completed a major study of the Ramsar Wetlands Convention.

Alan E. Boyle is Professor of Public International Law in the University of Edinburgh, and was previously Reader in Public International Law in the University of London. He is co-author with Professor Patricia Birnie of *International Law and the Environment* (1992) and has written widely on law of the sea and international environmental law. He has recently edited *Environmental Regulation and Economic Growth* (1994) and (with M. Anderson), *Human Rights Approaches to Environmental Protection* (forthcoming).

Dr Robin Churchill is a reader at Cardiff Law School, University of Wales, where he has been teaching international law and EC law since 1977. He is the author of *EEC Fisheries Law* (1987) and *The Law of the Sea* (with A.V. Lowe, 2nd edn., 1988), and has published numerous articles on international and EC law, many of which have been concerned with environmental issues.

Professor David Freestone has held a personal chair of International Law at the University of Hull since 1991. He has established research and consultancy interests in international environmental law, EC law and marine and coastal law. He is founder and editor-in-chief of the *International Journal of Marine and Coastal Law*, and his recent books include *The North Sea: Perspectives on Regional Environmental Co-operation* (1990) and *Basic Legal Documents on Regional Environmental Co-operation*

(1991) (both with Ton IJlstra), *International Law and Global Climate Change* (with Robin Churchill, 1991) and most recently *The Precautionary Principle in International Law: The Challenge of Implementation* with Ellen Hey, 1995).

Kristina Gjerde, JD, is currently a research associate at the University of Hull Law School. She was previously a maritime lawyer in the New York City law firm of Lord, Day and Lord (1984-1988) and then a Research Fellow at the Marine Policy Centre of the Woods Hole Oceanographic Institution (1988–1990). She has also acted as a consultant to various US, UK and international environmental organizations. Recently, with David Freestone, she edited a special issue of the International Journal of Marine and Coastal Law on *Particularly Sensitive Sea Areas and International Law*. In addition to various other reports and publications, she was a major contributor to *The Oceans and Environmental Security: Shared US and Russian Perspectives* (1994). Ms Gjerde is admitted to the practice of law in New York State.

Sam Johnston is the Jaques & Lewis Research Associate in the Department of Land Economy at the University of Cambridge. Educated in Australia, where he qualified as a solicitor, and in the United Kingdom, his specialism lies in the area of environmental law and policy. In addition to advising a wide range of governmental and non-governmental organisations, he has published extensively in the international environmental law field. With Professor Malcolm Grant he is currently preparing *Cases and Materials on Environmental Law and Policy* for publication by Sweet and Maxwell and, with Dr Timothy Swanson, an analysis of the negotiation of the Biodiversity Convention for publication by Edward Elgar.

R. Jayakumar Nayar is a lecturer in the School of Law, University of Warwick where he teaches in the postgraduate Law in Development programme and in International Law. He is currently working on the completion of his doctoral dissertation, *Human Welfare and International Law: Reconceptualising Human Rights as Resistance*.

David Mohan Ong is a lecturer in law at the University of Hull Law School where he teaches on the International Environmental Law, Law of the Sea and Marine Environmental Law Master's courses. He has published recently on South-east Asian State practice on the joint development of offshore oil and gas deposits and the relationship between Chapter 17 of Agenda 21 and the 1982 UN Convention on the Law of the Sea. He acted as a UN consultant (along with Professors David Freestone and Alan Boyle and Dr Katharina Kummer) to jointly prepare a survey of existing international agreements on marine environmental protection for the 1992 UN Conference on Environment and Development in Rio. Recently he has also acted as a consultant on various South-east Asian maritime boundary disputes and has prepared a working paper on this subject for the Council on Security Co-operation in the Asia-Pacific region as a Research Associate of the Malaysian Institute of Maritime Affairs.

Catherine Redgwell is a senior lecturer and Director of the Centre for Environmental Law at the University of Nottingham. Educated in Canada and the United Kingdom, her chief research interest lies in the field of public international law, particularly international environmental law, with subsidiary interests in natural resources law and EC environmental law. She qualified as a barrister and solicitor in British Columbia in 1985, and in 1992 was seconded to the Legal Advisers, Foreign and Commonwealth Office for a six-month period. Her publications range from an analysis of the Antarctic Environmental Protocol to a study of the application of the trust concept to environmental protection.

Gregory Rose is the Head of the Environmental Law Unit in the Australian Department of Foreign Affairs and Trade. He was previously with the Foundation for International Environmental Law and Development (FIELD) at the University of London, where he was Director of the Marine Resources Programme, Editor-in-Chief of the *Review of the European Community and International Environmental Law* (RECIEL) and lecturer in international environmental law (1990–1993). He has worked for the Australian Attorney-General's Department, consulted for the UN and for international environmental organisations and published on a range of topics in international environmental law, most recently on the Asia-Pacific region.

Dr Ian Walden is the Tarlo Lyons Senior Research Fellow in Information Technology Law in the Centre for Commercial Law Studies, Queen Mary and Westfield College, University of London. He is editor of *EDI and the Law* (1989), joint editor of *Information Technology and the Law* (2nd edn., 1990) and *EDI Audit and Control* (1993) and is also on the editorial team for the journal *Computer Law and Practice*. He is Vice-Chairman of the UK EDI Association's Legal Advisory Group and a legal advisor to London solicitors, Tarlo Lyons, and is currently on secondment to the European Commission, DG-III (Industry), as a national expert in electronic commerce law, until the end of 1995.

Dr Lynda M. Warren is a lecturer at Cardiff Law School, University of Wales, where she has been teaching property law and planning law since 1989. Previously she was a lecturer in marine biology at the University of London, Goldsmiths' College. She has published widely on environmental law and policy, especially conservation, and is particularly interested in the interface between law and science. She is a member of the Countryside Council for Wales.

John Woodliffe is a senior lecturer in the Faculty of Law, University of Leicester. He has a long-standing interest in energy and natural resources law and has written widely on the civil and military aspects of nuclear power and the impact of environmental considerations upon the formulation and implementation of European and international energy policy. He is the author of *The Peacetime Use of Foreign Military Installations under Modern International Law* (1992) and is co-author of a forthcoming book on regionalism and world trade. His main research interests are in environmental issues that relate to the international law of development, world trade law and the law of the sea.

Abbreviations

AJIL	*American Journal of International Law*
ASEAN	Association of South-East Asian Nations
ASIL	American Society of International Law
ASMA	Antarctic Specially Managed Area
ASPA	Antarctic Specially Protected Area
AT	Antarctic Treaty
ATCM	Antarctic Treaty Consultative Meeting
ATCP	Antarctic Treaty Consultative Party
ATS	Antarctic Treaty System
BFSP	British & Foreign State Papers
BGCI	Botanic Gardens Conservation International
CCAMLR	Convention on the Conservation of Antarctic Marine Living Resources
CCAS	Convention on the Conservation of Antarctic Seals
CEP	Committee for Environmental Protection
CGIAR	Consultative Group on International Agricultural Research
CIDIE	Committee of International Development Institutions on the Environment
CITES	Convention on International Trade in Endangered Species of Wild Fauna and Flora
Cm/Cmnd	Command Paper
CNPPA	Commission of National Parks and Protected Areas (of the IUCN)
Col JTL	*Columbia Journal of Transnational Law*
COM	Communication from the Commission [of the EC] to the Council
CPGR	(FAO) Commission on Plant Genetic Resources
CRAMRA	Convention on the Regulation of Antarctic Mineral Resource Activities
CYIL	*Canadian Yearbook of International Law*
DAC	Development Assistance Committee, OECD
DOEM	Designated Officials for Environmental Matters of the United Nations
EC	European Community
ECE	(United Nations) Economic Commission for Europe
ECOSOC	Economic and Social Council (of the United Nations)
EFTA	European Free Trade Association
EIA	Environmental Impact Assessment
EIPR	*European Intellectual Property Review*
EIS	Environmental Impact Statement
EP	Environmental Protocol
EPA	Environmental Protection Agency (US)
EPL	Environmental Policy and Law
EPO	European Patent Office
ESA	Endangered Species Act of 1973 (US)
ESCAP	(United Nations) Economic and Social Commission for Asia and the Pacific
FAO	Food and Agriculture Organisation
GATT	General Agreement on Tariffs and Trade
GDP	Gross Domestic Product
GEF	Global Environment Facility
GNP	Gross National Product

HC	House of Commons Paper
HL	House of Lords Paper/Debates
IARC	International Agricultural Research Centre
IBP	International Biological Programme
IBRD	International Bank for Reconstruction and Development
ICAO	International Civil Aviation Organisation
ICJ Rep.	Reports of the International Court of Justice
ICLQ	*International and Comparative Law Quarterly*
ICSU	International Council of Scientific Unions
IJMCL	*International Journal of Marine and Coastal Law*
ILA	International Law Association
ILC	International Law Commission
ILM	International Legal Materials
ILO	International Labour Organisation
ILR	International Law Reports
IMO	International Maritime Organisation
INC	Intergovernmental Negotiating Committee
INSC	International North Sea Conference(s)
IOC	Intergovernmental Oceanographic Commission
IPCC	Intergovernmental Panel on Climate Change
IPGRI	International Plant Genetic Resources Institute
IPR	Intellectual Property Rights
ITTO	International Tropical Timber Organisation
IUCN	International Union for the Conservation of Nature and Natural Resources (The World Conservation Union)
IUPGR	International Undertaking on Plant Genetic Resources
IWC	International Whaling Convention
JPTSO	*Journal of the Patent and Trademark Society Office*
LDC	London Dumping Convention
LNTS	League of Nations Treaty Series
LOSC	Law of the Sea Convention
Misc	United Kingdom Command Papers, Miscellaneous Series
MSY	Maximum sustainable yield
NEPA	National Environmental Policy Act of 1969 (US)
NGO	Non-Governmental Organisation
NIEO	New International Economic Order
NOAA	National Oceanic and Atmospheric Administration (US)
OAU	Organisation of African Unity
ODA	Overseas Development Administration (of the UK)
OECD	Organisation for Economic Co-operation and Development
OJ	*Official Journal of the European Communities*
OJ EPO	*Official Journal of the European Patent Office*
PARCOM	Paris Commission (on Land-based Pollution)
PASIL	*Proceedings of the American Society of International Law*
PCIJ Rep.	Reports of the Permanent Court of International Justice
PGR	Plant Genetic Resource
RECIEL	*Review of European Community and International Environmental Law*
RGDIP	*Revue générale de droit international public*
RIAA	Reports of International Arbitral Awards
SCAR	Scientific Committee on Antarctic Research
SIPRI	Stockholm International Peace Research Institute
SPA	Specially Protected Area
SPAW	1990 Kingston Protocol to the Cartagena Convention on Specially Protected Areas and Wildlife
TEWG	Transitional Environmental Working Group
TFAP	Tropical Forestry Action Plan

TIAS	Treaties and Other International Acts Series
TRIP	Trade Related Intellectual Property
UKTS	United Kingdom Treaty Series
UNCED	United Nations Conference on Environment and Development
UNCHE	United Nations Conference on the Human Environment
UNCLOS	United Nations Convention on the Law of the Sea
UNCTAD	United Nations Conference on Trade and Development
UNDP	United Nations Development Programme
UNECE	See ECE (above)
UNEP	United Nations Environment Programme
UNESCO	United Nations Educational, Scientific and Cultural Organisation
UNGA	United Nations General Assembly
UNTS	United Nations Treaty Series
UPOV	International Union for the Protection of New Varieties of Plants
UST	United States Treaties and Other International Agreements
WCED	World Commission on Environment and Development
WCMC	World Conservation Monitoring Centre
WCP	World Climate Programme
WCS	World Conservation Strategy
WIPO	World Intellectual Property Organisation
WMO	World Meteorological Organisation
WRI	World Resources Institute
WWF	World Wide Fund for Nature

The International Environmental Law and Policy Series

Series General Editor
Stanley P. Johnson

Advisory Editor
Günther Handl

Pollution Insurance: International Survey of Coverages and Exclusions, W. Pfennigstorf (ed.)
(ISBN 1–85333–941–5)
Civil Liability for Transfrontier Pollution, G. Betlem
(ISBN 1–85333–951–2)
Transboundary Movements and Disposal of Hazardous Wastes in International Law, B. Kwiatkowska, A.H.A. Soons
(ISBN 0–7923–1667–3)
The Legal Regime for Transboundary Water Pollution: Between Discretion and Constraint, A. Nollkaemper
(ISBN 0–7923–2476–5)
The Environment after Rio: International Law and Economics, L. Campiglio, L. Pineschi, D. Siniscalco, T. Treves (eds.)
(ISBN 1–85333–949–0)
Overcoming National Barriers to International Waste Trade, E. Louka
(ISBN 0–7923–2850–7)
Precautionary Legal Duties and Principles in Modern International Environmental Law, H. Hohmann
(ISBN 1–85333–077–0)
Negotiating International Regimes: Lessons Learned from the UN Conference on Environment and Development, B. Spector, G. Sj/ostedt, I. Zartman (eds.)
(ISBN 1–85966–077–0)
Conserving Europe's Natural Heritage: Towards a European Ecological Network, G. Bennett (ed.)
(ISBN 1–85966–090–8)
US Environmental Liability Risks, J. T. O'Reilly
(ISBN 1–85966–093–2)
Environmental Liability and Privatization in Central and Eastern Europe, G. Goldenmann et al (eds.)
(ISBN 1–85966–094–0)
German Environmental Law, G. Winter (ed.)
(ISBN 0–7923–3055–2)
The Peaceful Management of Transboundary Resources, G. H. Blake et al (eds.)
(ISBN 1–85966–173–4)
Pollution from Offshore Installations, M. Gavouneli
(ISBN 1–85966–186–6)
Sustainable Development and International Law, W. Lang
(ISBN 1–85966–179–3)
Canada and Marine Environmental Protection, D. Vanderzuraag
ISBN 90–411–0856–4
The Environmental Policy of the European Communities, S. P. Johnson, G. Corcelle
ISBN 90–411–0862–9
European Environmental law, J. Salter (Looseleaf Service)
(Basic Work ISBN 1–85966–050–9)
(Please order by ISBN or title)

Introduction

Michael Bowman and Catherine Redgwell

The 1992 United Nations Conference on Environment and Development (UNCED) produced a number of instruments, of which only two comprised treaty texts opened for signature. One, the 1992 United Nations Framework Convention on Climate Change, addresses the serious problem of the impact of human activities on the earth's atmosphere and consequent global warming. This topic was the subject of a previous study by the Committee on Environmental Law of the British Branch of the International Law Association, which anticipated the conclusion of the Convention with publication by Graham and Trotman in 1991 of *International Law and Global Climate Change*. The other, the 1992 United Nations Convention on Biological Diversity ("the Biodiversity Convention"), is the focus of this present volume, which reflects the work of the Committee since 1991. The bulk of the project was completed by the late summer of 1994, though in some cases it has been possible to take account of developments subsequent to that time.

The Convention on Biological Diversity

The Convention is the first international treaty explicitly to address all aspects of biodiversity ranging from the conservation of biological diversity and sustainable use of biological resources to access to biotechnology and the safety of activities related to modified living organisms. Though any survey of international environmental law would reveal international wildlife treaties stretching back a century or more, and a wide range of instruments concerned with habitat protection and the control or reduction in activities which adversely affect species and habitat, there is no doubt that explicit attention to biodiversity conservation is a recent phenomenon. For example, the 1972 United Nations Conference on the Human Environment, considered by many to constitute the genesis of modern international environmental law, makes no reference to biological diversity as such in the resulting Stockholm Declaration. However, there are references to the need for conservation, particularly in Principle 4 which calls for the safeguarding and wise management of "the heritage of wildlife and its habitat which are now gravely imperilled by a combination of adverse factors".

International Law and the Conservation of Biological Diversity (C. Redgwell and M. Bowman, eds.: 90 411 0863 7: © Kluwer Law International: pub. Kluwer Law International, 1995: printed in Great Britain), pp. 1–4

It was only in 1980 in the World Conservation Strategy (WCS), formulated by the IUCN in collaboration with UNEP, WWF, FAO and UNESCO, that explicit reference to diversity of life forms is found. The WCS articulates one of the three fundamental objectives of living resource conservation as the preservation of genetic diversity. The 1982 World Charter for Nature, and the Second World Conservation Strategy, *Caring for the Earth* (1990), follow this new emphasis on the need to conserve the "vitality and diversity of the Earth". Hence the latter explicitly calls for the conservation of biodiversity (rather than its predecessor's reference exclusively to "genetic diversity"), including "not only all species of plants, animals and other organisms, but also the range of genetic stocks within each species, and the variety of ecosystems". "Keeping within the Earth's carrying capacity" is a related theme of *Caring for the Earth* as part of, and indeed an essential prerequisite to, sustainable development. *Our Common Future*, the 1987 Report of the World Commission on Environment and Development, explicitly views species and ecosystems as resources for (sustainable) development.

One of the main threats imperilling development – and the conservation of biodiversity – is the rapid destruction of the earth's most species-rich ecosystems. A 1990 Report by the IUCN, WWF, World Resources Institute, Conservation International and the World Bank, *Conserving the World's Biological Diversity*, states that nearly one quarter of the earth's biological diversity is estimated to be at risk of extinction during the next twenty to thirty years. More species than ever before are so threatened, many (particularly insects) before they have even been described or classified. The threat is exacerbated by the fact that at least half of the earth's species are located in tropical rain forests, thus reinforcing the link between species and habitat.

The World Commission on Environment and Development called for a new approach to species and ecosystem conservation based upon the notion of "anticipate and prevent" and for the negotiation of a properly funded "Species Convention" along the lines of the 1982 United Nations Convention on the Law of the Sea. This reflected growing international recognition of the need for a global convention to remedy the perceived defects of the previous piecemeal approach to conservation of global biodiversity. Although global conventions did exist, their focus was selective, covering only internationally important sites (e.g. the World Heritage Convention), particular types of ecosystem (e.g. the Ramsar Convention on Wetlands of International Importance) or particular species (e.g. the Convention on the Conservation of Migratory Species of Wild Animals) or regulating a particular threat to endangered species (e.g. the Convention on International Trade in Endangered Species of Wild Fauna and Flora). Regional conventions added to this patchwork quilt of protection, but did not provide for the comprehensive global regulation needed.

Thus, in 1987, UNEP's Governing Council resolved to establish an *ad hoc* working group to examine "the desirability and possible form of an umbrella convention to rationalize current activities in this field, and to address other areas which might fall under such a convention" (Resolution 14/26). By 1990 consensus had been achieved regarding the negotiation of a framework convention building on, rather than attempting to consolidate or replace, existing international treaties. Assisted by related activities in the IUCN and FAO, UNEP was able to produce a first draft of a convention to launch the formal negotiating process in February 1991 in the Intergovernmental Negotiating Committee for a Convention on Biological Diversity (INC).

Within the INC, Working Group I addressed general issues such as fundamental principles, general obligations, *in situ* and *ex situ* conservation measures, and the relationship of the Biodiversity Convention with other international agreements. Working Group II dealt with access to genetic resources and technologies, technology transfer, technical assistance, financial mechanisms and international cooperation. By and large, this division of responsibilities is reflected in the structure of the Convention itself, with most of Working Group II's responsibilities falling in the later articles of the Convention. Negotiations concluded with the adoption of the Convention at Nairobi on 22 May 1992. The Convention was opened for signature at UNCED in Rio de Janeiro on 5 June 1992 and entered into force a scant eighteen months later on 29 December 1993 following the thirtieth ratification (by Mongolia).

The Structure of the Book

In the case of the Biodiversity Convention, the approach of the Committee has necessarily differed from that taken to *International Law and Global Climate Change* since we were addressing a Convention already concluded and, as we were to witness, one which entered into force only eighteen months later at the end of 1993. We have retained the approach of extending beyond the ambit of the Convention itself to examine the conservation of biodiversity in international law. As a consequence, the following chapters include not only detailed analysis of the Convention (Chapter 2) but also of particular features such as sustainable use of biological resources (Chapter 3), *ex situ* conservation (Chapter 7) and plant genetic resources (Chapter 8). The controversial issue of intellectual property rights, which delayed United States' signature of the Convention, is comprehensively analysed in Chapter 9.

Property rights are a linking theme of several chapters, given the Biodiversity Convention's reaffirmation of sovereign rights over natural resources, including biological resources, and increasing enthusiasm for the use of property rights – whether held by the individual, a group (e.g. Plant Breeders' Rights) or by the State – as a means of providing an incentive for the management and sustainable use of biological resources. The Biodiversity Convention implicitly recognises the uneven global distribution of natural resources, and consequently the heavy burden which implementation of the Convention will inevitably impose on many developing States which happen to be the depository of rich, even "mega-diverse", biological resources. Differences between developing and developed States over a whole host of issues, and the role of indigenous peoples, form the focus of Chapters 12 and 13 respectively, followed by an examination of the critical issue of resourcing the estimated US$3.5 billion needed annually to implement measures necessary to conserve biodiversity (Chapter 14). The problems of implementation of the Convention in the European Union (Chapter 11), which has ratified the Convention along with all of its Member States, and the United States (Chapter 10) are also examined.

In order fully to appreciate the context in which the Biodiversity Convention was negotiated, its essentially framework character, and the wide range of existing international conventions which, directly or indirectly, conserve biodiversity, this volume also includes an analysis of existing measures for protection of the terrestrial (Chapter 4), marine (Chapter 5) and Antarctic (Chapter 6) environment. Existing measures in the European Union and the United States are also addressed in the relevant chapters, with the overt policy of the latter being to ensure that no new instrument was needed for implementation of the Biodiversity Convention.

As a preliminary to these studies, the book commences with an analysis of the nature, development and philosophical foundations of the biodiversity concept in international law (Chapter 1). Debate concerning access to and use of the components of biological diversity, and in particular their actual or potential commercial value, has tended to obscure the non-economic values of biodiversity. These are clearly recognised in, for example, the 1987 *Report of the World Commission on Environment and Development* which refers to aesthetic, ethical, cultural and scientific grounds for conservation. Indeed, non-economic values are explicitly recognised in the preamble to the Biodiversity Convention which refers to "the intrinsic value of biological diversity" and the "ecological, genetic, social, economic, scientific, educational, cultural, recreational and aesthetic value of biological diversity and its components". The ultimate success of the Convention may arguably lie in the extent to which the implementation of its substantive provisions pays due regard to all of these forms of value.

1

The Nature, Development and Philosophical Foundations of the Biodiversity Concept in International Law

Michael Bowman

The Nature of the Biodiversity Concept

Biological diversity, or biodiversity as it is commonly called, has been described as "an umbrella term for the degree of nature's variety".[1] More specifically, it is understood to be a three-fold concept embracing:

(a) the diversity of ecosystems;
(b) the diversity of species;
(c) genetic diversity within species.[2]

Of these three elements, the diversity of ecosystems might be regarded as the concept commanding the *highest* level of importance, since all living organisms exist and function not in isolation but as part of a wider environment, occupying a particular niche within their appropriate ecosystem, and it is through the preservation of entire ecosystems that diversity can most effectively be secured.[3] On the other hand, species diversity might be claimed to be the *central* concept, since the species has traditionally been regarded as the taxonomic starting point for the classification of living organisms,[4] and it is to the protection of particular species or groups of species

---

[1] J.A. McNeely et al., *Conserving the World's Biological Diversity* (1990), at p. 17.
[2] *Ibid.* Biodiversity can be described by reference to a greater number of levels of biological organization: e.g. ecosystem, community, guild, species, organism, gene, but this threefold approach is the one adopted in Article 2 of the 1992 Convention on Biological Diversity, Misc. 3 (1993), Cm 2127. For further text location references and current status information regarding this and other conventions referred to, see M.J. Bowman and D.J. Harris, *Multilateral Treaties: Index and Current Status* (1984) and latest cumulative supplement.
[3] McNeely, *op. cit.*, n. 1 at p. 57.
[4] E.O. Wilson, *The Diversity of Life* (1992), at pp. 35–45.

International Law and the Conservation of Biological Diversity (C. Redgwell and M. Bowman, eds.: 90 411 0863 7: © Kluwer Law International: pub. Kluwer Law International, 1995: printed in Great Britain), pp. 5–31

that most international conversation efforts have hitherto been directed.[5] Finally, however, a case could be made for regarding genetic diversity as the most *fundamental* element, since it is in the very variety of genetic material existing within and between species that the raw material for scientific, industrial and agricultural innovation and development can be found, as well as the resilience and adaptability which will be needed if the earth's biosphere is to survive and flourish in the face of current trends of continuing environmental degradation.[6] The better view is therefore to regard these three aspects as mutually interdependent and equally important, so reflecting the symbiotic relationships existing within the biosphere itself.

It is considered axiomatic that biodiversity protection can best be effected *in situ*, that is through the conservation of natural habitats in such a way as to preserve entire ecosystems and the species they contain.[7] It is not, however, considered adequate merely to protect wildlife species from known and immediate threats to their survival. Preferably, sufficient genetic variation will be preserved within each species to maximise its capacity for resilience in the face of potential future threats, whether in the form of disease, habitat loss, climate change or the introduction of alien and competitive species.[8] Such genetic diversity is a function partly of sheer numbers, but is also affected by a variety of other factors, such as the existence of geographically scattered population groups within species, which are likely to have evolved distinct genetic characteristics.[9] Hence the need for protection of each species in each of the parts of its natural range.

It is also recognised, however, that an important contribution may be played by *ex situ* methods of conservation.[10] These include collection and propagation by botanical gardens, captive breeding programmes in zoos and like institutions and the maintenance of gene banks and similar facilities. It seems unlikely at present, however, that *ex situ* techniques can play more than a limited role, not least because they tend to rely on *in situ* approaches to enable their genetic stocks to be replenished. IUCN therefore advocates that the two approaches be regarded as "opposite ends of the total spectrum required for effective conservation".[11]

Biodiversity and the International Legal System

It is only relatively recently that the international community has begun to recognise the full scale and seriousness of the threat posed to the natural environment by human activities, and the need to establish a clear, coherent and comprehensive legal framework within which to tackle that threat. The alarmingly high rate of extinction of wildlife species in recent times has constituted a particular focus of concern, and provided a significant challenge to the established legal order. Traditionally, the exploitation of natural resources has been regarded as an aspect of national sovereignty, to be pursued in accordance with the customary norms governing State

---

[5] See generally S. Lyster, *International Wildlife Law* (1985).

[6] IUCN, *World Conservation Strategy* (1980), sections 1 and 3.

[7] McNeely, *op. cit.*, n. 1, pp. 55–62; World Resources Institute (WRI) *et al., Global Biodiversity Strategy* (1992), at p. 19.

[8] McNeely, Chapter 3; WRI, *ibid.,* Chapter 2.

[9] R.B. Primack, *Essentials of Conservation Biology* (1993), at pp. 31–33, Chapters 11, 12.

[10] McNeely, *op. cit.*, n. 1, at pp. 62–6; WRI, *op. cit.*, n. 7, at pp. 137–46; Primack, *ibid.*, Chapter 17; Chapters 30–4 in E.O. Wilson (ed.), *Biodiversity* (1988). See also the Chapter by L. Warren in this work.

[11] McNeely, *op. cit.*, n. 1, at p. 62.

jurisdiction. Thus each State was in principle free to exploit the wildlife resources present within its territory, whether permanently or temporarily, as a consequence of natural patterns of migration.[12] On the high seas and in other areas beyond national jurisdiction, States were understood to enjoy equal rights of exploitation.

Naturally, the absence of formal legal constraints upon this process would be likely to contribute significantly to the over-exploitation of living resources and the degradation of natural systems. With the emergence of the biodiversity concept as a key element in scientific thinking and conservation policy, it has therefore become necessary to examine the extent to which that concept has found a place within the international legal system.

The Development of the Biodiversity Concept in International Law

Just as Monsieur Jourdain, Molière's *bourgeois gentilhomme*, was delighted to discover that he spoke prose without knowing it,[13] so a chronological survey of legal developments in the environmental field would readily reveal that the international community has long been concerned with the problem of the conservation of biological diversity, albeit not expressly under that particular conceptual rubric.[14] Lyster refers to the existence of forestry conservation laws in Babylon as long ago as 1900 BC, and to the setting aside of land in Egypt as a nature reserve in 1370 BC.[15] Even the use of international treaty arrangements for the protection of wildlife now has a pedigree stretching back over a hundred years.[16] However, the development of anything resembling a comprehensive and coherent biodiversity strategy can only be regarded as an extremely recent phenomenon.

The Intergovernmental Conference of Experts on the Scientific Basis for Rational Use and Conservation of the Resources of the Biosphere, convened by UNESCO in Paris during September 1968, constituted an important early landmark in this process through its establishment of the Man and the Biosphere Programme, emphasising humanity's place in the natural order of things as well as the importance of a holistic, ecosystemic approach to nature conservation.[17] Although the more widely renowned 1972 United Nations Conference on the Human Environment is treated by many as constituting the genesis of modern international environmental law,[18] there is relatively little sign in the resulting Stockholm Declaration[19] of the biodiversity concept as such. A clear recognition of the importance of conservation was, nevertheless, apparent in several of the Declaration's key provisions. Principle 2 emphasised the need to safeguard the "natural resources of the earth including the air, water, land, flora and

---

[12] *Behring Sea Fur Seals Arbitration* (1898) 1 *Moore's International Arbitration Awards* 755.

[13] Molière, *Le Bourgeois Gentilhomme*, Act IV, Scene I.

[14] See, e.g. the Chronological Table in A.C. Kiss and D. Shelton, *International Environmental Law* (1990), and the Table of Major Treaties and Instruments in P.W. Birnie and A.E. Boyle, *International Law and the Environment* (1992).

[15] Lyster, *op. cit.*, n. 5, at p. xxi.

[16] Birnie and Boyle, *op. cit.*, n. 14, state at p. 421 that "the first relevant treaty was the 1885 Convention for the Uniform Regulation of Fishing on the Rhine".

[17] See R. Boardman, *International Organization and the Conservation of Nature* (1981), at pp. 65–66.

[18] M. Pallemaerts, "International Environmental Law from Stockholm to Rio: Back to the Future" in P. Sands (ed.) *Greening International Law* (1993), at p. 2; Kiss and Shelton, *op. cit.*, n. 14, at Chapter 2; J. Brunnée, *Acid Rain and Ozone Layer Depletion* (1988), at pp. 83–84.

[19] Declaration of the United Nations Conference on the Human Environment (Stockholm), UN Doc.A/CONF.48/14/Rev.1. Many of the ideas underlying the concept can, however, be seen to have influenced the Action Plan also agreed at Stockholm; see especially Recommendations 38–45.

fauna and especially representative samples of natural ecosystems", while Principle 3 stated that the earth's capacity to produce vital renewable resources must be maintained and improved. Most importantly, Principle 4 proclaimed that

> Man has a special responsibility to safeguard and wisely manage the heritage of wildlife and its habitat which are now gravely imperilled by a combination of adverse factors. Nature conservation including wildlife must therefore receive importance in planning for economic development.

It has been pointed out that this last provision did not feature in early drafts of the Declaration at all, and was only inserted at the Conference itself upon the proposal of the Indian delegation, though the explanation of its original absence has been suggested to lie less in a failure to appreciate the importance of wildlife conservation than in an awareness of the fact that a number of legal instruments devoted to this objective were already in existence.[20]

It is perhaps in the 1980 World Conservation Strategy,[21] through its underlying theme of sustainable development, that the real foundations for the biodiversity concept were laid. The strategy, which was formulated by IUCN as part of a collaborative effort with UNEP, WWF, FAO and UNESCO, established its three fundamental objectives of living resource conservation as being:

(a) to maintain essential ecological processes and life-support systems;
(b) to preserve genetic diversity and
(c) to ensure the sustainable utilisation of species and ecosystems.

Genetic diversity was understood to mean the

> range of genetic material found in the world's organisms, on which depend the functioning of many of the above processes and life-support systems, the breeding programmes necessary for the protection and improvement of cultivated plants, domesticated animals and micro-organisms, as well as much scientific and medical advance, technical innovation and the security of the many industries that use living resources.[22]

When linked with the requirement of the conservation, in the form of sustainable utilisation, of species and ecosystems, the principle of preservation of genetic diversity clearly captures the essence of the later biodiversity concept.

The World Conservation Strategy was promptly followed by the 1982 World Charter for Nature,[23] which embraced much of its thinking, while adding some distinct features of its own. The fourth preambular paragraph stated that the General Assembly was persuaded that "lasting benefits from nature depend upon the maintenance of essential ecological processes and life support systems, and upon the diversity of life forms...". The first four paragraphs of the substantive General Principles of the Charter[24] accordingly provided:

---

[20] L.B. Sohn, "The Stockholm Declaration and the Human Environment" (1973) 14 *Harvard ILJ* 423.

[21] IUCN/UNEP/WWF, *World Conservation Strategy* (1980).

[22] *Ibid.*, section 1.

[23] UNGA Res. 37/7; UN Doc.A/37/51 (1982). For the text of the Charter, together with its legislative history and a commentary to its provisions, see W.E. Burhenne and W.A. Irwin, *World Charter for Nature* (2nd rev. edn., 1986).

[24] There are five general principles in all; the fifth states that "Nature shall be secured against degradation caused by warfare or other hostile activities". Note also that paragraph 9, in the section headed "Functions", states that "due account shall be taken ... of the biological productivity and diversity... of the areas concerned".

1. Nature shall be respected and its essential processes shall not be impaired.
2. The genetic viability of the earth shall not be compromised; the population levels of all life forms, wild and domesticated, must be at least sufficient for their survival, and to this end necessary habitats shall be safeguarded.
3. All areas of the earth, both land and sea, shall be subject to these principles of conservation; special protection shall be given to unique areas, to representative samples of all the different types of ecosystems and to the habitats of rare or endangered species.
4. Ecosystems and organisms, as well as the land, marine and atmospheric resources that are utilised by man, shall be managed to achieve and maintain the optimum sustainable productivity, but not in such a way as to endanger the integrity of those other ecosystems or species with which they coexist.

The unofficial Commentary to the Charter states:[25]

These principles, which are not legally binding in a formal sense, represent moral or social rules which, if they are to have the force of law, must be transformed by the international community into the terms of conventions or into customary international law ... In the long run, the principles which are contained in the Charter should be integrated into the conscience of the world.

Taking up this challenge at the time of the Brundtland Commission Report[26] in 1987, the Legal Experts Group appointed by that Commission established the following as the core elements of Principle 3 of their Proposed Legal Principles for Environmental Protection and Sustainable Development:[27]

States shall maintain ecosystems and ecological processes essential for the functioning of the biosphere, shall preserve biological diversity, and shall observe the principle of optimum sustainable yield in the use of living natural resources and ecosystems.

This principle appears to be the direct descendant of the World Conservation Strategy and the World Charter for Nature, though it will be noticed that the terminology of "biological diversity" has by now replaced that of "genetic diversity". The Experts Group did not offer much by way of analysis of this principle, contenting themselves with recalling the large number of existing conventions which provided concrete manifestations of adherence to it within the context of particular environmental problems.[28] The following year IUCN joined with UNEP to embark on preparatory work for an international convention on the conservation of biological diversity. The convention was finally opened for signature at the Rio Earth Summit in 1992.

Biodiversity and Existing Conservation Measures

As indicated above, it is clear that the international community has concerned itself with the problem of the conservation of biological diversity for many decades, albeit in a rather piecemeal and short-sighted fashion. Early conservation efforts tended to be applied with regard to wildlife species directly exploited by man, while, more

---

[25] Burhenne and Irwin, *op. cit.*, n. 23, at p. 142. The commentary was prepared by the International Council for Environmental Law on the initiative and under the leadership of the Conseil Européen du Droit de l'Environnement and reviewed by members of the IUCN Commission on Environmental Policy, Law and Administration.

[26] *Our Common Future* (1987).

[27] For the full text see R.D. Munro and J.G. Lammers (eds.), *Environmental Protection and Sustainable Development: Legal Principles and Recommendations adopted by the Experts Group on Environmental Law of the World Commission on Environment and Development* (1987).

[28] *Ibid.*, pp. 45–54.

recently, conventions with a broader focus have emerged.[29] As far as the protection of ecosystems is concerned, the 1971 Ramsar Wetlands Convention[30] is the principal example, and its very wide definition of what constitutes a wetland gives it a particularly expansive scope.[31] The recently adopted Environmental Protocol to the Antarctic Treaty[32] accords substantial protection to the Antarctic ecosystem, particularly when considered alongside the earlier environmental instruments governing that region,[33] of which the marine living resources convention was widely commended for its "ecosystem" approach.[34] The 1982 Law of the Sea Convention[35] devotes an entire chapter to the marine environment, and is supported by a broad array of other treaties dealing with more specific aspects of the problem.[36] In addition to those treaties which are specifically concerned with the protection of particular kinds of ecosystem, the vast network of protected areas established under the World Heritage Convention[37] and the various regional nature conservation conventions[38] offer some degree of protection to a variety of sites and areas embracing many types of habitat.

The bulk of this legislation concerns *in situ* conservation, but there are increasing indications of recognition, both directly and indirectly, of the importance of *ex situ* methods. The 1985 ASEAN Convention,[39] for example, makes specific provision for the establishment of gene banks, while the 1973 Convention on International Trade in Endangered Species (CITES)[40] accords special treatment to specimens resulting from programmes of artificial propagation or captive breeding, and the 1979 Berne Convention on the Conservation of European Wildlife envisages the reintroduction of native species of wild flora and fauna in certain circumstances.[41]

The clear view of IUCN, however, was that the existing network of conventions was failing to prevent the exploitation of species and ecosystems at rates which far exceed their sustainable yield, and that greater international co-operation was required in order to reverse this trend.[42] Hence the need for the conclusion of an all-embracing conservation treaty was recognised, a need which the Biodiversity Convention was intended to meet.

---

[29] For examples, see the 1979 Bonn Convention on the Conservation of Migratory Species of Wild Animals, Misc 11 (1980), Cmnd 7888; the 1979 Berne Convention on the Conservation of European Wildlife and Natural Habitats, 1284 UNTS 209 and the other regional conventions discussed in Part III of Lyster, *op. cit.*, n. 5.

[30] The 1971 Convention on Wetlands of International Importance, Especially as Waterfowl Habitat, 995 UNTS 245.

[31] Lyster, *op. cit.*, at p. 184; P.J. Dugan (ed.), *Wetland Conservation: A Review of Current Issues and Required Action* (1990), at p. 9.

[32] Misc. 6 (1992), Cm 1960.

[33] See L. Kimball, "Environmental Law and Policy in Antarctica" in Sands (ed.), *op. cit.*, n. 18, and the chapter by C.J. Redgwell.

[34] 1980 Convention on the Conservation of Antarctic Marine Living Resources, UKTS 48 (1982), Cmnd 8714. See especially Articles 2(3) and 15 and, for discussion, see Lyster, *op. cit.*, at pp. 157–158; Kimball, *op. cit.*, n. 33, at pp. 130–131.

[35] UN Doc. A/CONF.62/122.

[36] See Birnie and Boyle, *op. cit.*, n. 14, Chapters 7 and 13; and the chapter by D. Freestone in this work.

[37] 1972 Convention for the Protection of the World Cultural and Natural Heritage, 1972 UNJYB 89.

[38] See n. 29.

[39] 1985 ASEAN Agreement on the Conservation of Nature and Natural Resources, *ASEAN Documents Series, 1967–1986*, p. 203.

[40] 993 UNTS 243. See Article 7(4), (5).

[41] Article 11(2).

[42] McNeely, *op. cit.*, n. 1, at p. 68.

An important issue arises concerning the relationship between this mass of existing legislation and the Biodiversity Convention itself. It was apparently decided at an early stage that it would not be practicable to establish the Convention as an "umbrella" treaty under the aegis of which all existing agreements would automatically be brought.[43] Rather, Article 22(1) of the Convention provides that

> The provisions of the present Convention shall not affect the rights and obligations of any Contracting Party deriving from any existing international agreement, except where the exercise of those rights and obligations would cause serious damage or a threat to biological diversity.

This plainly leaves intact all the commitments and entitlements which derive from existing conservation, or indeed other, treaties,[44] except to the extent that they are inimical to the basic objectives of the Biodiversity Convention itself – a notion which seems pregnant with practical difficulties of interpretation.[45]

Genetic Diversity within Species

It will be apparent from the above survey that rather little has been said on the question of genetic diversity, for the simple reason that few existing conventions make explicit reference to it. Indeed, a recent work on biodiversity in international law[46] devotes an entire section each to the conservation of species and of ecosystems but not even a full chapter to the preservation of genetic diversity.

It may be, therefore, that it is in this area that the Biodiversity Convention has the most to offer, though it would be wrong to assume that there are no provisions in existing treaties which bear upon this question. Perhaps the clearest reference occurs in Article 3(1) of the ASEAN Convention, which requires the parties, wherever possible, to "maintain maximum genetic diversity by taking action aimed at ensuring the survival and promoting the conservation of all species under their jurisdiction and control".[47] Article 4(3) of CITES, furthermore, provides for the restriction of trade in specimens of Appendix II species where necessary

> to maintain that species throughout its range at a level consistent with its role in the ecosystems in which it occurs and well above the level at which that species might become eligible for inclusion in Appendix I

(i.e. as a species threatened with extinction). This reveals an indirect concern for the preservation of genetic diversity both through the requirement of maintaining populations at levels "well above" that at which the survival of the species in question could be regarded as under threat, and through the perceived need to achieve this objective "throughout its range". In the same vein, the CITES definition of a species is such as to enable protection to be offered to "any sub-species ... or geographically separate population thereof".[48] A similar definition is adopted for the purposes of the

---

[43] F. Burhenne-Guilmin and S. Casey-Lefkowitz, "The Convention on Biological Diversity: A Hard-Won Global Achievement" (1992) 3 *Yearbook of International Environmental Law* 43, at p. 45.

[44] For example, trade agreements such as GATT.

[45] Note that a provision incorporated in earlier drafts of the Biodiversity Convention to the effect that existing conservation agreements could be renegotiated as protocols to that treaty was not included in the final text.

[46] C. de Klemm and C. Shine, *Biological Diversity Conservation and the Law* (1993).

[47] Paras. 2 and 3 of Article 3 establish supporting obligations.

[48] Article 1(a).

Migratory Species Convention.[49] The preamble to the latter, furthermore, expressly recognises the genetic value of wild animals. It may be, however, that the current texts of wildlife conventions will need to be reviewed in order to determine whether they pay sufficient regard to the need to preserve genetic diversity. At both the 1989 and 1992 meetings of the Conference of the Parties to CITES, for example, Denmark sought to establish an exemption from CITES permit formalities for 2 ml aliquot samples of blood and tissue to be used for DNA sampling, on the grounds that such techniques, which were invaluable for biodiversity studies, had not been developed at the time the Convention was drafted and had therefore not been provided for in the exemption provisions. Despite some support from other European States, this proposal was rejected as being contrary to the terms of the Convention.[50]

The Beneficiaries of Biodiversity Conservation

International law has traditionally concerned itself primarily with the rights and interests of States.[51] During the post-war period, however, the growth of the human rights movement has significantly readjusted the entire focus of the system, not only by recognising the substantive rights of individual human beings, but also by according them procedural rights of enforcement in many cases.[52] From a slightly different perspective, the needs of environmental protection have also come to pose a major challenge to the established structure of the international legal system. It has become common to observe that the natural environment knows no political boundaries and that the traditional regime of resource exploitation, grounded primarily in the notion of national territorial sovereignty, requires to be replaced by more overtly collectivist approaches.[53] Hence the birth of such concepts as the "common heritage of mankind".

Yet it is pertinent to observe that such concepts have perhaps made a greater impact in the realms of political rhetoric and academic discourse than they have upon the substance of international law. In particular, few treaties have given meaningful effect to the common heritage notion,[54] and it has become customary to treat issues of environmental protection as falling within the scope of the more diluted concept of the "common concern of mankind".[55] This appears to connote little

---

[49] Article 1(1)a.

[50] See CITES DOC.7.46, Com.II 7.2 and PLEN.7.6(Rev.), and also DOC.8.41(Rev.), Com.II 8.10, 8.11 and PLEN.8.7. Note that Article 7(6) does provide an exemption in respect of certain non-commercial exchanges between scientific institutions, but it was argued that this was insufficient in scope for the purposes envisaged.

[51] Note that Oppenheim was of the view that "the Law of Nations is a law between States only and exclusively": L. Oppenheim, *International Law: A Treatise* (1905), section 289.

[52] See, e.g. the 1966 International Covenant on Civil and Political Rights, 999 UNTS 171 and its Optional Protocol, 999 UNTS 302; the 1950 European Convention for the Protection of Human Rights and Fundamental Freedoms, 213 UNTS 221; and the 1969 American Convention on Human Rights, 1144 UNTS 123.

[53] For recent developments of this theme, see, for example, J.A. McNeely, "Common Property Resource Management or Government Ownership: Improving the Conservation of Biological Resources" (1991) *International Relations* 211; G. Handl, "Environmental Security and Global Change: the Challenge to International Law" (1990) 1 *Yearbook of International Environmental Law* 3.

[54] Birnie and Boyle, *op. cit.,* n. 14, at pp. 120–121. The only real examples are the 1979 Agreement Concerning the Activities of States on the Moon and Other Celestial Bodies, 18 ILM 1434 and the 1982 UN Convention on the Law of the Sea, UN Doc.A/CONF.62/122; as to the latter see also the 1994 Agreement Relating to the Implementation of Part XI, UN Doc.A/RES/48/263, 33 ILM 1309.

[55] See, for example, the preamble to the 1992 United Nations Framework Convention on Climate Change, Misc. 6 (1993), Cm 2137.

more than that such issues are the legitimate concern of all States and should not be regarded as being matters of purely domestic jurisdiction.[56] The preamble to the Biodiversity Convention itself adopts the terminology of common concern.[57] At the same time, it is noticeable that the treaty's substantive provisions are very firmly grounded in traditional state-centred patterns of thinking.[58]

As indicated above, however, individual States are no longer the sole repositories of rights in international law, and the growth of the human rights movement in the post-war period has not been without impact upon the development of environmental law itself. In particular, the right to a decent environment is sometimes asserted to be one of the so-called "third generation" of human rights.[59] At the individual level, it may certainly be meaningful to think in terms of a right to a living or working environment free of unacceptable pollution risks, or to an adequate supply of potable drinking water, but it is more difficult to translate such thinking into the realms of biological diversity. While many individuals might justly feel themselves to be the beneficiaries of a natural environment rich and diverse in wildlife, it would seem impossible to quantify the particular degree of diversity to which any one person could be *entitled*. It may therefore be more plausible to think in terms of collective rights or interests in this context, and particularly to have regard to the interests of future generations in receiving a wildlife heritage no less rich and diverse than that which the present generation enjoys. Many conservation treaties, including the Biodiversity Convention itself, accordingly make reference in their preambles to the need to safeguard the interests of future generations.[60] Once again, however, it is hard to see that much of substance is really added by the incorporation of this human rights dimension. The detailed, substantive content of wildlife conservation commitments is certain to continue to be determined by reference to environmental, rather than human rights, treaties and it is therefore only in the procedural context that any significant advantage is likely to be gained. This would most probably take the form of "securing ... rights of access to information, to participation in decision-making processes and to administrative and judicial remedies"[61] and could conceivably be manifest at the national or international, individual or collective levels. It would seem that relatively few developments of this kind are yet discernible at the international level, however, and it may be that real progress in this context would require the creation of some wholly novel institution, such as an ombudsman to represent the interests of future generations.[62] For the time being, therefore, preambular

---

[56] See A.E. Boyle, "International Law and the Protection of the Global Atmosphere: Concepts, Categories and Principles", in R. Churchill and D. Freestone (eds.), *International Law and Global Climate Change* (1991), at p. 11. See also, by the same author, Chapter 2 of this work.

[57] "Affirming that the conservation of biological diversity is a common concern of humankind".

[58] See especially Articles 3–9, 12, 15, 16, 19–21, 31.

[59] On rights regarding the environment, see Principle 1 of the Stockholm Declaration; Article 24 of the 1982 African Charter on Human and Peoples' Rights, 21 ILM 58; Birnie and Boyle, *op. cit.*, n. 14, at pp. 190–197; and on "third-generation" rights in general, see P. Alston, "A Third Generation of Solidarity Rights: Progressive Development or Obfuscation of International Human Rights Law" (1982) 29 *Netherlands International Law Review* 307.

[60] See, for example, the 1946 International Convention for the Regulation of Whaling, 161 UNTS 72; the 1968 African Convention on the Conservation of Nature and Natural Resources, 1001 UNTS 3; CITES, the Bonn Convention and the Berne Convention. Note also Article 4 of the World Heritage Convention.

[61] Birnie and Boyle, *op. cit.*, n. 14, at p. 194.

[62] Goa Guidelines on Intergenerational Equity, (1988) 18 *EPL* 190–191; E. Brown Weiss, *In Fairness to Future Generations* (1989). For an interesting domestic law application of the intergenerational equity concept, see *Minors Oposa v. Secretary of the Department of Environment and Natural Resources* (1994) 33 ILM 173.

references to the interests of future generations amount to little more than an affirmation of one aspect of the motivations underlying the treaties in question.[63]

It is also important to recognise that a heavy emphasis upon human rights perspectives in this context may give rise to concern that an unduly anthropocentric approach to biodiversity conservation is being adopted. The question whether nature itself is to any extent seen as being an intended beneficiary of international legal measures for the conservation of biological diversity is a complex one which cannot really be answered without an investigation of the philosophical foundations of the concept within the international legal system.

Philosophical Foundations of the Biodiversity Concept

It is, perhaps, characteristic of lawyers to direct their energies primarily towards the analysis of the substance and structure of the legal principles they create and apply, and to display rather little interest in any discussion of the underlying philosophical motivations which have generated the need for such principles in the first place. In the field of human rights, for example, it is sometimes argued that, as there is plainly a consensus in favour of a regime of protection, little purpose is served by investigation of the theoretical basis for the recognition of such rights.[64] This argument can be convincingly refuted, however, not least on the grounds that the precise nature and content of the protective regime must of necessity be shaped and informed by the underlying philosophical considerations which prompted its adoption. Indeed, in the absence of broad agreement concerning such fundamentals, the apparent consensus may ultimately collapse once the process of practical implementation of the regime is embarked upon. Such arguments apply with equal force in the context of environmental protection.

It is, admittedly, notoriously difficult to isolate with precision the political motivations underlying the adoption by governments of international legal measures in any context, and still more difficult to relate these motivations to any coherent, underlying philosophical theory. Where protection of the environment is concerned, it may be particularly unwise to expect these factors to be clear, consistent or uniform.[65] Nevertheless, it should be possible to identify particular instincts and attitudes which have played their part in spurring governments to action, and to relate these to broad currents of thought within the evolving fields of environmental ethics and conservation theory. In the case of international instruments for the protection of human rights, despite the wide variety of approaches to questions of underlying philosophy,[66] it is possible to detect as a unifying basic theme the recognition of the inherent dignity, in the sense of worthiness, of the individual human

---

[63] That should not be thought to be without significance, however, since the interpretation of the treaty will be undertaken in the light of its object and purpose: Article 31, 1969 Vienna Convention on the Law of Treaties. See also the text accompanying n. 166.

[64] For a discussion of such issues, see J. Donnelly, *Universal Human Rights in Theory and Practice* (1989) and, in reply, M. Freeman, "The Philosophical Foundations of Human Rights", (1994) 16 *Human Rights Quarterly* 491.

[65] International legal measures in the field of environmental protection are, perhaps, particularly likely to constitute compromise solutions, not least because of the sharply divergent interests of States, the technical complexity and scientific uncertainty surrounding many environmental problems, and the wide range of interest groups seeking to influence governmental opinion.

[66] For a discussion of these issues, see J.J. Shestack, "The Jurisprudence of Human Rights" in T. Meron (ed.), *Human Rights in International Law* (1984).

being.[67] Plainly, if action is to be taken to protect or conserve elements of the natural world, that must equally be based upon the recognition of some form of value in their continued existence. In order to be able to identify and assess the value of nature, however, it is first necessary to analyse and understand the nature of value.

The Nature of Value

Fortunately, the discipline of environmental ethics is by now sufficiently well developed to offer a workable theoretical framework for the analysis of questions of environmental value. Although usage of terminology in this respect is far from being absolutely uniform, philosophers commonly categorise the value of elements of the natural world as being instrumental, inherent or intrinsic.[68] The instrumental value of a particular entity lies in the use to which it may be put, as in the case of a fish being consumed for food or a tree cut down to provide timber. Inherent value, by contrast, is the value that an entity possesses on account of being prized for itself, rather than for its utility. In the world of human artefacts, a work of art would represent the classic example, and, where nature is concerned, inherent value might be derived not only from aesthetic, but equally from cultural or religious, considerations.[69] Both instrumental and inherent value, it will be noted, depend upon the existence of an external valuer or beneficiary. Intrinsic value, by contrast, is understood to be the value that entities have of themselves, for themselves, and therefore does not presuppose the existence of any external valuer at all.[70] It is in fact possible to detect a recognition of each of these distinct notions of value in the developing body of international law concerning environmental protection.

Early attempts at nature conservation can be seen to be primarily motivated by narrowly utilitarian and material considerations; that is, the value of wildlife was

---

[67] See, e.g. the preambles to the 1948 Universal Declaration on Human Rights, UNGA Res.217A (III); the 1966 International Covenant on Civil and Political Rights; the 1966 International Covenant on Economic Social and Cultural Rights, 993 UNTS 3; and the 1966 International Convention on the Elimination of All Forms of Racial Discrimination, 660 UNTS 195. See also Part 7 of the Declaration on Principles Guiding Relations Between Participating States, Final Act of the 1975 Helsinki Conference on Security and Cooperation in Europe, 14 ILM 1292, at p. 1295. Note that the concept of the "inherent dignity" of the human person referred to in these instruments should, in accordance with the terminology discussed below, be understood to involve a recognition of the intrinsic value of the individual human being.

[68] For an instructive account, see F. Matthews, *The Ecological Self* (1991), Chapters 3, 4. On the question of terminology, note that philosophers such as Tom Regan and Paul Taylor have been criticised for using the term "inherent" in the sense which is more commonly conveyed by the term "intrinsic" – see E.C. Hargrove, Preface, pp. xvii–xviii, in E.C. Hargrove (ed.), *The Animal Rights/Environmental Ethics Debate* (1992). Hargrove himself dispenses with the need for the term "inherent" by referring to anthropocentric and non-anthropocentric forms of both instrumental and intrinsic value – see his "Foundations of Wildlife Protection Attitudes", especially at pp. 169–175, *ibid.* Alternative terminological schemes are frequently employed to reflect broadly similar ideas, e.g. the notions of commodity, amenity and moral values discussed by B. Norton, "Commodity, Amenity and Morality: the Limits of Quantification in Valuing Biodiversity" in E.O. Wilson (ed.), *Biodiversity* (1988).

[69] Values derived from cultural or religious considerations may, of course, be either instrumental, as where particular societies are in some material way dependent upon certain species for nutritional or other requirements, or inherent, as where particular creatures are accorded a symbolic or totemic significance. Different creeds and cultures may also recognise elements of the natural world as possessing intrinsic value.

[70] Note that some philosophers dispute that value can exist independently of conscious valuers – see, e.g. J. Baird Callicott, "Animal Liberation, A Triangular Affair", in Hargrove (ed.), *op. cit.,* n. 68 at p. 48.

seen to be largely, if not exclusively, instrumental in character. This is apparent from the very title of such agreements as the 1902 Convention for the Protection of Birds Useful to Agriculture,[71] as well as from the preamble to the 1900 Convention for the Preservation of Wild Animals, Birds and Fish in Africa,[72] which spoke of the desire to protect species which were "useful to man or inoffensive". Firmly in keeping with this essentially anthropocentric perspective, these conventions excluded from protection, and even provided for the destruction of, species which were judged harmful (*nuisibles*) to human interests. These included, under the 1900 Convention, lions, leopards, crocodiles and various species of snakes and birds of prey.[73] Similarly treated under the 1902 Convention were eagles, hawks, pelicans, herons and pigeons.[74] Human interests, of course, also lay at the heart of early attempts at conservation of directly exploited wildlife species, such as whales, seals and fish.[75]

It is perhaps appropriate to observe that few of the conventions in question could be said to have achieved conspicuous success, even when judged by reference to their own rather unenlightened objectives. The lack of effective enforcement mechanisms and other structural weaknesses were no doubt a major factor, but the very narrowness of their focus and their lack of adequate grounding in environmental science must also have played their part. The infamous "blue whale unit", for example, adopted under the International Convention for the Regulation of Whaling for the purpose of determining permissible catch levels, may well have seemed a perfectly reasonable commercial currency through which to regulate the market in whale products,[76] but it bore no relation to any form of ecological reality and wholly failed to prevent the continued depletion of whale stocks.[77] It is also ironic to note that some of the species excluded from the protective beneficence of the earlier conventions are now numbered amongst the most strictly protected wildlife taxa.[78]

A second consideration, less material though equally anthropocentric in character, which came to achieve recognition in international conservation treaties of a slightly later era, was the aesthetic value of the natural world. From this perspective, nature was to be regarded, in much the same way as works of art, as having value of an inherent kind. This notion of the aesthetic value of nature has been accorded a pivotal place in the writings of certain North American conservation theorists[79] and featured particularly prominently in the 1940 Western Hemisphere Convention,[80]

---

[71] 102 BFSP 63.

[72] 94 BFSP 715. Note that this source indicates that the Convention was never ratified but that effect was given to many of its provisions in various British overseas territories.

[73] Article II(13), (15) and Table 5.

[74] Article 9 and List 2.

[75] For discussion of the treaties in question see Birnie and Boyle, *op. cit.*, n. 14, Chapters 11 and 13; Lyster, *op. cit.*, n. 5, Chapters 2, 3.

[76] Particularly whale oil – the system was, of course, based on the oil-producing potential of different cetacean species.

[77] Lyster, *op. cit.*, Chapter 2; G. Rose and S. Crane, "The Evolution of International Whaling Law", in P. Sands (ed.), *Greening International Law* (1993).

[78] See, e.g. CITES, Appendix I.

[79] See especially, Hargrove, *op. cit.* n. 68 and also in "An Overview of Conservation and Human Values", in D. Western and M.C. Pearl, *Conservation for the Twenty-First Century* (1989).

[80] 1940 Convention on Nature Protection and Wildlife Preservation in the Western Hemisphere, 161 UNTS 229.

which articulated the desire to "protect and preserve scenery of extraordinary beauty ... and natural objects of aesthetic ... value".[81] Similarly, the 1972 World Heritage Convention[82] defined the world natural heritage to include "natural features consisting of physical and biological formations ... of outstanding universal value from the aesthetic ... point of view" and "natural sites ... of outstanding value from the point of view of ... natural beauty".

The "aesthetic value" formula has been repeated in the preambles to a large number of later conventions, though it is noticeable that it usually appears in harness with a number of other justifications of an instrumental character, amongst which "economic" and "scientific" are usually prominent.[83] Although the aesthetic value of nature has undoubtedly played a significant part in the evolution of conservation theory and is still regarded by some philosophers as the central plank of environmental ethics,[84] it is difficult to regard it as an independently sufficient foundation for modern, global measures of biodiversity protection, given its rather cosy, middle-class, Northern hemisphere orientation. For those whose daily struggle is for survival, the aesthetic appeal of the natural world is unlikely to loom large on the list of life's priorities. It is undeniably the case that developing States are becoming increasingly aware of the economic potential of ecotourism, which plainly depends to a considerable extent upon the aesthetic appeal of the natural world to wealthy foreigners, but the real interest of such States is likely to lie in the extent to which this appeal can effectively be capitalised, and should therefore be taken to reflect an instrumental value of wildlife, albeit possibly of a non-consumptive kind.[85]

It was doubtless for such reasons that the World Conservation Strategy chose to emphasise once again the practical, utilitarian values of wildlife conservation, seeking to overturn the received wisdom that environmental protection and economic development are mutually antithetical – indeed the whole thrust of the document was to insist that the two were not merely compatible, but that the former was in fact a prerequisite to the latter. In the long term, economic development would be dependent upon the continued viability of natural systems and the ongoing availability of consumable natural resources. Hence "sustainable development" became the pivotal concept in the new thinking.[86]

This modern form of utilitarianism was, however, to be much broader, better informed and more far-sighted than that which had inspired the conventions adopted at the beginning of the century. Species-rich ecosystems, such as the tropical forests, were to be preserved as much for their potential as their current value to man, with the possible pharmaceutical value of hitherto unrecorded plant species being commonly cited as an example.[87] Equally, there was now a much greater awareness of the functional interdependence of natural communities and the need to preserve an appropriate ecological balance. Thus the desire to conserve a particular

---

[81] Second preambular paragraph. Note also Articles 1(1), 1(3) and 7.
[82] Article 2.
[83] See, e.g. CITES, the Bonn and Berne Conventions and the Biodiversity Convention itself.
[84] Hargrove, *op. cit.*, n. 79.
[85] Of course, even non-consumptive enjoyment of wildlife, such as bird or whale watching, can be damaging if pursued without due regard to the interests of the species under observation.
[86] World Conservation Strategy, Introduction.
[87] *Ibid.*, Section 3; Wilson, *op. cit.*, n. 4, Chapter 13; N.R. Farnsworth, "Screening Plants for New Medicines", in Wilson (ed.), *op. cit.*, n. 10.

species directly exploited by man might be seen to require also the protection of other species or of habitats upon which it was ecologically dependent.[88]

Considerable doubts must remain, however, as to the extent to which this modified conservation ethic will be sufficient to enlist the governments of developing States as enthusiastic recruits to the cause of environmental protection. To the contrary, there have already been attempts to dilute or disregard the language of existing treaty obligations on the grounds that it is unnecessarily strict for sustainable development purposes.[89] To some governments, indeed, the sustainable development justification for conservation (itself sometimes described as the "new imperialism") may appear to be little more than a cloak for cultural or ethical preferences which they do not necessarily share. It is certainly true that there are many, particularly perhaps in the developed world, who would argue that the real, ultimate justification for conservation does indeed derive from an ethical argument which would regard elements of the natural world as possessing intrinsic value, and therefore as falling within the scope of "moral considerability" in their own right.[90]

Such considerations have clearly played some part, moreover, in the process leading to the adoption of international conservation instruments. The preamble to the Berne Convention, for example, expressly recognises that European wildlife constitutes a natural heritage possessing intrinsic, as well as other kinds of, value. It is, of course, to be remembered that the Berne Convention is a treaty concluded principally[91] between the developed States of the European region; but acknowledgment of the intrinsic value of the natural world is also found in instruments of a global scope. Indeed, the point is brought out with particular clarity in the World Charter for Nature,[92] which states in its preamble that

> Every form of life is unique, warranting respect regardless of its worth to man, and, to accord other organisms such recognition, man must be guided by a moral code of action.

The Charter therefore seeks through its substantive principles to combine a recognition of the intrinsic value of wildlife with that of its functional utility to humans.

The World Charter for Nature certainly cannot be regarded as the creation of the developed States exclusively. It is significant to recall in this context that the project to draft the Charter had originally been proposed by President Mobutu of Zaire at the 12th General Assembly of IUCN in 1975.[93] Developing States were prominent in sponsoring the various resolutions proposing adoption of the Charter, which was ultimately approved by 111 votes to 1, with only 18 abstentions.[94] Furthermore, in

---

[88] B.G. Norton, "The Cultural Approach to Conservation Biology", in Western and Pearl, *op. cit.*, n. 79. Note particularly in this context the notion of "keystone" species: Wilson, *op. cit.*, n. 4, Chapter 9; Primack, *op. cit.*, n. 9, at pp. 43-48.

[89] See, for example, the (unsuccessful) attempt by a group of African states to circumvent the language of Article 3(3) (c) of CITES, which establishes that trade in Appendix I specimens may not be permitted if "for primarily commercial purposes". It was argued that the real issue should be whether trade was sustainable, rather than commercial: CITES, DOC.8.49.

[90] The term "moral considerability" is commonly attributed to Kenneth Goodpaster: see "On Being Morally Considerable" (1978) 75 *Journal of Philosophy* 308.

[91] But not exclusively. The Convention is open for acceptance by Member States of the Council of Europe, non-Member States which participated in its elaboration and the EEC. After its entry into force, other States might be invited to accede: Articles 19, 20. Among non-European states, Burkina Faso and Senegal are currently parties.

[92] See note 23.

[93] See Burhenne and Irwin, *op. cit.*, n. 23, at p. 14.

[94] *Ibid.*, p. 16. The United States was the sole dissentient and the Amazonian nations formed a significant bloc among the abstaining group. For the full voting roster, see p. 95.

all the many comments formulated by States with respect either to the document as a whole, or to its individual provisions, there appears to be no trace of any specific objection to the notion of the intrinsic value of wildlife.[95] Some reservations may be implicit, however, in a somewhat cryptic statement issued by Brazil on behalf of eight Amazonian nations[96] to the effect that

> the preamble contains philosophical and doctrinal principles which do not enjoy unanimous support, since from different points of view they are and will be considered heterodox, unfounded or simply irrelevant. They are therefore not likely to contribute to the protection of nature, which can only be founded on pragmatism if it is to have any real or practical effect.

This comment could equally have been directed at other aspects of the preamble, however.[97] Certainly, there can be little doubt that the Charter as a whole embraces a clear recognition both of anthropocentric and non-anthropocentric values in nature.

It has nevertheless been argued that the international community has subsequently backtracked from this pluralist approach with the forceful re-emergence of utilitarian priorities at the Rio Earth Summit.[98] Indeed, it is has been suggested that Principle 1 of the Rio Declaration,[99] which states that human beings are at the centre of concerns for sustainable development,

> represents a triumph of unrestrained anthropocentricity ... The word "nature" appears nowhere else in the text and there is no recognition of the intrinsic value of natural ecosystems and wild species.[100]

While there is doubtless force in these observations, it would be unwise to draw unwarranted conclusions from them. It is to be remembered in this context that the Rio Conference was not solely concerned with the conservation of nature. Rather, as its official title plainly indicates,[101] it had as its primary focus the twin themes of environment and development and the interrelationship between them. Though this may in itself have been a cause for regret to some, it was inevitable that the statement of general principles ultimately articulated by such a Conference would be similarly oriented. Since the very concept of development signifies the economic and social advancement of human beings,[102] it is self-evident, indeed almost tautologous, to suggest that human beings are at the centre of developmental concerns, whether sustainable or otherwise. Principle 1 therefore tells us little or nothing about the place of nature in the international system as a whole.

---

[95] *Ibid.*, Part I, sections C and D.

[96] *Ibid.*, p. 38. Brazil was speaking on behalf of itself, Bolivia, Colombia, Ecuador, Guyana, Peru, Suriname and Venezuela.

[97] Colombia, for example, had previously commented adversely upon preambular references to the theory of evolution and other matters: *ibid.*, pp. 41, 43. Note also the observation of the French government, couched in very similar terms to the later Brazilian statement but apparently directed at the 1980 draft of the seventh preambular paragraph: p. 48.

[98] M. Pallemaerts, "International Environmental Law from Stockholm to Rio: Back to the Future", in P. Sands (ed.), *Greening International Law* (1993).

[99] The Rio Declaration on Environment and Development, UN Doc.A/CONF.151/5 (7 May 1992).

[100] Pallemaerts, *op. cit.*, n. 98, at pp. 12–13. The single reference is also in Principle 1, to the effect that human beings are "entitled to a healthy and productive life in harmony with nature".

[101] Namely, the United Nations Conference on Environment and Development.

[102] Article 1 of the 1986 Declaration on the Right to Development, UNGA Res. 41/128, in fact refers to economic, social, cultural and political development. Note also that Article 2 states that "the human person is the central subject of development" and the Rio Declaration is therefore simply restating established principles in this context.

More worrying at first sight is Principle 4, which states:

> In order to achieve sustainable development, environmental protection shall constitute an integral part of the development process and cannot be considered in isolation from it.

On one reading, this might be taken to subordinate environmental protection entirely to the needs of the development process, but a more careful consideration suggests that this is unlikely to be the intention. To assert that environmental protection is an integral part of the development process, and cannot be considered in isolation from it, is not to deny that the former may also be an end itself, or that it should not equally be an integral part of many other governmental functions. One of the fundamental objectives of the environmental movement, recognised in the World Charter for Nature itself,[103] has in fact been to secure the integration of environmental considerations into the entire range of policy making processes, and Principle 4 of the Rio Declaration should therefore be regarded as an affirmation of that objective in the developmental context. Far from denying the wider significance of environmental protection, the Declaration actually goes on to confirm it in one particular context through its recognition that "Peace, development and environmental protection are interdependent and indivisible."[104]

It would indeed be surprising if the Rio Conference were thought to have undermined or abandoned the notion of intrinsic value, given that the concept was expressly reaffirmed in the very first paragraph of the preamble to the Biodiversity Convention, which was signed by some 150 nations at that very conference. The precise wording of the paragraph is worthy of note. It states,

> The Contracting Parties,

> Conscious of the intrinsic value of biological diversity and of the ecological, genetic, social, economic, educational, cultural, recreational and aesthetic values of biological diversity and its components...

It may be significant that intrinsic value is not merely listed here as one in a long list of values, but rather set apart as if to rank equally with all the various forms of instrumental and inherent value which follow. While the main emphasis of the convention is undoubtedly upon the instrumental value of nature, this bold recognition of its intrinsic value is a particularly striking feature.[105]

There is, of course, no reason why multiple justifications for the conservation of biodiversity should not be offered, provided that there is no element of internal tension between them. Although it may be going too far to suggest that the intrinsic

---

[103] This is expressly articulated in paragraph 7, but in fact underlies the whole of Part II of the Charter. Note also paragraph 16.

[104] Principle 25. It is important to recall that the principles of construction in accordance with which the relationship between successive legal instruments is to be determined are not reducible to a simple proposition that the later in time prevails. Article 30 of the Vienna Convention on the Law of Treaties, for example, treats that rule as applicable only where the two instruments relate to the same subject matter and are substantively incompatible. The World Charter for Nature and the Rio Declaration do not satisfy these criteria. While the Charter focuses upon the specific question of man's responsibilities with regard to nature, the Declaration deals with the wider issue of the place of environmental protection in the context of developmental concerns. In the absence of a specific and inescapable conflict between the provisions of the two instruments, mere differences of emphasis are to be resolved in accordance with the principle *specialia generalibus derogant, generalia specialibus non derogant*.

[105] Note that the *Explanatory Guide* to the Convention, prepared in 1994 by the IUCN Environmental Law Centre, describes this a "very important innovation", which "may be seen as acknowledging the right of all components of biodiversity to exist independent of their value to humankind": see L. Glowka et al., *A Guide to the Conservation on Biological Diversity* (1994), at p. 9.

and instrumental value justifications for conservation are inherently incompatible, it is undeniable that there is at least the potential for conflict between them. This becomes much more clearly apparent when the complex question of the locus of value – that is, the precise identification of those entities that can be considered to be the subjects or repositories of these various forms of value – is explored in more detail.

The Locus of Value

Although it is relatively common to speak loosely in terms of the exploitation of wildlife species or of natural ecosystems, it is plain that this is often little more than a form of shorthand. In reality, neither species nor ecosystems can be eaten, burned for fuel or converted into garments. Rather, it is individual animals or plants that are used for such purposes. This point is brought out with particular clarity in the CITES treaty, which, although its title incorporates the phrase "Trade in Endangered Species", refers more precisely throughout its substantive provisions to "trade in specimens" of such species.[106] A specimen is defined to mean an individual animal or plant, whether alive or dead, and also to include certain parts or derivatives thereof.[107] This involves the recognition that tortoiseshell, rhino horn, elephant ivory and similar derivatives may possess a commercial value of their own. With the benefit of advances in scientific knowledge and techniques, even genetic material can now be put to a wide variety of valued uses.[108] Instrumental value should therefore be seen to reside primarily in individual plants and animals, as well as in their parts and derivatives in many cases. The main relevance of the species in this context is that it represents the total current population of individuals and, through their shared genetic inheritance and resultant breeding capacity, the prospect of a continuing supply of further individuals into the future. Similarly, as far as the ecosystem is concerned, it constitutes both the physical location and the functional unit within which this process will occur. Thus, from a purely instrumental perspective, it is species and ecosystems which are to be conserved and protected in order that the value of individual wildlife specimens may be realised. Finally, it should not be overlooked that instrumental value can be found not only in the living but also in the abiotic elements of the natural world – indeed the importance of soil, air and water both for direct human utilisation and with regard to the maintenance of natural processes and life support systems can hardly be overstated.[109]

The locus of inherent value is perhaps a more complex question. From an aesthetic point of view, inherent value would appear to reside in any entity which is capable of being the subject of aesthetic appreciation. Individual plants and animals would certainly qualify, as might more complex assemblages of the same – though it is to be emphasised that we are probably thinking here in terms of aesthetically appreciable assemblages, such as forests, flocks, herds, vistas and landscapes, rather than more ecologically meaningful collectivities like species, communities and ecosystems. Of course, a knowledge and understanding of the latter will again be required in order to adopt the conservation measures necessary to guarantee a continuing supply of aesthetically valued entities, but species and ecosystems are

---

[106] Such references occur throughout Articles 1–8 of the treaty.
[107] Article 1.
[108] For examples, see WRI, *op. cit.*, n. 7, at pp. 4–5.
[109] World Conservation Strategy, Section 2.

arguably not the subject of inherent value in themselves.[110] It is, however, important to note once again that non-living elements, such as mountains, rivers and lakes, together with less permanent and substantial, or even wholly intangible, manifestations, such as clouds, sunsets and rainbows, may also be judged to possess inherent value.

Most complex of all, however, is the question of the precise locus of intrinsic value. If this concept is defined to mean the value which an entity has of itself for itself, it is plain that we are envisaging a "self" of a rather special kind. Whilst a mountain or a painting might be said to be valuable in or of itself, it is unlikely to be suggested to have value for itself. The task is therefore to identify the kinds of entity capable of possessing this form of value.

The question has been considered in a number of recent works, most notably perhaps in Freya Matthews's The Ecological Self.[111] In Matthews's view, if intrinsic value is defined as the value which an entity possesses of itself for itself, then the possession of such value must be limited to entities that are "self-realising" – that is, entities which possess self-interest and function in pursuit of that self-interest. This self-interest derives from the special purpose, or *telos*, of self-perpetuation exhibited by the entities in question:

> A self-realizing being ... is one which, through its nature, defines a self interest. What happens to it matters because it is actively seeking to preserve its own integrity, its identity.[112]

On this view, the paradigmatic example of self-realisation is the individual organism, on the grounds that it "satisfies its own energy requirements, grows, repairs or renews its own tissues and reproduces itself".[113] According to Matthews, the categorical difference between organisms and other entities is that

> organisms, by their activity, define an interest, or a value, namely their own. Other systems may be designed and constructed for the purpose of producing a particular effect, but this purpose is definable only relative to the intentions of the designer. In a deserted post-holocaust world, mechanical devices such as typewriters and bicycles, and even cybernetic devices such as ... guided missiles, would return to the estate of mere objects, their purpose having vanished from the world with the agents who designed them. Such systems do not embody their purpose in themselves: their purpose is super-added to their existence, and can be subtracted from it without the things themselves ceasing to exist, or ceasing to be the things they intrinsically are. Organisms differ from such systems in as much as they do embody their purpose in themselves; for an organism, to exist is to possess self-interest. Unlike the machine, which can exist as a durable material structure independently of fulfilling the purpose for which it is made, the existence of the organism coincides with its purpose, for its purpose is to exist.[114]

---

[110] Note, however, that some philosophers whose writings have helped to shape the present author's approach take the view that the biosphere as a whole may be of inherent value: see R. Attfield, *The Ethics of Environmental Concern* (2nd edn. 1991), at p. 159.

[111] F. Matthews, *The Ecological Self* (1991).

[112] *Ibid.*, p. 103. There is, of course, no suggestion that this "purpose" need in any sense be conscious. For discussion of the various other modern philosophers who have utilised the notion of *telos*, see p. 175, and for comparisons between the concept of self-realisation and (i) the medieval concept of *conatus*, employed by Spinoza and others (see pp. 109–116) and the more modern concept of *autopoiesis*, see p. 173. Matthews herself also draws heavily upon the analysis of the biologist Jacques Monod, who characterised all life by reference to the three criteria of teleonomy, autonomous morphogenesis and reproductive invariance: see pp. 99–100.

[113] *Ibid.*, p. 98. Note, however, the view that it is genes, rather than organisms, which are the true replicators: R. Dawkins, *The Extended Phenotype* (1982), at pp. 97–117.

[114] Matthews, *ibid.*, pp. 100–101.

The notion that intrinsic value resides in organisms – that is, in individual plants and animals, has been accepted by many other environmental ethicists, including Rodman,[115] Goodpaster,[116] Taylor,[117] Rolston[118] and Attfield[119] and it is not easy to see on what basis it could be seriously challenged.[120]

The question remains however, as to whether other entities – species, ecosystems or perhaps the biosphere itself – should also be regarded as possessing intrinsic value. Certainly, the notion of the intrinsic value of species has its proponents. Rolston, possibly the leading advocate, invites us[121] to consider whether

> singular somatic identity conserved is the only process that is valuable. A species is another level of biological identity reasserted genetically over time ... The species line is the *vital* living system, the whole, of which individual organisms are the essential parts. The species defends a particular form of life, pursuing a pathway through the world, resisting death (extinction) , by regeneration maintaining a normative identity over time. The value resides in the dynamic form....

This view is plainly likely to commend itself to many scientists, not least because the species is often regarded as "the grail of systematic biology",[122] but it is open to serious objections. As Ferre has pointed out,[123]

> species do not literally 'do' the things that Rolston's language attributes to them; e.g. they do not 'use' individuals, 'defend' a form of life or 'resist' extinction. Particular organisms defend their lives, and in so doing enhance the possibility that future organisms like them will continue enjoying the good of their kind. Species, however, do not defend themselves. To say they do commits ... the "fallacy of misplaced concreteness".[124]

Matthews[125] agrees that species are classes rather than systems, abstract rather than concrete entities, and that "their continuity of form is the *outcome* of the strivings of individual organisms" rather than an aim of the species itself. Attfield is of a similar view.[126] These reservations regarding the attribution of intrinsic value to species are not diminished by widespread disagreement as to the precise nature and definition of the species

---

[115] J. Rodman, "Four Forms of Ecological Consciousness Reconsidered", in D. Scherer and T. Attig, *Ethics and the Environment* (1983).

[116] K. Goodpaster, "On Being Morally Considerable" (1978) 75 *Journal of Philosophy* 308.

[117] P. Taylor, "The Ethics of Respect for Nature", in Hargrove (ed.), *op. cit.*, n. 68. See also his *Respect for Nature* (1986).

[118] H. Rolston III, "Value in Nature and the Nature of Value", in R. Attfield and A. Belsey (eds.), *Philosophy and the Natural Environment* (1994).

[119] R. Attfield, *op. cit.*, n. 110.

[120] One possible argument might derive from the view which would downplay the significance of individual organisms as merely the repositories of genes or the temporary and dispensable representatives of species; thus R. Dawkins, *The Selfish Gene* (1976, paperback edn. 1989) states at p. 266 that "the individual body, so familiar to us on our planet, did not have to exist. The only kind of entity that has to exist in order for life to arise, anywhere in the universe, is the immortal replicator", that is, the gene. But other scientists have taken a quite different view of the balance of importance between genes and organisms, re-emphasising the importance of the latter – see, for example, A. Rayner's review of B. Goodwin, *How the Leopard Changed its Spots* (1994), in *New Scientist,* 7 January 1995, at p. 36 – and even Dawkins has conceded that there "really *is* something pretty impressive about individual organisms": *op. cit.*, n. 113, p. 250.

[121] *Op. cit.*, n. 118, at pp. 20-21. See also Taylor, *op. cit.*, n. 117.

[122] Wilson, *op. cit.*, n. 4, at p. 38.

[123] F. Ferre, "Highlights and Connections" in Attfield and Belsey (eds.), *op. cit.*, n. 118, at p. 229.

[124] This concept is attributed to A.N. Whitehead, *The Function of Reason* (1929).

[125] Matthews, *op. cit.*, n. 111, at pp. 179–180.

[126] Attfield, *op. cit.*, n. 110, at pp. 150–151.

[127] For discussion, see D.S. Woodruff, "The Problem of Conserving Genes and Species", in Western and Pearl (eds.), *op. cit.*, n. 79, at pp. 80–82.

concept, leading some to doubt its very validity,[127] or by the fact that even enthusiasts for the biological species concept recognise that it has significant limitations.[128]

Where ecosystems are concerned, the picture is also unclear. Here Matthews[129] joins Rolston[130] in believing that ecosystems may "qualify, up to a certain point, as selves" and so possess intrinsic value, largely on the grounds that "*by their own efforts*, they procure the energy for their self maintenance". The main reason for the qualified form in which this view is articulated appears to relate to the inevitable problems of identification and demarcation of ecosystems – how is it to be determined where one ends and another begins? Such problems presumably cannot arise with regard to the biosphere as a whole, however, and supporters of the Deep Ecology movement would indeed seek to ascribe intrinsic value at such a level.[131] Others remain unconvinced, however. As Attfield[132] points out,

> Certainly everything which is of value ... is located in the biosphere, and the systems of the biosphere are necessary for the preservation of all these creatures. But that does not give the biosphere or its systems intrinsic value. Rather, it shows them to have instrumental value, since what is of value in its own right is causally dependent on them. As to the biosphere as a whole, with all its richness and beauty, those features which it has but its components lack suggest that its value is inherent.

If this argument is accepted, it would seem to follow that abiotic elements of the natural world cannot possess intrinsic value.

Not surprisingly, legal instruments which have made reference to the intrinsic value concept have not explored the question of the precise locus of value in any depth. The Berne Convention rather vaguely attributes intrinsic value to "wild flora and fauna".[133] The World Charter for Nature, a little more helpfully, states that "every form of life" warrants respect, and adds that, "to accord other *organisms* such recognition, man must be guided by a moral code of action".[134] This seems plainly to call for respect to be accorded at the level of the individual organism, in keeping with the philosophical arguments outlined above. Unfortunately, the matter is confused, rather than clarified, by the unofficial Commentary to the Charter, which states[135] that:

> respect for nature must above all take the form of respect for all forms of life. The words "all forms of life" should be understood to express the idea that all living species should be conserved and to envisage the preservation of genetic diversity. It does not mean that the life of each individual animal or plant should be protected.

This passage seems open to two distinct criticisms. The first is that it fails to address the fact that the text actually refers to respect for organisms rather than the conserva-

---

[128] See Wilson, *op. cit.*, n. 4, who concedes at p. 45 that "the biological-species concept has chronic deep problems". Note that it could equally be argued, however, that "the organism is a concept of dubious utility, precisely because it is so difficult to define satisfactorily": Dawkins, *op. cit.*, n. 113, at p. 253.

[129] *Op. cit.*, n. 111, at pp. 129–135.

[130] *Op. cit.*, n. 118, at pp. 22–30.

[131] See generally A. Naess, "The Shallow and the Deep, Long-Range Ecology Movement" (1973) 16 *Inquiry* 95; B. Devall and G. Sessions, *Deep Ecology: Living as if Nature Mattered* (1985). See also A. Leopold, *A Sand County Almanac* (1949); J. Lovelock, *Gaia, a New Look at Life on Earth* (1979). For an approach which would extend intrinsic value beyond Nature as it is manifest on the planet earth, see K. Lee, "Awe and Humility: Intrinsic Value in Nature. Beyond an Earthbound Environmental Ethics", in Attfield and Belsey (eds.), *op. cit.*, n. 118, at p. 89.

[132] *Op. cit.*, n. 110, at p. 159.

[133] Fourth preambular paragraph.

[134] Fifth preambular paragraph: see the text accompanying n. 192.

[135] Burhenne and Irwin, *op. cit.*, n. 23, at p. 136.

tion of species. The second problem is that it appears to rest upon an assumption that, if the concept of respect for nature were to apply at the level of the individual plant or animal, that would inevitably involve the consequence that their individual lives would in every case have to be protected. This assumption seems entirely unwarranted. It is perfectly true, as Matthews has pointed out,[136] that

> If something is characterised as intrinsically valuable then it is simply analytic that, other things being equal, it should not be destroyed or prevented from existing. It has a prima facie claim to our moral consideration.

The claim, however, is merely a prima facie claim and applies only when "other things" are "equal". It can therefore clearly be overridden by other factors, which may include a consideration of the instrumental value of the entity concerned for human or other purposes.[137]

The idea that individual organisms are the repositories of intrinsic value may derive further support from the Biodiversity Convention itself. It is worthy of note that the preamble actually ascribes intrinsic value to "biological diversity", rather than to either species or organisms as such. However, "biological diversity" is itself defined[138] to mean "the variability among living organisms from all sources", and it is not easy to see how such variability can be *intrinsically* valuable unless the individual organisms themselves possess that quality.[139] Furthermore, the definition goes on to state that it "includes diversity within species" and this inevitably places the spotlight upon individual organisms, since "every individual of a species that originates from sexual reproduction has a slightly different combination of genes".[140] Individuals, therefore, in a very real sense constitute the "repositories of genetic diversity".[141] It could be the case that the notion of diversity within a species might be intended to refer only to the possible existence of subspecies, varieties and distinct populations,[142] but as E.O. Wilson has pointed out,[143] such categories are essentially arbitrary, since, unlike the species taxon, they lack a single, defining criterion. Subspecies and races really amount to little more than groups of individuals displaying certain randomly selected similarities. It would seem to follow that once intrinsic value is seen to exist in diversity at levels below that of the species, then the individual organism becomes the prime focus of attention. Finally, if the Convention, through its express reference to the intrinsic value of diversity, is to be

---

[136] *Op. cit.*, n. 111, at p. 118.

[137] See the section headed "The Measure of Value", below.

[138] Article 2.

[139] Indeed, it is not easy to see how "diversity" as such can possess intrinsic value at all, since it is not a "self". It may possess inherent value, as well as instrumental value for other "selves", particularly ecosystems, if they are capable of being so regarded: see generally Attfield, *op. cit.*, n. 110, at pp. 149–150.

[140] Draft Explanatory Guide, *op. cit.*, n. 105, at p. 24, Box 5. Each individual is, therefore, unique. This genetic variation commonly stems from mutations in chromosome structure; for accessible accounts of this process and its consequences, see C. Tudge, *Last Animals at the Zoo* (1991), pp. 76–80 and Wilson, *op. cit.*, n. 4, at pp. 157–162.

[141] Wilson, *ibid.*, p. 159.

[142] Note, however, that it has been suggested that, by analogy with CITES and other conventions, subspecies, varieties, etc., might be understood to be included within the term "species", which is not defined in the Biodiversity Convention and may not necessarily be used in a strictly scientific sense: see the Draft Explanatory Guide, *op. cit.*, n. 105, p. 1 Box 1. If this is correct, then the diversity exhibited by the existence of subspecies is already catered for within the notion of "diversity between species".

[143] *Op. cit.*, n. 4, at pp. 64–69.

interpreted as recognising the intrinsic value of individual organisms, then by parity of reasoning it might equally be taken to recognise that of species and ecosystems, despite the philosophical differences of opinion on this issue.[144]

While the conclusion that species and ecosystems possess intrinsic value may be regarded as tolerable, if inconvenient, in certain quarters, the view that each individual organism possesses similar value is likely to achieve less ready acceptance, since it might be thought to lead to a much more serious practical limitation upon the right of States to exploit their natural resources, which is one of the most highly cherished entitlements under international law. It is important to understand, however, that this view does not in fact preclude the exploitation of individual animals or plants at all, since it does not in any sense involve treating them as sacrosanct. Rather, as already indicated, it gives them a prima facie claim to our consideration. Nor does the attribution of intrinsic value to all organisms necessarily imply the possession of such value in equal measure; for as Attfield has observed,[145]

> the question of the scope and limits of moral consideration is not to be confused with that of moral significance. An answer to the first question commits nobody to any particular view about the relative importance of one set of claims or interests over another.

Clearly, the strength of each claim will depend upon the degree of value each organism is judged to possess. The more difficult issue concerns the establishment of the criteria in accordance with which the measure of such value may be determined.

The Measure of Value

Once again, the key to this particular mystery can most readily be found in Matthews's analysis. As she points out,[146]

> If the value of a being resides in its self-maintaining activity, then the measure of its intrinsic value would seem to be the degree of its power of self-maintenance.

This power of self-maintenance is, she suggests, in general linked with the degree of complexity of the organism in question, since greater complexity confers greater autonomy, in the sense of a capacity to take stock of its surroundings and actively adapt them to its needs.[147] Thus the blue whale, for example, a highly complex creature, would possess significantly greater intrinsic value than a much simpler organism such as the krill upon which it feeds.

Yet having advanced this hypothesis, Matthews appears almost at once to qualify it, if not to resile from it completely, on the grounds of the whale's total ecological dependence upon krill. If the two are "ecologically interconnected", she argues,[148] then the value gap between them begins to close as

---

[144] Nevertheless, the Convention appears, through its definition, to suggest that individual organisms have pride of place where intrinsic value in diversity is concerned. One treaty which appears to accord express recognition to the intrinsic value of a particular *ecosystem* is the 1991 Environmental Protocol to the Antarctic Treaty, which refers in Article 2 to the "intrinsic value of Antarctica, including its wilderness and aesthetic values and its value as an area for the conduct of scientific research". Unfortunately, this appears so to confuse the various forms of value as to make it impossible to know what to make of it.

[145] *Op. cit.*, n. 110, at p. 140.

[146] *Op. cit.*, n. 111, at p. 122.

[147] For an account of the nature and significance of complexity from a scientific perspective, see R. Dawkins, *The Blind Watch-maker* (1986), Chapter 1.

[148] At p. 125.

a flow of intrinsic value, from one self to others, commences ... The intrinsic value of a given self may ultimately be a function not only of its own power of self-maintenance, but also of the intrinsic value of the many other species of organism on which that power depends. In this way the original hierarchy of selves, in respect of intrinsic value, is broken down.

The cogency of this part of her analysis may be doubted, however. In particular, if intrinsic value is defined to mean the value that an entity has of itself, for itself, it is difficult to accept that there can ever be a "flow of intrinsic value from one self to others", because the concept would lose its meaning as soon as the flow began. The value that krill possess *for themselves* is worthless, indeed without rational meaning, in relation to whales. Whales are, by definition, only capable of benefiting from the value that krill have *for whales*, and that must be a form of instrumental, rather than intrinsic, value. What Matthews's argument does confirm is that if we wish to ensure the conservation of whales, on account of their high intrinsic value, it will be equally important to ensure the conservation of krill, on account of their crucial instrumental value (and regardless of their lesser intrinsic value); but this simply demonstrates that different kinds of value may have to be worked into the same equation.

While ecological interdependence is, therefore, undeniably a key issue in the formulation of conservation policy it does not seem to present any obstacle to acceptance of the notion of degrees of intrinsic value, and the criterion of complexity seems the best by which to assess the quantum. The factors which contribute to complexity are, of course, too numerous and sophisticated for detailed discussion here, but the general process through which they have evolved has been conveniently summarised by Rolston:[149]

One-celled organisms evolved into many-celled, highly integrated organisms. Photosynthesis evolved and came to support locomotion – swimming, walking, running, flight. Stimulus response mechanisms became complex instructive acts. Warm-blooded animals followed cold-blooded ones. Neural complexity, conditioned behaviour and learning emerged. Sentience appeared – sight, smell, hearing, taste, pleasure, pain. Brains evolved, coupled with hands. Consciousness and self-consciousness arose. Persons appeared with intense concentrated unity.

Any approach which seeks to assess intrinsic value by reference to the complexity of organisms would almost certainly accord pride of place to the human animal, and it is this very consequence which has caused some philosophers to reject the idea of quantification of intrinsic value altogether and to support instead an ethic based upon "biocentric egalitarianism". Thus, Paul Taylor has argued that only by rejecting human claims of superiority over other living things can we adopt the proper attitude of respect for nature.[150] Yet for the purposes of developing a legal regime for nature conservation this view seems wholly unrealistic.[151] Nor would it seem to be necesary, for the very factors which place mankind at the top of the pecking order may themselves provide the source of the obligation to treat nature with respect. In that context, Ferré[152] has identified six major capacities which characterise full personhood, as expressed in mature, healthy human beings. These are the

[149] *Op. cit.,* n. 118, at p. 24.
[150] "The Ethics of Respect for Nature" in Hargrove (ed.), *op. cit.,* n. 68, at p. 111.
[151] Particularly bearing in mind the anthropocentric bias discernible at the Rio Summit and discussed above.
[152] F. Ferré, "Personalistic Organicism: Paradox or Paradigm?", in Attfield and Belsey, *op. cit.,* n. 118, at p. 65.

powers of enjoying consciously, thinking logically, remembering, planning, preferring/judging and acting with moral responsibility. It is these very qualities, and particularly the last, which give rise to the moral code of action which the World Charter for Nature treats as the basis of respect for other forms of life. Furthermore, to the extent that other life forms share these higher qualities (and surely few would seek to deny that *some* creatures share *some* of them), such creatures should be entitled to have them respected.[153]

Such an approach would clearly place mammals towards the upper end of the scale, and would, indeed, treat all sentient creatures as deserving of special consideration. In particular, while the instrumental value of animals may, in appropriate circumstances, continue to permit their killing and use to satisfy human needs, there can plainly be no justification for subjecting them to unnecessary suffering as part of this process. There may even come a point when their intrinsic value is judged to be so great as to render killing them morally unacceptable in most circumstances. Some might feel that a plausible case could already be made out for primates and cetaceans in that regard.[154] Where plants are concerned, however, instrumental values are likely to predominate. As Ferré states[155]

> Grass may have some modicum of intrinsic value, on my view, but not very much. Still, the instrumental value of grass may far outweigh the higher, more intense intrinsic values of particular mammals which graze upon it.

Similarly, the valued consumption uses of trees may justify their destruction notwithstanding their intrinsic value, though there would again be circumstances where the contemplated use might be too trivial to be permissible, and the wanton destruction of a tree should always be considered as indefensible.[156]

Naturally, the desire to preserve diversity is itself a further element to build in to the equation. It is widely accepted that organisms not only have a good of their *own,* but also a good of their *kind,*[157] and the possibility of the extinction of any particular kind thus becomes a relevant factor in any decision to destroy an individual plant or animal. Thus the value of an organism may be increased by reference to its rarity; in the case of an endangered species the fate of each individual specimen becomes a matter of heightened concern.[158]

Conclusions

The contention that the biodiversity concept in international law must logically be based upon a recognition of the intrinsic value of individual organisms is likely to be unpalatable to many and to be regarded as unconvincing or eccentric by others. From a philosophical or political perspective, it might be thought to involve an

---

[153] *Ibid.*, p. 67.
[154] As to the former, see P. Cavalieri and P. Singer, *The Great Ape Project* (1993) and, as to the latter, A. D'Amato and S.K. Chopra, "Whales: Their Emerging Right to Life" (1991) 85 AJIL 21. Some writers would extend such consideration much more widely, for example to all "subjects-of-a-life": see T. Regan, *The Case for Animal Rights* (1984).
[155] *Op. cit.,* n. 152, at p. 69.
[156] Attfield, *op. cit.,* n. 110, at p. 155; S. Clark, *The Moral Status of Animals* (1977), at p. 172.
[157] See, for example, Rolston, "Biology Without Conservation: An Environmental Misfit and Contradiction in Terms", in Western and Pearl, *op. cit.,* n. 79, at p. 234; Attfield, *op. cit.,* n. 110, at p. 155; Ferre, *op. cit.,* n. 123, at pp. 229–230.
[158] Attfield, *ibid.*; Matthews, *op. cit.,* n. 111, at 134.

acceptance of such controversial concepts as animal rights or, even worse, "plant rights", and also to run counter to the well-known arguments of such philosophers as J. Baird Callicott that "animal liberation" and environmental ethics are based on distinct and incompatible intellectual foundations.[159] It may be challenged as being emotive, or even emotional[160] or be said to represent a form of "cultural imperialism".[161] Finally, it might be thought to challenge man's position of dominance over the natural world. From the strictly legal point of view, it might appear to place excessive emphasis upon the preambular paragraphs of international legal instruments, and insufficient upon the dispositive provisions, as well as to take inadequate account of the fact that some of the instruments in question are at best of a "soft law" character.

Yet none of these objections would stand up to serious scrutiny. An acceptance of the moral considerability of organisms does not involve any commitment to a rights-based approach to recognition of that status.[162] Callicott has subsequently retreated from his original views[163] and both philosophers and scientists have demonstrated that there is in fact no necessary incompatibility between accepting the importance of biological communities as unified systems and according value to individual creatures.[164] Those who would label demands for the recognition of such value as emotional would do well to ponder the intellectual substance of their own positions in comparison with the carefully reasoned theses of the targets of their criticism. They might be forced to conclude that the matter was less one of cultural imperialism than of ethical imperatives. It is in any event the case that respect for individual elements of the natural world is a feature of many religions and cultures, and therefore the question is one of upholding the nobler aspects of many creeds, rather than seeking to impose one cultural tradition upon others.[165]

As regards the legal issues, it is true that the above discussion relies heavily upon inferences to be drawn from the preambles of international legal instruments, but that would seem to be entirely legitimate as a means of identifying the

---

[159] "Animal Liberation: A Triangular Affair" in Hargrove (ed.), *op. cit.*, n. 68. Note that the term "animal liberation" is used generically to cover various philosophical rationales for a humane ethic.

[160] It is almost impossible, it seems, to avoid such charges when any issue of animal welfare is raised; for an illustration of this point, check the Word Index to D. Paterson and M. Palmer (eds.), *The Status of Animals: Ethics, Education and Welfare* (1989) for references to "emotional", "emotive", etc. It is interesting to note that research in the United States has demonstrated that children appear to go through three distinct phases, in terms of their attitudes towards animals, as they mature; these various stages have been labelled "exploitative", "emotional" and "ethical": D. Paterson, "Assessing Children's Attitudes Towards Animals", *ibid.*, pp. 60–61. The international community is perhaps in the process of a similar transition.

[161] See, for example, the comments of the whaling nations at the 44th Annual Meeting of the IWC, (1992) 13 *Traffic Bulletin* 57.

[162] Attfield, *op. cit.*, n. 110, at p. 141, 167–168.

[163] "Animal Liberation and Environmental Ethics: Back Together Again" in Hargrove (ed.), *op. cit.*, n. 68.

[164] See the various essays collected in Hargrove, *ibid.*, especially M.A. Warren, "The Rights of the Non-Human World". See also R. Rodd, *Biology, Ethics and Animals* (1990) and R.D. Ryder (ed.), *Animal Welfare and the Environment* (1992), especially Chapters 3, 4, 8, 10, 11, 14 and 15.

[165] See, for example, the Muslim, Buddhist, Christian, Hindu and Jewish Declarations on Religion and Nature, formulated at the 1986 Assisi Conference, and reprinted in (1987) 17 *EPL* 47, at pp. 87–90, all of which stress at some point the importance of individual organisms. Note also B.A. Masri, *Islamic Concern for Animals* (1987).

underlying philosophy and essential object and purpose of those instruments.[166] It is undeniably the case that particular emphasis has been placed upon the World Charter for Nature, which is not, and was never intended to be, a legally binding document. Yet in many respects the Charter simply provides a theoretical underpinning for the substantive provisions of other conservation instruments which do possess legally binding force. Certainly, instances of recognition of the need for the protection of individual organisms can be found in many treaties. Article 5 of the 1950 International Convention for the Protection of Birds,[167] for example, requires the parties to prohibit methods of hunting or trapping which are likely "to result in the mass killing or capture of birds or to cause them unnecessary suffering", whilst the power under Article 5(f) of the Whaling Convention to regulate "appliances which may be used" has been implemented so as to preclude the employment of certain devices which are judged to be particularly inhumane.[168] The control of methods of capture of wild animals is in fact a feature of many conservation treaties[169] and, while most of the provisions in question are aimed principally at controlling the use of techniques which are indiscriminate, many also have the effect of proscribing cruelty. Indeed, some are only really explicable on the basis that the elimination of cruelty is their primary objective.[170]

Most significantly of all, perhaps, a concern for the protection of individual specimens pervades the entire text of CITES,[171] one of the most widely ratified of all conservation treaties. Although the Biodiversity Convention itself does not appear

---

[166] Note in this context the views of G. Fitzmaurice, in "The Law and Procedure of the International Court of Justice 1951–4: Treaty Interpretation and Other Points" (1957) 33 BYIL 203, at pp. 228: "Although the objects of a treaty may be gathered from its operative clauses as a whole, the preamble is the normal place in which to embody, and the natural place in which to look for, an express or explicit general statement of the treaty's objects and purposes. Where these are stated in the preamble, the latter will, to that extent, govern the whole treaty". The significance of the preamble in that regard is admittedly diminished where the treaty contains an operative provision detailing its objectives, which is true of the Biodiversity Convention itself (Article 1). Even there, however, the preamble may shed light upon the motives which underlie those objectives. In that regard, Fitzmaurice states at p. 229 that "recitals or statements of the motives of the parties ... in entering into the treaty would, in general, be conclusive as to those motives ... and would normally be conclusive (unless contradicted elsewhere, in the operative part of the text) as to the intentions of the parties in entering into the treaty" and that "the interpretational conclusions to be drawn from the preamble are as binding on the parties as those to be drawn from any other part of the treaty". Thus the primary objective of the Biodiversity Convention, as explained in Article 1, is the conservation of biological diversity; the intrinsic value of such diversity, which is recognised in the opening words of the preamble, constitutes a significant motivation underlying this objective.

[167] 638 UNTS 186.

[168] Specifically, the cold-grenade harpoon, which was outlawed by para. 6 of the Schedule. See also Lyster, op. cit., n. 5, at pp. 26–27.

[169] In addition to those already mentioned, note, for example, the 1968 African Convention, Article 7(2) and its predecessor, the 1933 Convention relative to the Preservation of Flora and Fauna in their Natural State, Article 10; the 1972 Convention for the Conservation of Antarctic Seals, UKTS No. 45 (1978), Cmnd. 7209, Article 3(h); the 1979 Berne Convention, Article 8 and the 1980 Antarctic Marine Living Resources Convention, Article 9(2)(h).

[170] Note, for example, the ban on the use of blinded or mutilated decoys in Appendix IV of the Berne Convention. Article 3 of the Antarctic Seals Convention makes express reference to both the "rational" and the "humane" use of seal resources.

[171] See Articles 3, paras. (2)(c), (3)(b), (4)(b), (5)(b), 4 paras. (2)(c), (5)(b), 6(b), 5 para. (2)(b), 7 para. (7)(c), 8 paras. (3), (5) and 12 para. (2)(b).

to contain any such provisions, it leaves intact, as seen above, all these obligations deriving from existing treaties, except where they would damage or threaten biological diversity itself, which they are in practice most unlikely to do.[172] Furthermore, given the apparent recognition in the preamble of the Convention of the intrinsic value of individual organisms, even the negotiation at some future time of a Protocol concerning methods of capture or related matters cannot be ruled out entirely. Indeed, the extent of public concern over humane issues is already such that the establishment of ethically defensible means of exploitation might ultimately prove to be a prerequisite, in a practical sense, to the Convention's stated objectives regarding the sustainable use of the various components of biodiversity and the utilisation of genetic resources.[173]

In truth, the real difficulty in this area appears to lie in the current unwillingness of many governments to treat ethical, humane issues with due seriousness, whether or not they are embodied in formal legal obligations. To that extent, the World Charter for Nature has simply illuminated an area of tension which lies at the heart of practice under many international treaties. There could, perhaps, be no clearer demonstration of the problem than in the dialogue of the deaf which has recently been pursued through the International Whaling Commission. To the whaling nations, the relatively plentiful stocks of certain cetacean species has meant that there could be no legitimate justification for refusing to sanction the resumption of commercial whaling.[174] To others in the international community, however, the real need is to prevent not only the extinction of cetacean species, but also the death and suffering of individual sentient and highly intelligent creatures.

This entire analysis suggests that, while any element of tension between the principal justifications for the conservation of biological diversity is in fact capable of rational resolution, this may not be possible to achieve in ways that all states will find easy to accept. To that extent, therefore, there is an unresolved internal conflict at the heart of the biodiversity concept itself. It is not easy to see how the concept can prove capable of bearing the substantial load it is likely to be asked to carry in the field of global conservation in the absence of some hard thinking about fundamental philosophical questions and their practical legal consequences.

---

[172] It is certainly possible that humane and conservation considerations may conflict, as in the case of the culling of overabundant species or the eradication of alien or exotic species, but it is less easy to see any conflict arising from existing treaty obligations.

[173] The objectives of the Convention are expressly indicated in Article 1.

[174] The force of this argument was, however, significantly undermined by revelations in January 1994 of illegal and unrecorded Soviet whaling since the 1960s: see *New Scientist*, 5 March 1994, at p. 4.

2

The Rio Convention on Biological Diversity

Alan E. Boyle

Introduction

The adoption of the Convention on Biological Diversity, and its entry into force on 29 December 1993, marks an important new development in the protection of the natural environment. Previous conservation conventions had mainly been concerned with the rational use of common property or shared resources such as fish, with the protection of migratory animals and their habitat, and with the suppression of international trade in endangered species.[1] A more recent development has been the negotiation of treaties concerned with particular ecosystems, notably in Antarctica, in regions such as South-East Asia, the Caribbean and the Western Indian Ocean or under the World Heritage Convention.[2] These agreements, together with the International Court's decision on conservation of common property resources in the *Icelandic Fisheries* cases,[3] have gradually extended the role of international law in promoting the conservation and sustainable use of natural resources, and in themselves contribute significantly to the protection of biological diversity. Nevertheless they fall short of establishing a comprehensive global regime for the protection of nature, and largely leave untouched resources located wholly within a State's own national boundaries. The Convention on Biological Diversity represents, at least in principle, an attempt to internationalise, in a more comprehensive and inclusive way, the conservation and sustainable use of nature, based on the concept of biological diversity. Biological diversity has been described as "the total

---

[1] 1982 UN Convention on the Law of the Sea; 1971 Ramsar Convention on Wetlands of International Importance; 1979 Bonn Convention on the Conservation of Migratory Species of Wild Animals; 1973 Convention on International Trade in Endangered Species.

[2] 1991 Protocol to the Antarctic Treaty on Environmental Protection; 1972 World Heritage Convention; 1990 Kingston Protocol Concerning Specially Protected Areas and Wildlife in the Wider Caribbean; 1985 Nairobi Protocol Concerning Protected Areas and Wild Flora and Fauna in the Eastern African Region; 1985 ASEAN Agreement on the Conservation of Nature and Natural Resources.

[3] (*UK* v. *Iceland*) (1974) ICJ Rep. 3; (*Germany* v. *Iceland*) ibid., p. 175.

International Law and the Conservation of Biological Diversity (C. Redgwell and M. Bowman, eds.: 90 411 0863 7: © Kluwer Law International: pub. Kluwer Law International, 1995: printed in Great Britain), pp. 33–49

variety of genetic strains, species and ecosystems".[4] It is defined in Article 2 of the Biological Diversity Convention as

> the variability among living organisms from all sources including, inter alia, terrestrial, marine and other aquatic ecosystems and the ecological complexes of which they are part; this includes diversity within species, between species and of ecosystems.

It should be conserved, it has been argued, on moral grounds, "because all species deserve respect regardless of their use to humanity", and on utilitarian grounds, because they are all components of our life support system, and because biological wealth supplies food, raw material and genetic material for agriculture, medicine and industry.[5] There is some evidence that the biological diversity of nature is increasingly threatened, due to pollution, deforestation, overuse of resources, harmful land use and development practices and other factors. One estimate predicts that up to 25% of the world's species could be extinct by the middle of the next century.[6]

If, as we shall see, the Convention displays serious flaws, and would have benefited from a more considered negotiating process, its goals are nevertheless ambitious. The main purpose of this chapter is to consider in a very tentative and preliminary way the problems it poses, and how much further work will be needed to convert it from an instrument of political symbolism into an effective conservation regime.

The Negotiation of the Convention

The genesis of a convention on biodiversity can be found in the work of several international bodies. The World Conservation Strategy, adopted by IUCN, UNEP and the World Wildlife Fund in 1980 included among its main objectives the preservation of genetic diversity and the sustainable utilization of species and ecosystems. The revised strategy adopted in 1991 also calls for the protection of biodiversity as an important element in sustainable living. In particular the report urges governments to protect the remaining natural ecosystems unless there are overwhelming reasons for change, to use these ecosystems sustainably, to maintain as large an area as possible for supporting a diversity of sustainable uses and species, and to restore and rehabilitate degraded ecosystems.

The 1982 World Charter for Nature, adopted formally as a resolution of the UN General Assembly,[7] provides *inter alia* for the maintenance of the "genetic viability" of the earth, the conservation of unique areas, representative samples of ecosystems and habitats of rare or endangered species. It stresses that ecosystems used by man must be managed to achieve and maintain optimum sustainable productivity, that all planning should include strategies to conserve nature, establish inventories of ecosystems and assess the effects on nature of proposed policies and activities. In 1983 an FAO conference adopted an Undertaking on Plant Genetic Resources[8] whose

---

[4] IUCN/WWF/UNEP, *Caring for the Earth* (1991), at p. 28.

[5] *Ibid.* See also the 1982 World Charter for Nature, UNGA Res.37/7 which declares that "Every form of life is unique, warranting respect regardless of its worth to man, and to accord other organisms such recognition, man must be guided by a moral code of action". See generally R.F. Nash, *The Rights of Nature* (1989).

[6] *Caring for the Earth*, p. 28, and see generally World Conservation Monitoring Centre, *Global Biodiversity* (1992).

[7] See n. 5, *supra*

[8] FAO Res.8/83 (1983). Most developed States reserved their position on this resolution, however.

objective was to ensure that these resources should be explored, preserved, evaluated and made available for plant breeding and scientific purposes. The resolution was non-binding but laid out a series of articles defining the measures states would take to protect and preserve these resources. A Commission on Plant Genetic Resources was also created.

The first body to propose specific legal principles, however, was the World Commission on Environment and Development. Its legal principles required states to maintain ecosystems essential for the functioning of the biosphere "in all its diversity", to maintain "maximum biological diversity" by ensuring the survival and promoting the conservation of all species of flora and fauna in their natural habitat, and to observe the principle of "optimum sustainable yield" with regard to the exploitation of living resources.[9]

It was this proposal and a report from the Executive Director on the rationalisation of international conventions on biological diversity which led UNEP's Governing Council to initiate the process of drafting a convention in 1989. Resolution 15/34 recognised the need for co-ordinated and effective implementation of existing legal instruments and agreements but also endorsed the adoption of a further legal instrument, possibly to be in the form of a framework convention, whose purpose would be to ensure global conservation of biodiversity. An Expert Working Group was established, and subsequently converted into an Intergovernmental Negotiating Committee, in which some seventy states participated. Finally, Resolution 44/228 of the UN General Assembly brought the conservation of biological diversity within the mandate of the UN Conference on Environment and Development. This made it essential that the negotiations in UNEP should result in a convention ready for signature at Rio in June 1992. These constraints of timing, and the complexity of the issues, go far to explaining the rushed and unsatisfactory nature of the Convention that was eventually adopted.

At the second session of the Working Group in 1990 the Executive Director of UNEP stressed that the agreement should not infringe the sovereignty of States over their resources, that it must protect the interest of States in which the resources are located and must provide incentives for conservation of biological diversity without inhibiting growth or sustainable development. These considerations reflect the considerable gulf which separated developed and developing States in their attitude to the negotiations. There was early consensus among participants that those who most enjoy the economic benefits of biological diversity (i.e. developed States) should contribute equitably to the cost of conservation and sustainable management, that genetic resources should in some form be accessible to all and that technology and information on their use should be transferable to all. What the meeting sought was a spirit of co-operation on these issues between gene-rich developing counties and technology-rich developed countries. Emphasising that this would not simply be a conservation convention, and in response to the demands of developing states, UNEP's Governing Council further defined the basic objectives of the negotiations in 1991 by resolving that "both the conservation of biological diversity and the rational use of biological resources

---

[9] Article 3, Draft Convention on Environmental Protection and Sustainable Development, adopted by the Experts Group on Environmental Law, 1986, in R.D. Munro and J.G. Lammers, *Environmental Protection and Sustainable Development* (1986). IUCN also produced draft articles for a convention in 1987, and a revised text in 1989; an FAO draft appeared in 1990.

shall be integral and inseparable elements of the convention".[10] Negotiations made very slow progress, however, and there were serious disagreements between developed and developing States.[11]

The Group of 77 developing States saw the convention as part of a broader agenda to use UNCED for the purpose of restructuring global economic relations to obtain the required resources, technology and access to markets to enable the south to undertake development that was both environmentally sound and rapid enough to meet the needs of its population.[12] For developing States, therefore, the key issues in the biodiversity negotiations were:[13]

(a) the establishment of special systems of intellectual property rights and appropriate mechanisms for compensating the South for the biological resources provided by it;
(b) the establishment of mechanisms giving the South access to the biotechniques that are developed through the use of the genetic resources that it provides;
(c) additional funding to facilitate implementation of the convention and access to technology.

In many respects they were successful in meeting these objectives.

For the developed States, the economic issues were no less important but their perspective was rather different. The United States was predictably opposed to the Convention's provisions on technology transfer, financing and access to resources, and, as with the Law of the Sea Convention, it initially refused to sign the final text. President Bush claimed that the treaty "threatened to retard biotechnology and undermine the protection of ideas".[14] But the US objections went beyond the economic philosophy on which the convention was based. In its public declaration, the US government drew attention to the convention's weaknesses, particularly as regards environmental impact assessments, the legal relationship with other treaties, and the scope of obligations affecting the marine environment. It observed that "the hasty and disjointed approach to the preparation of this convention has deprived delegations of the ability to consider the text as a whole before adoption" and concluded that the text did not reflect well on the treaty-making process.[15] However, President Clinton's new administration did eventually sign the convention, although the prospects for US ratification and the conditions which might be attached remain uncertain.

The European Community expressed regret at the weakness of the convention in contrast to its own directives, but it did not share to the same degree the US objec-

---

[10] (1991) 21 EPL 171.
[11] See *Ad Hoc* Working Group of Experts on Biological Diversity, 2nd Session, Geneva, 19–23 February, 1990; Intergovernmental Negotiating Committee on Biological Diversity, 5th Session, Nairobi, February 1992, and F. Burhenne-Guilmin and S. Casey-Lefkowitz, "The New Law of Biodiversity", (1992) 3 *Yearbook of International Environmental Law* 43.
[12] South Centre, *Environment and Development: Towards a Common Strategy of the South in the UNCED Negotiations and Beyond* (1991), at p. 3.
[13] *Ibid*, pp. 23–24.
[14] Remarks by the President of the USA in Address to the UN Conference on Environment and Development, Rio de Janeiro, 12 June 1992. See M.D. Coughlin, "Using the Merck-INBio Agreement to Clarify the Convention on Biological Diversity" 31 Col JTL 1993, p. 337.
[15] Declaration made at the UNEP Conference for the Adoption of the Agreed Text of the Convention on Biological Diversity, 22 May 1992, (1992) 31 ILM 848. See Coughlin, *ibid*.

tions to transfer of technology and access to biotechnology, and did sign the convention. The British were concerned at the financial commitments placed on developed states in Articles 20 and 21, and signed subject to an interpretative declaration in which nineteen other States joined.[16]

Nevertheless, although negotiations on substantive issues were still taking place at the same time as the Rio Conference, it was possible to adopt a convention, but only by incorporating some significant changes proposed by developing States and by deleting a number of substantive provisions on which no agreement could be reached and whose omission renders the final text markedly different from the fifth negotiating draft.[17] In particular, three important articles were removed, either from the body of the treaty, or entirely:

(a) The precautionary principle: the fifth draft had stressed the need to "anticipate, prevent and attack the causes of reduction or loss of biodiversity at source", and provided that "lack of full scientific certainty shall not be used as a reason for postponing [action or measures] to avoid or minimise a threat to biodiversity".[18] This wording reflected the increasingly widespread adoption of the precautionary principle, which, in modified form, appears in Principle 15 of the Rio Declaration. In the final text these elements are transferred to the preamble, leaving their ultimate status less clear.

(b) Responsibility for damage to biodiversity: the fifth draft had provided that those responsible for activities which damaged or threatened biodiversity would be responsible for the costs of avoiding or remedying the damage.[19] Such explicit attribution of environmental responsibility is exceptional in international agreements; its deletion in the final text conforms to the prevailing reluctance of states to address the issue.

(c) Global lists of protected areas and species: the draft had proposed that such lists be adopted by the Conference of the Parties,[20] following the model of the World Heritage Convention. The final text makes no provision for global lists or for areas to be selected by the Conference of the Parties and instead leaves the selection and management of protected areas or species and the setting of priorities to each party.[21]

The Convention's Objectives

Reflecting the differing interests of developed and developing states, Article 1 of the Convention sets out two main objectives:

(a) the conservation of biodiversity and the sustainable use of its components;

(b) the fair and equitable sharing of the benefits arising out of the utilisation of genetic resources.

---

[16] See text accompanying n. 57.

[17] Intergovernmental Negotiating Committee For a Convention on Biological Diversity, Fifth Session, Nairobi, May 1992, UNEP/Bio.Div/N7-ING.5/2 (1992).

[18] Draft Article 3(5) and (6).

[19] Draft Article 14.

[20] Draft Article 15.

[21] Convention, Article 8. See Burhenne-Guilmin and Lefkowitz, *op. cit.*, n. 11, at p. 52.

The first part of Article 1 reflects the policy laid down by UNEP and referred to earlier of stressing the interdependent character of conservation and rational use of resources. Despite the preamble's recognition of the "intrinsic value" of biodiversity, including its ecological, cultural and aesthetic aspects, this is not a "preservationist" convention: it assumes human use and benefit as the fundamental purpose for conserving biodiversity, limited only by the requirement of sustainability and the need to benefit future generations. Thus references to conservation of biodiversity must be read in conjunction with the sustainable use of its components, a qualification insisted on by developing countries. To that extent it is philosophically closer to the fisheries or migratory animals treaties, where the purpose of conservation is sustainable human use, than to the Protocol to the Antarctic Treaty on Environmental Protection or the Whaling Moratorium adopted under the International Convention for the Regulation of Whaling, which are the best examples of a preservationist concept of conservation not intended to serve human economic needs.

"Conservation" is deliberately not defined at all, while "sustainable use", is defined only in very general terms in Article 2:

> the use of components of biological diversity in a way and at a rate that does not lead to the long term decline of biological diversity, thereby maintaining its potential to meet the needs and aspirations of present and future generations.

A definition of this generality, like the comparable formulation adopted by the WCED, gives little or no guidance on how specific resources should be managed, or how the priorities of the present and the future are to be reconciled or valued.[22] In practice, since so much is left to the discretion of individual States, the avoidance of long-term decline will depend on how well the provisions on environmental impact assessment and *in situ* and *ex situ* conservation operate.[23] With the removal to the preamble of references to scientific uncertainty and the need for anticipatory action it is unlikely that these will do much to deter States from pursuing development policies which are imprudent and ultimately unsustainable, as has frequently been the case in the management of fisheries resources.

Emphasising that this is not simply a conservation convention, the second part of Article 1 calls for "appropriate" access to genetic resources, "appropriate" transfer of relevant technologies, and "appropriate" funding. These issues are dealt with in subsequent articles in more detail and are considered below.[24] The intended beneficiaries are resource-providing countries in the developing world which lack the finance and technology to benefit from exploitation of the genetic resources they are committed to conserving. Thus a trade off between conservation and economic equity is at the heart of the convention and makes it unusual amongst environmental agreements. Although the Ozone Convention provides for transfer of technology and access to funding, it does so only for the limited purpose of ensuring compliance with that convention's control regime, and not with any view to conferring economic benefits on developing countries. The success of the Biological Diversity Convention is likely to be heavily dependent on the extent to which the anticipated benefits do flow to developing countries. As with the Law

---

[22.] On sustainable development, see generally M. Redclift, *Sustainable Development*, (1987); M. Jacobs, *The Green Economy* (1991); D. Pearce *et al.*, *Blueprint for a Green Economy* (1989).

[23.] See the Section on Conservation, *infra*.

[24.] See the Section on Measures to Benefit Developing States, *infra*.

of the Sea Convention, there is a risk that expectations in this regard may be unrealistic.

The Legal Status of Biodiversity Resources

The resources which represent the biological diversity of nature are not in most cases like migratory animals or high seas fish: they will legally be neither "shared natural resources" nor "common property" available to anyone. For the most part they are located within the territory of individual States and to that extent are subject to the sovereignty of the State concerned. General Assembly resolutions have in the past supported the principle of "permanent sovereignty" over such natural resources and the right of every State to possess, use and dispose of them freely.[25]

The Convention on Biological Diversity reaffirms the sovereignty of States over their own biological resources, and their sovereign right to exploit these resources pursuant to their own environmental policies.[26] Moreover, although Article 15 requires States to "endeavour to create conditions to facilitate access to genetic resources" by other parties, it does not create or recognise any right of "free" access for these States. On the contrary, it notes that "the authority to determine access to genetic resources rests with the national governments and is subject to national legislation", and goes on to provide explicitly that "Access to genetic resources shall be subject to prior informed consent of the Contracting Party providing such resources". This provision has its analogy in the 1982 UNCLOS Articles on scientific research in the Exclusive Economic Zone,[27] and although the method of implementation is uncertain and potentially problematic in practice,[28] it does reflect the underlying principle, firmly supported by developing countries, that sovereignty gives each state the right to control access to its own resources.

That sovereignty is not unlimited or absolute, however. It is subject to the requirements of conservation and sustainable use set out in Articles 6–9 of the Convention, and to the customary obligation not to cause damage to other states or to areas beyond national jurisdiction,[29] reflecting Principle 2 of the Rio Declaration.

None of this suggests any redefinition of the legal status of biological resources. The Convention's treatment of these resources remains fully within the existing rules of international law, and is consistent with the principle of permanent sovereignty. The only novelty in this respect is in the preamble, which describes the conservation of biological diversity as a "common concern of humankind". The fifth draft had referred to the "common concern of all peoples", but was amended following Brazilian opposition to the possibility that this might be seen as conferring rights on indigenous peoples.

---

[25] UNGA Res.1803 XVII (1966); UNGA Res.3201 (S-VI) (1974); UNGA Res.3281 XXIX (1974); 1982 World Charter for Nature, para. 22. On the legal status of natural resources see I. Brownlie, "Legal Status of Natural Resources in International Law", (1979/I) 162 *Recueil des Cours* 267ff.

[26] Preamble, Articles 3, 15.

[27] Articles 245–252. See also 1989 Basel Convention on the Control of Transboundary Movements of Hazardous Wastes and their Disposal, Article 6 of which institutes a regime of prior informed consent for hazardous waste trade.

[28] Compare the much more detailed requirements for prior consent set out in the 1989 Basel Convention on the Control of Transboundary Movements of Hazardous Wastes.

[29] See Article 3. This is the first time this principle has been put in the body of a treaty, rather than in the Preamble.

The term "common concern" was first used by UNGA Res. 43/53 to avoid attributing the status of "common heritage" to the global climate.[30] Early references to biological diversity as the "common heritage of mankind"[31] were altered to "common concern" for the same reason. It is evident that the Convention does not seek to internationalise ownership of biological resources in the way that the 1982 UNCLOS treats ownership of deep seabed mineral resources.[32] The terms of access to the resource are fundamentally different in these two regimes, and in the case of biological diversity do not involve international management by any institution comparable to the International Seabed Authority. What the two regimes do share is a common philosophy that the benefits of access to the resource must be shared equitably, as we have seen, but on a much more limited basis in the Biological Diversity Convention which does not affect the status of the resource itself.

But, if the convention does not internationalise ownership of biological resources, its reference to "common concern" is significant in the much more limited sense of legitimising international interest in the conservation and use of biological resources otherwise within the territorial sovereignty of other States. Like concern for human rights, it acknowledges that the management of a State's own environment and resources is a matter in respect of which all States, or all parties to the convention, have standing, even if they are not directly injured by any specific misuse of resources.[33] In this sense, permanent sovereignty over biological resources is no longer a basis for exclusion of others, but entails instead a "commitment to co-operate for the good of the international community at large".[34]

General Environmental Provisions

Environmental Impact Assessment

Failure to refer explicitly to the precautionary principle does not mean that States are free to disregard the possible consequences of their actions. Article 14 of the convention requires parties "as far as possible and appropriate" to "introduce appropriate procedures" requiring environmental impact assessment of proposed projects that are "likely to have significant adverse impacts on biological diversity". This wording compares unfavourably with comparable provisions in other environmental treaties. The 1991 Espoo Convention on Environmental Impact Assessment in a Transboundary Context, for example, is much more specific about the type of activities which must be assessed and the content of the EIA documentation which must be produced, and leaves less room for diverse judgements by different States on what is required. Moreover, the qualifying words "as far as possible and as appropriate" in Article 14 may enable parties to avoid an EIA altogether where they find it

---

[30] See A.E. Boyle, "International Law and the Protection of the Global Atmosphere: Concepts, Categories and Principles" in R. Churchill and D. Freestone, *International Law and Global Climate Change* (1991), at pp. 11–13, and UNEP, *The Meeting of the Group of Legal Experts to Examine the Concept of the Common Concern of Mankind in Relation to Global Environmental Issues* (1990).

[31] *Ad Hoc* Working Group of Experts on Biological Diversity, 2nd Session, Geneva, February 1990, para. 11. See also the FAO International Undertaking on Plant Genetic Resources, Res.8/83, which referred to plant genetic resources as "a heritage of mankind".

[32] Compare the 1982 UNCLOS, Articles 133–191.

[33] See Boyle, *op. cit.* n. 30.

[34] G. Handl, "Environmental Security and Global Change: The Challenge to International Law", (1990) 1 YIEL 32.

inconvenient to conduct one. The Rio Convention is also weaker in its provision for public participation; whereas this is an essential part of the EIA procedure under the Espoo Convention, in Article 14 it is merely called for "where appropriate". Finally, "programmes and policies" likely to have significant adverse impacts on bio-diversity do not have to be the subject of an EIA; the Convention merely requires parties to ensure that the environmental consequences are taken into account.

Implementation of these provisions requires a judgement that projects or pro-grammes are "likely" to have a "significant" adverse impact. Yet it will often be difficult to know what the likelihood and scale of any adverse impact will be until an EIA has been carried out. Although the convention is not unique in adopting this formulation,[35] adverse impacts will tend to be less easily predictable in the context of biological diversity, where the effects may be subtle, long term, and surprising, than, say, in the case of pollution from smelters. It is regrettable that the drafters did not prefer a text which took greater account of this inherent uncertainty by deleting the word "likely" and referring instead to "possible adverse impacts".

Weak though this provision for EIA may be, however, it is far reaching in one respect. Most of the earlier agreements which require an EIA do so solely in the context of transboundary harm. Their purpose is to protect other states, or areas beyond national jurisdiction.[36] The Rio Convention, in contrast, deals primarily with the assessment of effects *within* the territory of the party conducting it.[37] Thus it will apply in circumstances to which the Espoo Convention is wholly inapplicable, and to that extent Article 14 is important in emphasising the responsibilities now undertaken by States for the management of their own national environment. From this perspective, and taking account of the intrusion on national sovereignty which it represents, the weakness of the article may not be so surprising.

Good Neighbourliness

Another indication of the Convention's primary concern with the internal environ-ment of states is its perfunctory treatment of transboundary issues. Although custom-ary international law and many environmental agreements require notification and consultation between States with a view to minimising the harmful extraterritorial effects of activities within their jurisdiction or control,[38] this is scarcely reflected in the Biological Diversity Convention. Article 14 merely calls for parties to "promote, on a reciprocal basis" notification, exchange of information and consultation, by "encouraging the conclusion of bilateral, regional or multilateral arrangements". This compares very unfavourably even with Principle 19 of the Rio Declaration.[39]

---

[35] See e.g. UNEP, Principles of Environmental Impact Assessment, 1987 ("likely to significantly affect the environment"). Compare Articles 206 of the 1982 UNCLOS: ("reasonable grounds for believing that activities … may cause substantial pollution or significant and harmful changes").

[36] See e.g. 1983 Convention for the Protection and Development of the Marine Environment of the Wider Caribbean Region, Article 12; 1986 Convention for the Protection of the Natural Resources and Environment of the South Pacific Region, Article 16; 1991 Espoo Convention on Environmental Impact Assessment in a Transboundary Context.

[37] See also Principle 17 of the Rio Declaration on Environment and Development; 1985 ASEAN Agreement on the Conservation of Nature and Natural Resources, Article 14; 1990 Kingston Protocol Concerning Specially Protected Areas and Wildlife, Article 13; 1991 Protocol to the Antarctic Treaty on Environmental Protection, Article 8.

[38] See P. Birnie and A. Boyle, *International Law and the Environment* (1992), at pp. 102–109.

[39] And compare also 1985 ASEAN Agreement on the Conservation of Nature and Natural Resources, Article 20(3).

It may be argued that the customary rule represented in Principle 19 has no or only limited application to biological diversity, and that in this sense Article 14 is a welcome advance; but given that harm to biological diversity will generally involve loss of species, loss of habitat, or destruction of natural resources through pollution or development, it seems more than likely that these would fall within Principle 19's reference to "adverse transboundary environmental effect".[40] Thus Principle 19 of the Rio Declaration and Article 14 of the Rio Convention are potentially applicable simultaneously; that such different formulations can be adopted at the same conference displays the lack of co-ordination among those involved. A more satisfactory reflection of customary law and other treaty formulations is Article 14(d), which requires immediate notification of potentially affected States in the case of imminent or grave danger or damage to biological diversity. There seems little reason to doubt that this is already international law.[41]

Similarly, there is no doubt that States are responsible in international law for ensuring that activities within their jurisdiction or control do not cause damage to the environment of other states or of areas beyond the limits of national jurisdiction. Article 3, as we have seen, reiterates Principle 2 of the Rio Declaration, which itself follows Principle 21 of the Stockholm Declaration, a principle widely regarded as representing international law on the matter.[42] But as noted earlier, the convention makes no explicit provision for payment of compensation or restoration costs, and merely provides in Article 14(2) for the Conference of the Parties to examine the issue of liability and redress. Other conventions which have similarly left State responsibility for further study have not led to any agreement on the issue.[43] Moreover, whereas the fifth draft had specifically referred to harm to "biological diversity" in this context, the Convention merely talks of harm to the "environment". While it may be assumed that this includes biological diversity, it is now no longer entirely clear that this is the case, since the term "environment" is itself not defined.[44]

Conservation

Articles 5–9 are intended to give effect to the convention's first objective of conserving biological diversity. Article 5 deals with areas beyond national jurisdiction and calls for co-operation between the parties directly or through competent international organisations. This is likely to involve FAO and regional institutions with responsibility for marine resources, and may be linked to efforts endorsed by UNCED to improve the conservation and management of world fishing stocks. Article 6 is concerned with developing national strategies, plans or programmes reflecting the objectives of the convention and integrating conservation and sustain-

---

[40] Compare also the 1985 ASEAN Agreement, Article 20, which applies to damage to "the environment or the natural resources" of another party or beyond national jurisdiction.

[41] Birnie and Boyle, *op. cit.*, n. 38, at pp. 108–109; and see also 1985 ASEAN Agreement, Article 20(3).

[42] Birnie and Boyle, *op. cit.*, at pp. 89–102. See also UNGA Res.2996 XXVII (1972), and 1985 ASEAN Agreement, Article 20(1).

[43] See e.g. 1972 London Dumping Convention and contrast 1972 Convention on International Liability for Damage Caused by Space Objects.

[44] Compare 1988 Convention on the Regulation of Antarctic Mineral Resource Activities, Article 1(15); 1991 Protocol to the Antarctic Treaty on Environmental Protection, Article 3(2)(b); 1985 Vienna Convention for the Protection of the Ozone Layer, Article 1(2), which define the environment in terms of the impact on the ecosystem and its components.

able use of biological resources into other relevant plans, etc. Article 7 calls for states to identify what biological components are important for conservation and sustainable use, to monitor these, and to identify and monitor processes and activities likely to have a significant adverse effect. Little more need to be said about the provisions except that they are likely to require significant resources and technical expertise to implement.

Articles 8 and 9 deal with *in situ* and *ex situ* conservation. The former is defined as "the conservation of ecosystems and natural habitats and the maintenance and recovery of viable populations of species in their natural surroundings and, in the case of domesticated or cultivated species, in the surroundings where they have developed their distinctive properties". The latter means simply the "conservation of components of biological diversity outside their natural habitats". As the preliminary paragraph of Article 9 makes clear, *ex situ* conservation, for example in zoos, is "predominantly for the purpose of complementing *in situ* measures".

In situ conservation, as envisaged in Article 8, involves measures to protect a wide range of interests and components which collectively constitute the essential elements of biological diversity. These include:

(a) protected areas;
(b) regulation and management of biological resources both inside and outside protected areas;
(c) protection of ecosystems, natural habitats, and viable populations of species;
(d) environmentally sound and sustainable development in areas adjacent to protected areas;
(e) rehabilitation of degraded areas and recovery of species;
(f) control of use and release of modified living organisms where there are likely to have adverse environmental impacts;
(g) protection of threatened species and populations;
(h) regulation or management of processes and activities which threaten biodiversity.

Much of this is already reflected in existing conservation treaties, including the CITES Convention, the Ramsar, Berne and Bonn Conventions, and regional conventions such as the ASEAN Convention, the South Pacific Convention, the Kingston and East African Protocols and the Antarctic Environment Protocol.[45] The measures to be taken do reflect a broad and comprehensive view of what constitutes conservation, but they are left to individual states to implement "as far as possible and appropriate". While many countries already take measures of the kind indicated, much will depend on how willing States prove to be in practice to extend what is already done. It is worth emphasising again that the Convention leaves the setting of priorities and the choice of methods to individual parties; it does not compel uniformity or give the parties collectively the power to mandate measures. Much remains to be agreed if co-ordinated implementation is to occur.

An important question arises, however, over the relationship between the convention and other existing treaty regimes and how these may be affected by

[45] For further discussion of these treaties see S. Lyster, *International Wildlife Law*, (1985) and P. Birnie and A. Boyle, *op. cit.*, n. 38, Chapter 12. See also the 1992 Helsinki Convention on the Protection of the Marine Environment of the Baltic Sea Area, Article 15 of which specifically calls for States to take "appropriate" measures to conserve natural habitats, biological diversity and ecological processes, and the "Forest Principles" adopted at the Rio Conference in 1992.

Article 8. Article 22 provides that the Convention "shall not affect the rights and obligations of a Contracting Party deriving from any existing international agreement, *except where the exercise of those rights and obligations would cause a serious damage or threat to biological diversity*".[46] Conservation agreements are unlikely to be inconsistent with the Convention in this way; more problematic, however, are fisheries agreements under which excessive exploitation of stocks takes place, or the resumption of whaling under or outside the International Convention for the Regulation of Whaling. Parties to the Biological Diversity Convention could find themselves forced to reconsider the operation of these other regimes. Apart from this possibility, the convention is intended to operate as a loose framework for other more specific conservation regimes and problems of incompatibility are unlikely. But it remains unclear what the precise relationship between the convention and these other regimes will be, how they will be linked, and how action taken by supervisory commissions or meetings of parties will be satisfactory coordinated with the Biological Diversity Convention. These issues are explored more fully in later chapters.

Measures to Benefit Developing States

Ensuring the participation and co-operation of the developing states in whose territories many of the world's most significant biological resources and ecosystems are located was an essential political objective of the negotiations. Their interests are addressed in four different ways.

Common but Differentiated Responsibility

Principle 7 of the Rio Declaration refers to the "common but differentiated responsibilities" of States, and to the special responsibility of the developed States. Although the same phraseology is not employed in the Biological Diversity Convention, it is apparent that developed and developing States do not bear the same burdens. This accounts for the frequent references throughout the convention to what is "possible and appropriate", which will vary from State to State. A more explicit recognition of the differing standard expected from different States is found in Article 6, which requires each party to take general measures for conservation and sustainable use "in accordance with its particular conditions and capabilities". The Articles of the Convention which deal with transfer of technology and financial resources are, of course, meant to enhance the capabilities of developing States,[47] but failure to bring about these anticipated benefits may result in far-reaching limitations on the implementation of the convention by these parties. Article 20(4) states that:

> The extent to which developing country Parties will effectively implement their commitments under this Convention will depend on the effective implementation by developed country Parties of their commitments under this convention related to financial resources and transfer of technology and will take fully into account the fact that economic and social development and eradication of poverty are the first and overriding priorities of the developing country Parties.

---

[46] Emphasis added. Compare 1969 Vienna Convention on the Law of Treaties, Article 30, which provides, *inter alia*, that for parties to both treaties the "earlier applies only to the extent that its provisions are compatible with those of the latter treaty".

[47] See below, and see also Chapters 33–37 of Agenda 21, which deal with aspects of "capacity building".

This article is modelled on a comparable provision introduced into the Montreal Protocol to the Ozone Convention;[48] it could simply be a statement of the obvious, but it could also be read as making implementation by developing countries conditional on the assistance they receive from developed States. In effect it affords developing States the power to put pressure on developed States to ensure they have the necessary means to comply with the convention. Article 10 of the Montreal Protocol gives the meeting of the parties explicit competence to deal with complaints of non-transfer of resources, however; there is no such arrangement in the Biological Diversity Convention.

Fair and Equitable Sharing of the Benefits of Utilisation of Genetic Resources

As we have seen this is the Convention's second objective. Article 15(7) requires each party to ensure the sharing "in a fair and equitable way" of the benefits arising from the commercial and other utilisation of genetic resources with the party providing access to such resources. Since access is of course determined solely by the party with sovereignty over the resources, this does little more than state the obvious: that access can be granted on terms agreed by both sides. In these circumstances the party in possession of the resource is well placed to secure whatever benefits it seeks. The Convention may, however, serve the modest purpose of restraining excessive and "inequitable" demands for transfer of benefits, but if this is the intention the absence of any compulsory, binding third party dispute settlement machinery will give little opportunity for challenging the terms on which access is offered, even if they "run counter to the objectives of this Convention".

Access to and Transfer of Technology

Articles 16–19 deal with transfer of technology in several different senses. First, the parties undertake in Article 16(1) to provide or facilitate access and transfer to other parties of technologies "that are relevant to the conservation and sustainable use of biological diversity or make use of genetic resources and do not cause significant damage to the environment". Second, parties must take measures "with the aim that" parties which provide genetic resources have access to and transfer of technology which makes use of those resources.[49] Third, parties must take measures to provide for the "effective" participation in biotechnology research of those providing genetic resources, and to "promote and advance priority access on a fair and equitable basis" to the results and benefits of biotechnologies based on the provision of genetic resources.[50] Transfer of technology provisions in earlier treaties, such as the 1982 UNCLOS, have usually been controversial, on several grounds. There is first the reluctance of governments to compel companies and private parties to transfer technologies that may not be commercially available; second there have been objections to the terms on which any transfer will take place, particularly if this is not at market prices, and, third, there is the question of intellectual property rights which may be lost if transfer is required. The Biological Diversity Convention does

---

[48] Article 10. See also 1992 Rio Convention on Climate Change, Article 4(7).
[49] Article 16(3).
[50] Article 19(1), (2). See Coughlin, *op. cit.*, n. 14, for an example of an access agreement which provides for technology transfer.

attempt to deal with some of these issues, although how satisfactorily remains to be seen.

Transfers under Article 16(1) must be on "fair and most favourable terms", and in other cases on "mutually agreed" terms. Governments are specifically required by Article 16(4) to ensure that the private sector facilitates access to, and joint development and transfer of, technology. These provisions are likely to be easiest to implement in the case of countries providing access to genetic resources since they will again be in a position to bargain for the benefits they will receive, but for some governments, such as the United States, the suggestion of compulsion placed on industry will undoubtedly be objectionable.

Intellectual property issues are important because the transfer of patented technology is specifically envisaged. Article 16(2) provides that access and transfer "shall be provided on terms which recognise and are consistent with the adequate and effective protection of intellectual property rights", while Article 16(5) calls for the parties to co-operate to ensure that intellectual property rights "are supportive of and do not run counter to" the objectives of the convention. This appears to be an attempt to satisfy both sides; intellectual property rights are to be respected but only insofar as they assist rather than hinder implementation of the convention.

However, behind these references there remain unresolved questions about the scope of intellectual property rights and whether they benefit the providers of genetic resources or only those who make use of them. Natural genetic resources, or genetically altered organisms which result from experimentation, are not necessarily always patentable or a source of legally protectable rights. Discovery of a new species of fish, for example, could not be patented; like most natural resources it is simply a commodity which can be bought and sold by anyone. Patentable rights may arise either in respect of a new process for isolating and developing substances, or for new uses for existing substances or possibly, in respect of a substance which had no previous known existence. The extent to which these principles enable the products of biotechnology to be protected will vary, and remains controversial in national patent systems. It is, for example, still uncertain whether genetically altered organisms can be patented as such, or how far patent law will always protect new uses for existing substances.[51] How far this part of the Convention will be important thus depends in part on how far intellectual property law itself is prepared to go in protecting the products of biotechnology and the original natural genetic resource. For the United States it is clear that the risk of losing protection for genetic engineering is thought to be too high to support the Convention; this is not a problem which has deterred other developed States, such as the EEC countries, however.

Financial Benefits

For the first time in any global environmental treaty, "developed" States are required to provide "new and additional" financial resources to enable "developing" States to meet the costs of implementation, and to benefit from provisions on

---

[51] See *Diamond v. Chakrabarty* 65 L. Ed. 2d. 144 (1980); *John Moore v. Regents of the University of California* (1988) 15 USPQ 2d. 1753 (1990); *Mobil/Friction Reduction Application G02/88* (1989) 5 *European Patent Office Reports* (1990), 73; 1973 European Patent Convention, Articles 52(2)(a) and 53(b), and G. Winter, "Patent Law Policy in Biotechnology", (1992) 2 *Journal of Environmental Law* 167. For more detailed consideration, see Chapter 9 by I. Walden.

technology transfer and access to benefits.[52] Which States fall into each group is not identified in the convention, but will be agreed by the Conference of the Parties.[53] The provision of resources is not an open-ended commitment, however; under Article 21 the costs to be covered must be agreed between the developing State in question and the financial mechanism created by Article 21. This body is intended to manage the funds provided under Article 20 and any other funds given on a voluntary basis. Its policy and operation will be determined by the Conference of the Parties.[54] The Global Environmental Facility[55] will act as the financial mechanism on an "interim basis" pending the first meeting of the parties.[56]

A commitment to the principle of "additionality" – that environmental aid should be additional to other forms of development aid – has been an objective of developing States since Stockholm in 1972. It is remarkable that after prolonged resistance they now have agreement on the inclusion of this obligation in the convention. But much remains unsettled, including what constitute the incremental costs of implementation, how much developed States must contribute, and how this question is to be decided. That there is probably no agreement on these points is indicated by a British declaration made on signing the convention, which states that:

> the decisions to be taken by the Conference of the Parties under paragraph 1 of Article 21 concern "the amount of resources needed" by the financial mechanism, and that nothing in Article 20 or Article 21 authorises the Conference of the Parties to take decisions concerning the amount, nature, frequency or size of the contributions of the Parties under the Convention.[57]

Presumably this is meant to indicate that contributions are in effect voluntary and are determined by each party. An alternative reading of Article 21(1) more likely to be preferred by developing States is, of course, that the Conference of the Parties determines not only the resources needed to ensure the "predictability, adequacy and timely flow of funds" but also sets the level and frequency of contributions required; however, the Article does not say so explicitly. This is the sort of issue for which adequate dispute settlement machinery may prove essential.[58] As we saw earlier, however, implementation of the convention by developing States "will depend on" the transfer of resources by developed States, so the issue of additional funding is undoubtedly central to the convention's prospects of success, whatever Article 21 may mean.

Institutional Supervision and Dispute Settlement

The Conference of the Parties

The supervisory role given to the Conference of the Parties is typical of many modern environmental treaties. The parties are required, *inter alia*, to keep implementation of

[52] But see also 1992 Convention on Climate Change. Compare 1987 Montreal Protocol to the Ozone Convention, Article 10, which makes no reference to "additionality".

[53] Article 20(2).

[54] Articles 20(2), 21(1). See also Burhenne-Guilmin *et al.*, *op. cit.* n. 11, at p. 55.

[55] On the establishment of the GEF, see IBRD, Resolution 91–5, (1991) 30 ILM 1758 and "Establishment of the GEF", *ibid.*, at p. 1739.

[56] Resolution 1 of the Nairobi Conference for the Adoption of the Agreed Text, 1992, (1992) 31 ILM 843. On the restructuring of the GEF agreed in 1993 see below, Chapter 14.

[57] 12 June 1992, text provided by FCO. Nineteen developed states issued a joint interpretative declaration in similar terms when adopting the final text of the convention.

[58] See below.

the convention under periodic review, to adopt amendments, annexes, and protocols, to receive and consider information and reports, to review scientific, technical and technological advice provided by the subsidiary body established under Article 25, and to consider and undertake any additional action.[59] There is provision for UN agencies and NGOs to have observer status,[60] but the convention lacks the explicit provision for openness and publicity of information found, for example, in the Antarctic Environment Protocol.[61] As in most supervisory bodies, the effectiveness of these arrangements will depend partly on political will and partly on the provision and receipt of adequate information. Here too the convention is weaker than some other treaties in that it lacks any provision for independent monitoring or inspection. The Conference is thus dependent on the good faith of reporting States[62] and on such other sources of information − such as NGOs − as are to hand. In this respect the Antarctic Environmental Protocol is a better model for effective supervision since it does provide for independent inspection of sites and facilities.[63]

Moreover, the Biological Diversity Convention also lacks a formal non-compliance procedure comparable to the one created by Article 8 of the Montreal Protocol to the Ozone Convention, even if in practice the Conference of Parties may well view its role in this way. Nor, finally, does the Conference have the same power to adopt binding amendments to annexes by majority vote that the parties to the Montreal Protocol possess in certain cases.[64] Instead, parties may opt out of new or amended annexes by express objection within one year. On the spectrum of supervisory bodies this Conference of Parties thus falls closer to the weaker end.

Dispute Settlement

There are several issues raised by the convention's wording which are likely to require resort to effective dispute settlement machinery. Article 27 does provide for disputes concerning "interpretation and application" of the convention, but the only compulsory method of settlement is negotiation. All else, including resort to arbitration or the ICJ, is optional, although States may declare acceptance of one or both of these methods as compulsory. This is the typical clause found in most environmental treaties; it offers little or no assurance that unresolved matters of interpretation, or alleged excess of power by the Conference of the Parties or the financial mechanism can be settled by any third party process.[65]

Conclusions

There is much sense in the US objections to the weakness and unsatisfactory nature of the treaty text:

(a) It is a poor reflection of customary law and treaty developments concerning transboundary environmental effects.

---

[59] Article 23.
[60] Article 23(5).
[61] Articles 11(5); 14(4).
[62] Article 26.
[63] Article 14.
[64] Article 2(9), as amended 1990.
[65] Compare the *ICAO Council Case* (1972) ICJ Rep. 6, at pp. 56–60.

(b) It is weaker than other contemporary environmental treaties in omitting explicit reference to the precautionary principle, in the limited supervisory role of the Conference of the Parties, and in its decision-making structure.

(c) Its central obligations of conservation and sustainable use are weak, potentially contradictory and may prove difficult to operate in practice. Moreover the heavily qualified wording of the Convention's central articles, including the frequent use of the words "as far as possible" and "appropriate" leaves open to question how far the parties are in reality committed to anything. It will not be clear for some time whether the Convention provides a viable framework for real progress or is merely an exercise in political symbolism.

(d) Its driving force is as much the allocation of economic benefits to the developing world and a reorientation of the world economy as it is a concern with conservation and sustainable use. It remains to be seen whether the transfer of technology provisions and funding mechanisms will be of real help given the poor record of other treaties with comparable objectives, such as the 1982 UNCLOS. Overall therefore the omens are uncertain, but much will depend on who eventually ratifies and on what then happens in practice to a convention whose implementation and elaboration will be a considerable challenge for many States and for the Conference of the Parties.[66]

---

[66] As of 31 January 1995, 179 states had signed the convention, and 102 states and the European Union had ratified, approved or acceded to it. The first meeting of the Conference of the Parties took place in Nassau in November 1994.

Resolution 2 of the Nairobi Final Act of the Conference to adopt the Convention sets out priority issues for consideration by the Intergovernmental Committee on the Convention on Biological Diversity. See also the Note by the Executive Director of UNEP on "Issues before the IGC of the Convention on Biological Diversity", UNEP/CGD/IC/1/3(1993). The first session of the IGC was held in October 1993. A series of expert panels have also been established to consider issues such as a protocol on the safe transfer and handling of living biotechnology modified organs: UNEP/Bio.Div./Panels/Inf.1 (1993). For an informative study of the problems of protecting biodiversity in the USA, see R. Tobin, *The Expendable Future* (1990).

3

Sustainability, Biodiversity and International Law

Sam Johnston

Introduction

Although the origins of the concept can be traced back to early civilisation, earliest examples of international law adopting some form of sustainable use or management of living resources began appearing in the sixteenth century[1] and first appeared in modern treaty form in the 1885 Convention for the Uniform Regulation of Fishing in the Rhine. The concept first appeared as a guiding philosophy, generally applicable, as opposed to a technique which might be used for a particular resource, in the 1980 World Conservation Strategy ("WCS"). By the UNCED in 1992, sustainable use had become universally accepted as the basis upon which all living resources should be managed.[2]

The key legal definition of the concept is the interpretation of the term in the Convention on Biological Diversity ("CBD"), which declares in Article 2 that:

> "*Sustainable use*" means the use of components of biological diversity in a way and at a rate that does not lead to the long-term decline of biological diversity, thereby maintaining its potential to meet the needs and aspirations of present and future generations.

This definition is elaborated in more detail in the following articles of the Convention. The concept adopted by the Convention owes its origins to the WCS, which described sustainable use as "analogous to spending the interest whilst keeping the capital".[3] Whereas the concept once simply referred to controlling

---

[1] P.W. Birnie, and A.E. Boyle, *International Law and the Environment* (1992), at p. 421, n. 2.

[2] Although numerous examples can be cited to support this notion, we need go no further than the CBD, signed by 156 States and the EU entering into force on 29 December 1993, to confirm the idea that it is now an accepted principle of international law, which if not legally binding upon all States as customary international law is certainly binding upon all those States which have ratified the Convention and binding, to the extent of Article 18 of the Vienna Convention on the Law of Treaties, on those States which have signed it.

[3] See IUCN, *World Conservation Strategy* (1980), at p. 19.

International Law and the Conservation of Biological Diversity (C. Redgwell and M. Bowman, eds.: 90 411 0863 7: © Kluwer Law International: pub. Kluwer Law International, 1995: printed in Great Britain), pp. 51–69

exploitation by setting quotas, establishing hunting seasons or some other regulatory prohibition, contemporary understanding of the concept now implies a number of other key elements. For example:

(a) it implies a duty to *preserve* biodiversity to the extent that the resource has to be maintained in order to ensure that there is no long-term decline;

(b) given that the resource is biological diversity, it implies that it must be managed on a *biological basis* as opposed to a political one;

(c) due to the interdependence of biological systems, management of living resources cannot simply focus on the particular species being used, it must also consider the impact upon other species and the *ecosystem* as a whole;

(d) a further result of this interdependence and the fact that some aspects of bio-diversity which are essential for sustainable use have already been exploited to such a level that they are no longer viable, means that it implies a duty to *rehabil-itate* or restore those denuded aspects of biodiversity which are essential for future sustainable use;

(e) often the use of biological resources is driven to unsustainable levels by factors other than the number of users or the absence of proper quotas and more funda-mental causes are responsible. As a result, the use of biological resources needs to be considered in the widest possible context and needs to be *integrated* into general policy development procedures;

(f) as it requires consideration of the needs and aspirations of future generations, it implies the principle of *intergenerational equity*;

(g) as current understanding of biodiversity is such that in most cases it is not known what level of use is sustainable, it also implies a duty to undertake *research* to develop a better understanding of biological systems;

(h) given that biological systems are dynamic it implies a duty to *monitor* both the use and the ecosystem itself, so as to ensure that the use remains sustainable;

(i) this also implies that management systems must be *flexible*;

(j) the dynamic nature of the resource also, arguably, implies that a *precautionary approach* to management is needed to avoid long-term decline as a result of some unusual perturbation.

This chapter examines the extent to which international law reflects these ideas.

Sustainable Use as a Legal Concept

Different contexts in which sustainable use is applied will emphasise different facets. Whereas in the context of high seas fisheries, the primary emphasis of the concept is on setting quotas, the emphasis in the context of biotechnology and the use of genetic resources will be more on ensuring that the release or application of genetic-ally modified organisms will not cause irreversible damage through exotic species pollution. In the context of utilisation of natural terrestrial habitats, the emphasis will be on integrated management and on ensuring that the decision-making processes are transparent and democratic. In relation to sites of special global significance, the emphasis will be on the intergenerational duty to preserve these exceptional places for future generations.

Different measures are also demanded simply by different situations. For example, ensuring sustainable utilisation of natural habitats in the developed world will require quite different measures from those required in the developing world. Such a variety of examples, all invoking the term sustainable use of biodiversity, has meant that the term is rather enigmatic, with no one accepted meaning.

The Convention on Biological Diversity (CBD)

Nevertheless, due to the universal applicability of the CBD, the substantial attention which the concept receives as one of the central objectives of the Convention, the pivotal role envisaged for the Convention and the contemporary nature of the Convention, the concept as adopted by the Convention is central to any discussion about sustainable use.

The Convention can be interpreted as containing provisions calling for: quotas;[4] preservation;[5] management on the basis of biological unity;[6] holistic ecosystem approach to management;[7] an integrated approach;[8] rehabilitation of denuded aspects of biodiversity;[9] intergenerational equity;[10] research efforts;[11] monitoring of the effects of use;[12] establishment of management systems which are flexible and

---

[4] Apart from the definition in Article 2, Article 8(c) requires that the parties regulate or manage use of "biological resources". Article 10(b) also generally requires controlling of use in that it calls upon parties to adopt measures that "avoid or minimise adverse impacts" on biodiversity. This call is effectively repeated by Article 8(l) which calls upon parties to regulate activities which have significant adverse impact upon the conservation and sustainable use of biodiversity. Further control on use is called for in Article 9(d) and the products of biotechnology in Article 8(g). Some further support can be gained from Article 8(k) which deals with protection of threatened species and the overall ecosystem approach or philosophy adopted by the Convention.

[5] Although the Convention does not use the term preservation in this sense (it does use it in Article 8(j) in another sense), it is none the less implicit in the meaning of "conservation" as used by the Convention. For example, Article 9(c) calls for the rehabilitation of threatened species which is, in effect, preservation of biodiversity. Indeed, the definition of "sustainable use" refers to maintenance of biodiversity which must entail a degree of preservation.

[6] Although the Convention does not contain an explicit call for biodiversity to be managed on the basis of biological unity, it is implicit in many of the provisions. For instance, Article 8(d) (which calls for the promotion of the protection of ecosystems) can only be achieved if it is carried out on a biological basis as opposed to a political one.

[7] Consideration of the effects of use upon more than the target species is required by Article 10(b). This requirement is specifically mentioned for biotechnological use in Article 8(g) and (h). Furthermore, to the extent that the Convention reflects modern thinking about the need to incorporate the local people into the management of protected areas, to ensure that protected areas are seen by the local society as an asset instead of a resource lost or locked up, it also supports an ecosystem approach to management. Thus, calls for "environmentally sound and sustainable development in areas adjacent to protected areas with a view to furthering protection of these areas" (Article 8(e)), which, for instance, provides legal support for the buffer zoning concept pioneered by Unesco's Man and the Biosphere Programme can be considered as requiring an ecosystem approach. The calls to protect and promote the indigenous people rights, to the extent that it ensures consideration of these peoples in management of natural habitats, can also be considered as support for the ecosystem or holistic approach to managing use.

[8] Articles 6(b) and 10(a) call upon the parties to integrate consideration of the sustainable use of biodiversity into the wider context of overall national plans and policy.

[9] Articles 8(f), 9(c) and 10(d).

[10] The definition of sustainable use itself implies intergenerational equity. Article 8(i) indirectly calls for intergenerational equity by exhorting parties to provide conditions which balance present uses with conservation. Para. 23 of the Preamble could also be construed as supporting this with its emphasis on "future generations".

[11] Articles 7, particularly (a) and (b), 9(b), 12, 13, 18, the "clearing-house mechanism" proposed in Article 18(3) and the Subsidiary Body on Scientific, Technical and Technological Advice established under Article 25, all recognise the necessity to develop our understanding of biodiversity.

[12] Articles 7(c) and 25.

able to respond to change;[13] and the precautionary approach.[14] Significantly, the Convention also calls upon the parties to use economic instruments to encourage sustainable use of biodiversity at the private,[15] national[16] and international levels.[17]

Similar interpretations of the principle of sustainable use exist in other UNCED instruments such as Agenda 21, the Forestry Principles, the Programme of Action for the Sustainable Development of Small Island States and in the current draft of the document being negotiated for the UN Conference on Straddling Fish Stocks and Highly Migratory Species of Fish.[18] Arguably, support for this interpretation can also be found in several earlier instruments such as the African Convention, the United Nations Convention on the Law of the Sea (UNCLOS), the ASEAN Agreement, the Convention on the Regulation of Antarctic Marine Living Resources, the Convention for the Protection, Management and Development of the Marine and Coastal Environment of Eastern African Region and its Protocol Concerning Protected Areas and Wild Fauna and Flora in the Eastern African Region and the Convention for the Protection and Development of the Marine Environment of the Wider Caribbean Region and its Protocol Concerning Specially Protected Areas and Wildlife.

The concept as elaborated in the Convention, however, suffers from a number of important deficiencies. Even within the context of framework conventions, the normative content of the obligations are relatively weak and most of its provisions are little more than purely exhortatory. For instance, compared with UNCLOS, the CBD provisions are vague and almost devoid of commitment and from this viewpoint disappointing, given that the CBD was adopted ten years later.[19] As a result, every aspect of the concept adopted by the Convention lacks the necessary detail to be capable of effecting specific activities or even in some respects providing any

---

[13] See Articles 8(l) and (i).

[14] Although the precautionary principle was removed from the penultimate draft it is still referred to indirectly in the Preamble (8th and 9th paragraph).

[15] Article 10(e).

[16] Article 11.

[17] Under Article 20 each contracting party undertakes to provide "incentives" and "financial support" to help other contracting parties undertake activities which are intended to achieve conservation, sustainable use of biodiversity and the equitable sharing of its benefits.

[18] A/CONF.164/13/22.

[19] For instance, Article 61 of UNCLOS obliges States to control use in their EEZ and provides:

> 2. The coastal states, taking into account the best scientific evidence available to it, shall ensure through proper conservation and management measures that the maintenance of the living resources in the exclusive economic zone is not endangered by over exploitation...
>
> 3. Such measures shall also be designed to maintain or restore populations of harvested species at levels which can produce the maximum sustainable yield, as qualified by relevant environmental and economic factors, including the economic needs of coastal fishing communities and the special requirements of developing States, and taking into account fishing patterns, the interdependence of stocks and any generally recommended international minimum standards, whether subregional, regional or global.
>
> 4. In taking such measures the coastal States shall take into consideration the effects on species associated with or dependant upon harvested species with a view to maintaining or restoring populations of such associated or dependant species above levels at which their reproduction may become seriously threatened.

This is a far stronger commitment than Article 8(c) of the CBD which simply provides: "Each Contracting Party shall, as far as possible and as appropriate ... Regulate or manage biological resources important for the conservation of biological diversity whether within or outside protected areas, with a view to ensuring their conservation and sustainable use."

guidance. For example, nothing in the Convention provides much indication as to how quotas for any particular resource should be established, nor what criteria are important in setting quotas nor what general principles of management should be relied upon to establish quotas. Should quotas for fisheries be established on concepts used in other instruments? If so, which one; maximum sustainable yield, optimum sustainable yield, optimum yield, optimum population, optimum sustainable population, optimum economic yield, optimum ecological resource management or the ecosystem approach of CCAMLR?[20] What basis should be used for forestry quotas? Should there be a precautionary management approach? Nor is there any indication as to how to reconcile the interests of future generations with those of the present in calculating quotas for the use of any resource, or over what time frame to measure "decline by" or even what is meant by "adverse impacts".[21] Similar problems regarding detail exist for all the other aspects of the concept.

There are a number of more fundamental problems with the terms used in the Convention which may hinder efforts to develop a more binding regime to ensure sustainable use of biodiversity. A number of the key features of the concept are not referred to directly in the Convention and are present by implication only. The concepts of intergenerational equity, the precautionary approach, a biological basis for management and ecosystem management are only implied in the Convention. Their absence is noteworthy, even in the context of a framework convention, and has the consequence that the presence of these elements, although essential for sustainable use, is not assured in any binding regime which is developed under the auspices of the CBD.

Another problem is that several important sectors which use the components of biodiversity are not covered by the CBD. One such sector is *ex situ* collections gathered before the Convention came into force, which probably represent the single most valuable aspect of biodiversity and which are specifically excluded from the obligations of the Convention.[22] Even though efforts are currently being made partially to rectify this deficiency,[23] it will be many years before even this partial solution is implemented.[24]

A further problem arises from the definition of biological diversity in the Convention to include "diversity within species" as well as between species and of ecosystems. To use the components of biological diversity in a way that does not lead to the long-term decline of diversity within species means not only retaining the number of different species but also the population size of each species. A consequence of this is that countries are obliged to not only retain representative samples of all natural habitats, but the same amount of natural habitat as well. Laudable

---

[20] Explanation of these terms can be found in Birnie and Boyle, *op. cit.*, n. 1, p. 435. Article 22 provides little help in resolving this dilemma because they all represent different alternatives none of which are inherently in conflict with one another nor for that matter contrary to "the law of the sea".

[21] The complexity of the issues are well described in a FAO report to the United Nations Conference on Straddling Fish Stocks and Highly Migratory Fish Stocks, the Precautionary Approach to Fisheries with Reference to Straddling Fish Stocks and Highly Migratory Fish Stocks (A/CONF.164/INF/8).

[22] Article 15.3.

[23] Resolution 3 of the Nairobi Final Act requested that the management of the *ex situ* collections of the CGIARC be brought into line with the aims of the CBD. It is, however, only partial because these holdings only constitute 12% of the world plant germplasm accession and do not include any animal or microbial holdings.

[24] See further below for discussion of the current position of these measures and the problems which need to be overcome before management of the network can be brought into line with the principles of the CBD.

though this may be, it effectively means that countries which, for instance, use living resources which cannot be readily replaced, such as tropical rainforests, in order to abide by this obligation of sustainable use must stop *all* use, or conversion of their tropical rainforest: an obligation which for many countries is not only unacceptable but also impossible to fulfil. Although the obligation is qualified by "appropriateness", adopting provisions which automatically require the use of this qualification indicates poor drafting and more generally makes it even harder to ensure the observance of other provisions.

Another drafting problem with the CBD is the inconsistent way in which "sustainable use" is used. In some places, it speaks of the "conservation and sustainable use of biological diversity". In others it speaks of the "conservation of biological diversity and the sustainable use of its components". In still other places it speaks of the "sustainable use of biological resources". In contrast Chapter 15 of Agenda 21 speaks of the "conservation of biological diversity and the sustainable use of biological resources" and does this consistently throughout. With regard to the first usage it is technically not possible to use biological diversity itself, in that the term biological diversity refers to no more than the variability of life and what is used by society is only its tangible manifestations such as genetic material, organisms and ecosystems. The varying terminology although of itself perhaps of little consequence does suggest a confusion or lack of clarity amongst those who negotiated the CBD and allows for a degree of legal obfuscation which can only hinder the development of a binding regime.[25]

Finally, insisting on emphasising the importance of sustainable use by referring to it separately when speaking of conservation does invite the understanding that conservation is a concept which is somehow independent of sustainable use, despite the fact that this is at odds with the widely accepted contemporary meaning of conservation. The possible detrimental effect of having such a limited definition of conservation in such a pivotal instrument and the tensions and conflicts that this may create with other instruments which use the term in the more conventional and wider sense are manifest.

Sustainable use depends upon notions of good governance and management. These are manifest in the CBD in the provisions calling for environmental impact assessments, research and training, information collecting, public participation in decision making, transparent and democratic governance. Analysis of these provisions is beyond the scope of this chapter;[26] however, in many respects these provisions are inadequate and in some instances actually detract from the accepted international norms relating to such provisions. The extent that these inadequacies obstruct sound principles of management being adopted may present problems for future regimes developed under the Convention.

Custom and Other Instruments

The extent to which customary international law and other instruments can be used to help overcome these deficiencies is limited. Although there is growing support for the modern concept of sustainable use, none of the duties which it implies, even if it is accepted that they are principles of customary international law, are well developed principles, involving specific duties, clearly defined and generally accepted. Prior to UNCED all that could be ascertained with absolute certainty about customary

---

[25] Indeed it may be one of the reasons behind the problems which arise from the use of "natural resources" in Article 15.1, which is raised later below.

[26] See chapter by Alan Boyle, *infra*.

international law in this respect was that, although there was widespread evidence that most States accept that it is their duty to co-operate in the protection and use of living resources, the content of this duty remained unclear. Birnie and Boyle conclude in their review that the concept is merely a general framework within which conservatory, economic and social goals can be balanced.[27] Since their review, UNCED and its aftermath have provided a plethora of other instruments invoking the clarion call of sustainable use or utilisation of living resources.[28] Although these instruments have amplified the concept so that it now refers to more than simply setting quotas based upon target species, and have even arguably developed customary law to the extent that the concept implies all the corollary duties outlined above, few, if any, of these instruments have significantly developed the concept beyond the general guiding terms found in the CBD, even within the parameters of the relevant instrument, let alone in any wider sense.

For example, continuing the theme of quotas for fisheries, Chapter 17[29] of Agenda 21 which deals with the use of marine living resources has two of its seven programme areas specifically devoted to sustainable use. The chapter contains many of the key features of the concept, with calls for management plans and quotas based upon on multi-species approach which also "take into account the relationships among species, especially in addressing depleted species".[30] This call for ecosystem type management is repeated several times throughout the text.[31] There are also commitments to maintain and restore populations of marine species at levels that can produce the maximum sustainable yield,[32] controls on fishing gear and practices that minimise waste[33] and for effective monitoring and enforcement with respect to fishing activities.[34] The chapter also calls for management approaches to be "precautionary and anticipatory in ambit".[35] It does not, however, develop the key concept in this context, namely, the basis upon which quotas should be established. Rather, it simply repeats the provisions of UNCLOS in this respect with calls for quotas to be established on the basis of maximum sustainable yield.[36] It does not explain what is meant by this term and, therefore, can hardly be considered as developing the concept in any significant way.

---

[27] Op. cit., n. 2, at pp. 435–440.

[28] Some of the more important instruments and conferences looking at various aspects of sustainable utilisation are: (a) the UNCED instruments themselves – the Rio Declaration, Forestry Principles and Agenda 21; (b) the follow-up to UNCED; the 1995 UN Agreement on Straddling Fish Stocks and Highly Migratory Species of Fish, the Programme of Action for the Sustainable Development of Small Island States adopted at the UN Global Conference on Sustainable Development of Small Island States (n.b. Chapter O), the Desertification Treaty and the 1995 session of the Commission on Sustainable Development ("CSD"), which reviewed Chapter 15 of Agenda 21; (c) other relevant work includes UNEP's Montevideo Program (n.b. Heading K: Conservation Management and Sustainable Development of Soil and Forests) and the renegotiated International Tropical Timber Agreement; and (d) the preparatory work of the ICCBD for the first COP, in particular the results of Working Group I.

[29] "Protection of the oceans, all kinds of seas, including enclosed and semi-enclosed seas, and coastal areas and the protection, rational use and development of their living resources".

[30] Para. 17.45.

[31] For example, paras. 17.70, 17.74(c) and 17.46(b).

[32] Para. 17.78(b).

[33] Para. 17.78(d).

[34] Para. 17.78(e).

[35] Para. 17.1.

[36] Paras. 17.49(b) and 17.74(c).

The document being negotiated at the UN Conference on Straddling Fish Stocks and Highly Migratory Fish Stocks[37] which is intended to build upon Chapter 17 of Agenda 21, also contains many of the features of the concept as developed by the CBD. For instance, it favours a precautionary approach for the management of fish stocks and intergenerational equity is an important consideration as well. It calls for biological unity as a basis of management. In some respects the current draft text[38] even contains material developments on some aspects of the concept. For instance, it casts some doubt about the sustainable basis of maximum sustainable yield and calls for a reconsideration of its role in setting quotas. The current draft suggests that maximum sustainable yield be used as a philosophical reference instead of a mathematical one, or that it be used as minimum standard for rebuilding depleted stocks. There is also a recognised need for a set of criteria to be developed to measure sustainable use. Despite the promise of this draft text, one suspects that as its aspirations come closer to reality these developments will probably disappear and the text will be brought more in line with UNCLOS and Agenda 21.

Thus, despite nearly four years of negotiations and the adoption of numerous codes, principles and agreements, little progress has been made in developing the legal understanding of sustainable use in the context of setting quotas for fisheries. Few, if any, of the substantive issues have been settled. The FAO, in a preparatory report for the UN Conference on Straddling Fish Stocks and Highly Migratory Fish Stocks in May, 1994, noted that "One of the major tasks for research and management is to develop agreement on standards, rules, reference points and critical thresholds on which to base decisions and meet the management requirements of the 1982 Convention [UNCLOS] and Agenda 21, for the various types of ecosystems and resources."[39] Management concepts "remain at the level of international rhetoric".[40] The failure to develop the legal understanding of this particular aspect of the concept in this context does not bode well for other aspects, given that the issues which need to be resolved in the context of high seas fisheries are relatively well defined and simple when compared to other uses of biodiversity.

Similar conclusions may be drawn about the extent to which international law defines the other aspects of sustainable use. Thus, although the concept now obliges countries to do more than simply set quotas, which represents an important step forward in the development of the concept from a guiding principle to a binding commitment, what exactly it implies or obliges is not entirely certain due to the ambiguity with which legal instruments, particularly the CBD, refer to a number of the essential features. Furthermore, much remains to be decided before any of these duties, and sustainable use as a legal principle, can be considered to be more than a guiding philosophy. Moreover, there seems little in the review processes proposed for UNCED or any other negotiations which might be the harbinger of great change in this regard.

---

[37]   At the time of writing, there is still disagreement over what type of instrument the Conference will adopt. Many of the coastal States with EEZs are calling for a convention or some other form of legally binding document, while many of the distant water fishing States think that a recommendation or some other non-binding instrument is enough and others are calling for the whole issue to be developed within the UNCLOS framework.

[38]   A/CONF.164/13/22.

[39]   Para. 42.

[40]   Para. 57.

Economic Theory of Sustainable Use

Even though the traditional concept of sustainable use may have been little developed in recent instruments, these evidence for the first time comprehensive consideration of an economic approach to understanding the term. It is in this sense that the CBD and UNCED documents have made the most significant legal contribution to the development of the sustainable use concept.

The importance of this development is hard to underestimate. As 95% of biodiversity is now found within the jurisdictional control of States, management of biodiversity is now an issue which largely deals with sovereign assets, as opposed to unowned international resources. As a result, an approach based upon exhortation to implement traditional regulatory techniques such as quotas on use, or obligations to establish protected areas, will probably not be as effective as one based upon persuasion through incentives. The general failure of more traditional regulatory type approaches to controlling use in the context of living resources not only supports this supposition but demands that alternative approaches are developed. Examples of the inadequacies of the regulatory type of approach abound. One instructive example, are the attempts to regulate whaling. Under the 1948 International Convention for the Regulation of Whaling, the International Whaling Commission has tried for nearly fifty years to implement a regulatory regime which managed whaling on a sustainable basis. Yet, despite the numerous attempts to develop quite sophisticated management plans, none has so far been successful. The only approach which has had some success in conserving whales is the moratorium on all commercial whaling which has been in force since 1985: an approach which can hardly be described as sustainable use of a resource. Indeed, even at the national level, where quotas and other regulatory techniques are capable of being enforced in a manner unlike the consensual nature of international regulation, the purely regulatory approach to sustainable use has rarely been successful.[41]

Economic theory holds that in order for any resource to be properly managed, the price of that resource needs to reflect all the values that society places upon it, social and economic, including the costs of the external effects associated with exploiting, transforming and using the resource together with the costs of future uses forgone. Put simply this means internalising the external benefits and costs, known as externalities, associated with using a resource. Failure properly to value these resources means that incorrect signals are sent to decision-makers, conveying, in turn, misleading information about the resource scarcity and thus providing inadequate incentives for management, efficient utilisation and enhancement of living resources. Sustainable use as an economic concept has, therefore, very little to do with setting quotas or other regulatory controls over the exploitation of living resources, rather it is about creating the right incentives so that those who manage biodiversity, the stakeholders, will be motivated to conserve it.[42]

Internalising externalities is not a new concept in international law and has been introduced to many lawyers in the guise of the polluter pays principle. This represents the reverse of what is necessary in the context of biodiversity conservation, in that the objective of the polluter pays principle is to internalise the environmental costs of an activity, thereby providing a means whereby producers are motivated to look after the negative environmental effects of their product. What is more important in the context of biodiversity conservation, given the current situation, is to internalise the

---

[41] National attempts to ban the taking of endangered species represents an example of such a policy failure which is relevant for almost every country in the world.

[42] See T.M. Swanson, *The International Regulation of Extinction* (1994).

benefits that biodiversity generates. Complete internalisation of biodiversity's costs and benefits is commonly known as full cost resource pricing or the user pays principle. Such an approach to sustainable use has several advantages compared with traditional regulatory approaches; it is more self-enforcing in that it relies upon motivation and self-interest rather than punishment, and it also addresses the underlying causes of non-sustainable use of biodiversity such as poverty and debt.

The most important type of instrument to internalise benefits and motivate proper management of biodiversity is the development of some form of property right over the components of biodiversity. The allocation and enforcement of property rights as an instrument for internalisation can be accomplished through the use of land titles, water rights, user rights (licensing fees, concession bidding), intellectual property rights ("IPRs"), discovery rights, farmers' rights, development rights, discovery rights, or reassignment rights. In addition, some customary regimes for the management of common property resources may have an equivalent effect.[43] Property rights are used in this context in the widest possible manner and are not restricted simply to the notion of absolute title to the property but also include legal rights which could be considered as less than absolute ownership. A property right can be manifested in ways other than the simple assertion of title. For instance, it might simply be the controlling of access through licensing, contracting, or simply requiring notification of use, or even declaring patrimony or sovereignty over a species. As many of the benefits of biodiversity manifest themselves internationally, to ensure full internalisation the property rights need to be recognised internationally as well.

Other measures which help to internalise international benefits and costs are attenuation or elimination of policies that distort resource allocation such as agricultural subsidies; correction of underpriced primary products through readjustment of international terms of trade or international commodity agreements; use of economic incentives such as subsidising the setting aside of agricultural land, or using disincentives such as taxing activities which degrade the environment; provision of assistance, financial or technical, which is somehow hypothecated to biodiversity conservation on the basis of notions of equity or responsibility; "side payments" or lump sums transferred directly to compensate a country for taking measures such as not developing a natural habitat; calls for some form of tradeable permits, whereby developed countries are obliged to purchase "biodiversity credits" from developing countries; or international taxation schemes where the revenue is raised on the basis of biodiversity use and hypothecated for biodiversity management.

In order for property rights of any kind to be able to achieve internalisation they need to posses certain features. Property rights can only be assigned over a resource if the condition of "excludability" can be met. This is largely dependant upon institutional arrangements. Air is often cited as the archetypal "non-excludable" good, but measures like the Montreal Protocol and the proposals under the UN Framework Convention on Climate Change illustrate that exclusion can be arranged. Once property rights are assigned it must be possible for the holders to enforce their entitlement. This depends not only upon the capacity of the holders but also upon the administrative capacity of the system and on the cost of enforcement. Property rights must also be exclusive, transferable, secure and conferred for a sufficiently long time period. Another requirement is that those assigned the rights should be able to both take the decision concerning the use of the resource and to implement those decisions. Moreover they will need to understand how the

---

[43] For a more detailed explanation of economic instruments and their role in environmental protection see OECD, *Economic Instruments for Environmental Protection* (1989) and UNCTAD, *The Effect of the Internalization of External Costs on Sustainable Development* (1994) (GE.94–50541).

resource relates to its environment so that decisions are likely to be followed by the appropriate effect. Administrative practicality, however, argues in favour of assigning the rights to national governments, even though the understanding essential to effective management of the resource is likely to be situated at a more local level.

The Principle of Full Cost Resource Pricing

Prior to UNCED

As a principle of international law, the principle of full cost resource pricing has its origins in the notion that the principle of equity obliges countries, particularly developed ones, to pay for the environmental services which they receive from other countries, particularly developing ones. This idea first emerged in several bilateral treaties early this century which used substitution to secure the agreement of all parties to measures designed to ensure sustainable use of a particular resource. One early example of the substitution method was the 1911 Treaty for the Preservation and Protection of Fur Seals between Japan, the UK (on behalf of Canada), the USA and the USSR. The Treaty did not set any quotas on the number of seals that could be taken but instead banned pelagic (taking at sea) sealing. This ban was of obvious benefit to the USA and the USSR, where the breeding grounds for the seals were located. Under the terms of the Treaty, in return for Japan and Canada agreeing to effectively give up sealing, the USA and USSR were required to share a proportion of their harvests with them.[44]

Similar notions of equity are also evident in the early treaties which called for protected areas such as the 1940 Washington Convention or the 1968 African Convention. Under these conventions, payment was again in kind: parties were called upon to reciprocally set aside natural habitats for the benefit of all. The 1972 World Heritage Convention developed this concept of reciprocity by linking it with material assistance. Although the financial assistance provided by the World Heritage Convention was principally motivated by the pragmatic notion that conservation was only possible in many developing countries if assistance was forthcoming, equitable notions of compensation for global services rendered were also a motivating factor. The idea was developed further under the EEZ concept contained in UNCLOS, whereby prudent management was encouraged not by merely exhorting parties to undertake certain measures, but by allocating control and in a sense ownership, of the marine resources of the EEZ to the coastal States. UNCLOS further encouraged prudent management by acknowledging the right to sell any spare capacity in the fishery stocks within the EEZ to other countries[45] which, in theory, therefore, rewarded countries who properly managed their EEZ.

Attempts to create some system of rewarding countries which manage their resources in a sustainable way have also been made in a number of other conventions. The sustainable utilisation exceptions to the trade prohibitions of CITES such as the "ranching exemptions", species based quotas and the ill-fated African Elephant Management Quota System are one example. Others include: the "wise use" exception in the Ramsar Convention; the attempts by the FAO to develop the concept of Farmers' Rights and its accompanying international fund on plant genetic resources within its Global System for the Conservation and Utilisation of Plant Genetic

---

[44] Another example of substitution is the 1987 US–Canadian Agreement on Caribou of the Porcupine Caribou Herd (IUCN ELC TR 2863).

[45] Article 62(4)(a).

Resources;[46] attempts by the International Tropical Timber Organisation ("ITTO") to implement its Sustainable 2000 Programme; attempts of seven Meso-American States to develop property rights over their living resources;[47] and the Andean Pact's attempts to develop a common approach to regulating access to their genetic resources.

To a limited extent the Global Environment Facility ("GEF") and the funds made available under it are in recognition of services rendered and as such can be considered as an example of full cost resource pricing. The compensation that is promised under the GEF is, however, limited to the "incremental" cost of implementing measures for the conservation of biodiversity which provide global environmental benefits and in no way attempts to reward conservation with the benefits that biodiversity provides. Other examples which indirectly seek to incorporate principles of full cost resource pricing at an international level include: debt-for-nature swaps; attempts by commodity cartels to increase prices so as to better reflect true production costs; efforts to make international trade instruments more environmentally orientated; and efforts to eliminate subsidies in trade instruments.

These pre-UNCED examples are, however, only isolated attempts to incorporate some basic economic incentives into international efforts to conserve biodiversity. They do not represent a unified or co-ordinated attempt to incorporate the principle of full cost resource pricing into international law and it is not possible to argue that they represent the basis for suggesting that there is a customary international law principle to this effect. At best, they may used by some to suggest that they amount to an emerging custom. Apart from the fact that they do not comprehensively cover use of living resources, certain fundamental features of the principle have not been universally accepted. For example, prior to UNCED the idea that species could be "owned", or that there was a general duty to compensate States for conservation of natural habitat, were, at the very least, contentious ideas. Furthermore, almost without exception, they have failed in implementation.[48]

UNCED

The Convention on Biological Diversity (CBD)

The CBD, at a general level at least, embraces the idea of the principle of full cost resource pricing. There is, however, little in the way of detail in the Convention and normatively it goes no further than merely exhorting parties to adopt appropriate measures. How this might be achieved is not elaborated in the Convention. None the less, the Convention does contain a number of provisions which will prove useful for any attempts to implement the principle.

A key aspect to implementing the principle is the creation of some form of internationally recognised property right in the components of biodiversity. In this regard the Convention contains a number of important developments. First and foremost is the explicit acknowledgement in Article 15.1 of sovereign rights over "natural resources"

---

[46] C89/REP: Res. C4/89, C5/89 and C3/91.

[47] See the Non-binding Agreement of Intent to Coordinate their Legislation regulating the use of Genetic Resources found in Central America (6 June 1992) signed by Belize, Costa Rica, El Salvador, Guatemala, Honduras, Nicaragua and Panama, IER 397, 17 June 1992, under which these nations intend to develop rules on access to biological materials in exchange for material benefits, transfer of technology and training which represents a first step towards developing a legal regime addressing biological resource concessions in the area.

[48] For example, under CITES's Elephant Quota Scheme, African countries were only capturing 5% of the value of their raw ivory exports (Barbier 1990).

and a State's right to determine access to the "genetic resources" located within the jurisdiction of the State. The definition of "genetic resources" adopted by the CBD is broad enough to effectively include all genetic material or, in other words, all the usable components of biodiversity. This is supported by the principle of permanent sovereignty over natural resources found in Article 3 which entitles States to "exploit their own resources". These provisions represent a significant development of international law, as prior to the CBD, due to efforts to declare genetic resources the "common heritage of mankind", principally by the FAO in relation to plant genetic resources,[49] it was questionable whether international law would accept the notion that all the usable components of biodiversity were appropriable. For instance, whilst it is now generally accepted that individuals of a species found solely within a State's territory are capable of being "owned", prior to the CBD, concepts of ownership over improved genetic resources were extremely limited[50] and never extended to species themselves or to genetic resources of wild species, which in an economic sense represent a vital element of biodiversity. The effect of the provisions of the CBD is arguably to allow these last remaining unownable usable components of biodiversity to be appropriable for the first time. This provides a legal basis which will not only allow more complete internalisation but will also provide for the first time the basis for the privatisation of an aspect of biodiversity which has the potential to make the greatest contribution, in a material sense, to internalisation at the international level: the resources and wealth of genetic diversity.

Notions of sovereignty or ownership over genetic resources are given further meaning in later Articles which provide that access to these resources be on the basis of prior informed consent[51] and upon mutually agreed terms:[52] provisions which provide further legitimacy for controls over access. The Convention requires that there be equitable sharing of the benefits of the use of genetic resources, both in a financial, scientific[53] and technological[54] sense, which means that not only are countries entitled to demand payment for use of their genetic resources, but that those countries which use them, are obliged to compensate for that use.[55]

More generally, calls for the equitable sharing of benefits also support the principle, as internalisation of biodiversity's externalities would go a long way to realising this goal.[56] The extent to which the rights of indigenous and local people in respect of their

---

[49] See Article 1 of the International Undertaking on Plant Genetic Resources ("IUPGR"). This is by no means the only instance of such a declaration. For example, the Preamble of the Ramsar Convention declares that "waterfowl ... should be regarded as an international resource" (para. 5).

[50] See further the chapter by Ian Walden in this volume.

[51] Article 15.5.

[52] Article 15.4.

[53] Article 15.7.

[54] Articles 16.3, 19.1 and 19.2. In this regard, Article 16 requires not only fair and equitable sharing of technology but also requires that in respect of developing countries, access and transfer of technology must be provided or facilitated on "fair and most favourable terms", including on concessional and preferential terms where mutually agreed.

[55] Article 15.7. Further support for the obligation to pay is also found in Article 20.1.

[56] Although there is a considerable degree of overlap between the principle of full cost resource pricing and ensuring that the benefits of the use of biodiversity are equitably distributed, they are distinct concepts which ultimately require different measures. Whereas full cost resource pricing is simply about ensuring that those who actually control biodiversity receive sufficient incentives for conserving it, equitable sharing is principally concerned with ensuring that everybody is fairly rewarded and receives just compensation for their past, present and future contributions to the management of biodiversity. Thus, whereas the former is primarily concerned with hypothecation of the benefits of use and has little to say about existing systems of ownership, the latter is principally concerned with the inequities of existing form of rights over biodiversity.

knowledge and use of the components of biodiversity are developed as a result of Articles 8(j), 10(c) and 18.4, will also result in internalisation.[57] Further support for the principle arises as a possible consequence of the requirement to integrate consideration of sustainable use of biodiversity into the wider national decision-making process. An effective method for achieving integration of sustainable use into such processes is through "monetising" the costs and benefits that biodiversity provides, or even through internalising its externalities.

The CBD as a whole, therefore, seeks to create a new relationship between the providers and users of genetic resources where the provider is not only entitled to share in the benefits arising from the use of these resources but is also provided with the legal basis on which to demand their entitlement. This change has been described as a "fundamental paradigm shift" away from the common heritage principle. Although most commentary on this shift has focused upon the equitable issues which this raises in the context of the North/South debate, it also provides the foundation for proper implementation of the principle of full cost resource pricing.

The subjects of control over genetic resources, ownership and development of measures to ensure the equitable distribution of benefits, all to some extent manifestations of full cost resource pricing, have received extensive attention in the preparatory work for the first Conference of the Parties of the CBD. Intellectual Property Rights are seen by many as providing a way in which countries that manage and provide genetic resources for commercial uses can ensure that they receive their "fair" share of the benefits. At the last ICCBD, IPRs were "noted as possibly constituting a basis for fair and equitable sharing of benefits" and several delegates called for further study of the impact of IPRs on the objects of the Convention. This idea is explicitly recognised in Article 16.5, which also calls upon the parties to "cooperate in this regard subject to national legislation and international law in order to ensure that such rights are supportive of and do not run counter to its objectives". These calls have their basis in Articles 15.7, 16 and 19 of the CBD, which calls upon parties to take measures to ensure that the benefits of the use of genetic resources, and the benefits arising from the biotechnologies based upon genetic resources are shared in a "fair and equitable way" with the parties providing the genetic resources. Ensuring the equitable sharing of benefits would go a long to realising internalisation.

Ensuring access to the benefits of commercial use of plant genetic resources for countries of origin is a central aim of the efforts to bring the holdings of the Consultative Group of International Agricultural Research Centres ("CGIARC")[58] within the ambit of the CBD, the renegotiation of the International Undertaking on Plant Genetic resources ("IUPGR")[59] and its possible adoption as a protocol to the CBD and the efforts to develop and implement the FAO's Farmers' Rights.[60]

---

[57] Note that in this regard the ICCBD agreed to ask the first COP to initiate a study into how to implement Article 8(j), an important aspect of which will be the type of rights that indigenous people should be given over the biodiversity with which they interact in order that they are able to play their role in its conservation and sustainable use. Also see the background paper prepared for ICBBD/2 by the Interim Secretariat, "The rights of indigenous and local communities embodying traditional lifestyles: experience and potential for implementation of Article 8(j) of the Convention on Biological Diversity" (UNEP/CBD/IC/2/14).

[58] Item 4.2.3 of the Second ICCBD's Agenda. Also note Resolution 3 of the Nairobi Final Act and the recommendations of the ICCBD to the COP.

[59] Resolution 7/93 of the 27th FAO Conference, 1993.

[60] See Resolution 3 of the Nariobi Final Act, Resolution 3/93 of the FAO Conference, 27th Session, November 1993 and the planned Fourth International Technical Conference of the FAO's Commission on Plant Genetic Resources.

The attention that these issues received in the ICCBD process,[61] although not addressing anything of substance and someway off being resolved, do suggest that these matters may be one of the first substantive issue that the CBD and its administrative mechanisms consider. The Convention, therefore, not only provides the legal basis for internalisation of a key benefit of biodiversity, but will be one of the key fora in which such measures are developed.

As with the other features of sustainable use, the Convention does not provide much in the way of detail and leaves many vital issues unresolved. For instance, as noted previously, property rights can only be assigned over a resource if the condition of "excludability" can be met, yet the extent to which the Convention conveys control over the components of biodiversity to a particular country and allows biodiversity to be made an excludable resource is not certain. The Convention, rather than conferring sovereign rights over a country's "biological diversity", confers these rights over a country's "natural resources"; a term not defined in the Convention, nor for that matter in any of the other UNCED documents. This gives rise to two problems. First, it is arguable whether the term "natural" covers genetic resources which have been improved either through biotechnology or informally through more traditional and local breeding practices. These types of genetic resources are perhaps the most valuable type of genetic resources. To not be able to appropriate them would seriously limit the extent to which the full benefits of biodiversity may be internalised and thus represent a significant limitation to the principle of full cost resource pricing. The other problem arises from whether the use of "natural" allows ownership of the intellectual resource which genes represent. Though, there may be no doubt that the term includes individuals of a particular species and genetic material in a physical sense, some authors have questioned whether it can extend to the intellectual property of biodiversity.[62]

The attendant problems with using some form of IPRs as a method for internalisation are considerable as existing IPRs possesses few of the necessary features for internalisation described above and the CBD provides little guidance as to how these issues might be resolved. For example, existing types of IPRs, such as patents or plant variety rights, which require some element of novelty, cannot be used to establish ownership over wild species nor landraces used in traditional societies, nor for that matter for the knowledge of local and indigenous peoples. Furthermore, even if IPRs were extended to cover such species, there would still be serious problems with allocation of those rights, how the rights might be enforced, the administrative structure to support the rights and how the benefit of the rights might be delivered to the stakeholders. None of these issue has been formally considered at the international level and, therefore, it will be some time before IPRs will be able to make a meaningful contribution to implementing internalisation.

As a result of IPRs deficiencies, the FAO has for plant genetic resources which are useful for agriculture promoted an alternative intellectual property rights scheme known as Farmers' Rights. These are meant to be a mechanism through which the contribution that traditional farming communities make to the raw genetic material which modern agriculture relies upon is recognised[63] and are to be implemented

[61] See the report of Working Group I of the ICCBD (UNEP/CBD/IC/2/L.3) and the furore concerning the World Bank and the CGIARC network.
[62] See C. de Klemm, *Biological Diversity: Conservation and the Law* (1993), at p. 59.
[63] See Resolution 3 of the Nairobi Final Act, Resolution 7/93 of the FAO Conference, 27th session, November 1993 and the upcoming Fourth International Technical Conference of the FAO's Commission on Plant Genetic Resources.

through "an international fund for plant genetic resource", which is to be managed by the FAO's Commission on Plant Genetic Resources ("CPGR"). Despite the fact that Farmers' Rights to some extent overcome the problems of traditional IPRs in that they recognise wild and traditional varieties of plants and avoid some of the allocation and enforcement problems by vesting these responsibilities in the CPGR, none the less there remain a number of difficult legal and practical issues which need to be resolved before the scheme can be implemented.

For example, the CPGR, the principal promoter of the concept, recognises the nebulous state of Farmers' Rights. At their last biannual meeting, the 5th Session, the Commission

> agreed, however, that a number of questions remain open and would need to be addressed. These include the nature of the funding (voluntary or mandatory); the question of linkage between the financial responsibilities and the benefits derived from the use of PGR, and the question of who should bear financial responsibilities (countries, users or consumers). It also remained to be determined how the relative needs and entitlements of beneficiaries, especially developing countries, were to be estimated, and how farmers and local communities, would benefit from the funding.

The Commission also noted that despite the fact that there had been "significant debates on these and related issues in FAO, UNEP and UNCED, as well as a number of NGO fora ... nonetheless, more conceptual thinking was required, to answer these questions, and to determine appropriate mechanisms for the realisation of Farmers' Rights."[64]

Furthermore, the insistence on using an "international fund" managed by the CPGR not only has little chance of being acceptable politically but also has serious limitations in the extent to which the idea of full cost resource pricing can be fully incorporated. For example, neither the FAO nor other examples of these international funds, such as the GEF, have properly developed mechanisms for delivering global benefits to local holders. As public mechanisms they are neither an efficient nor a transparent means of delivery, in that they are controlled by national and international structures as opposed to local ones. They also retain an element of voluntariness, in the sense that payments through these mechanisms are often thought of as aid by many developed countries and not as payment for services rendered. In the long term, private mechanisms, whereby the actual user pays the actual owner will be superior and necessary to properly implement internalisation.

With such fundamental issues requiring further "conceptual thinking" it will be some time before Farmers' Rights are implemented.

Other Instruments

The extent to which other international instruments and international law generally can resolve some of these problems is limited. In this regard Agenda 21 provides a good illustration of the limitations found in other instruments. Even though the Plan calls for internalisation of costs and benefits and the use of economic incentives throughout its text, nowhere does it outline specific measures which might be taken to implement internalisation let alone actually require any particular step.

For example, Chapter 8 on decision-making procedures recognises the increasing use and importance of economic instruments and devotes an entire programme area

---

[64] Report of the Commission on Plant Genetic Resources, Fifth Session, Rome 19–23 April, 1993 (CPGR/93/REP), para. 18.

to the issue.[65] Paragraph 31(b) is particularly noteworthy and calls upon governments to "move more fully towards integration of social and environmental costs into economic activities, so that prices will appropriately reflect the relative scarcity and total value of resources and contribute towards the prevention of environmental degradation". The chapter goes on to to identify some activities which will help implement these goals such as reducing or eliminating "subsidies" and to "reform or recast existing structures of economic and fiscal incentives to meet environment and development objectives" and to "move towards pricing consistent with sustainable development objectives"[66] which is slightly more specific but still is some way off identifying concrete activities. The chapter, however, indirectly acknowledges that full cost resource pricing is far from being an accepted principle by its extensive calls for further research into the concept.

In Chapter 17, States commit themselves to "develop economic incentives, where appropriate, to apply ... means consistent with the internalization of environmental costs".[67] Internalisation is called for in general terms elsewhere in the chapter on the basis that coastal States "should obtain the full social and economic benefits from sustainable utilization of marine living resources within their exclusive economic zones and other areas under national jurisdiction".[68] The chapter acknowledges that the EEZ concept embodied in UNCLOS sets "forth rights and obligations of States with respect to conservation and utilization of those resources",[69] which is, to the extent that EEZs represent the principle of full cost resource pricing, further support for the concept.

Throughout Agenda 21, there are further calls for economic incentives for various sectors to use living resources in a sustainable manner and for the internalisation of the economic costs and benefits.[70] Specific examples include: altering unsustainable consumption of natural resources; deforestation; protection of mountains; promoting sustainable agriculture; and the use of freshwater resources. Chapter 26 also calls for the recognition and protection of indigenous peoples rights, which, as noted before, indirectly represents calls for internalisation. Although these calls for internalisation provide support for the universality of the principle, they are all rather vague and do not suggest specific measures which might be taken to implement internalisation. Furthermore, review of these issues at each session of the CSD have concluded with similarly vague exhortatory statements.[71]

The Forestry Principles, the Programme of Action for the Sustainable Development of Small Island Developing States, the current draft of the document being negotiated for the UN Conference on Straddling Fish Stocks and Highly

---

[65] See Programme Area (c) – Making effective use of economic instruments and market and other incentives, paras. 8.27–40.

[66] Para. 8.32.

[67] Para. 17.22(d).

[68] Para. 17.73 and also, indirectly, para. 17.79(c) which calls upon States to "implement, in particular in developing countries, mechanisms to develop mariculture, aquaculture and small-scale, deep-sea and oceanic fisheries within areas under national jurisdiction where assessments show that marine living resources are potentially available" and paragraph 17.79(g) which calls upon States to "Enhance the productivity and utilization of their marine living resources for food and income".

[69] Para. 17.69.

[70] Further examples can be found in paras. 2.34, 10.6(d), 11.20, 11.21(a), 11.22(a) and (j) and (l), 11.24, 11.27(d), 13.6(b) and (d) and (f), 14.2, 14.18(e), 14.46(b) and (d), 14.60(a) and (f), 18.12(e), 18.79, 30.9, 32.4, 32.5, 32.6(b), 35.7(d)(iii) and 35.20.

[71] For example, see para. 28 of the Report of the 1994 Session of the CSD.

Migratory Species of Fish,[72] the Desertification Treaty, the revised International Tropical Timber Agreement and several other regional initiatives also recognise the importance of using economic instruments to ensure sustainable use of biodiversity. None of these instruments, however, refers to specific measures or significantly develop the principle of full cost resource pricing.

In conclusion, although there exists ample support for the notion that States have the legal authority to take measures to implement internalisation, there exists no positive obligation to implement the concept. The calls for internalisation in the context of biodiversity thus far have been purely exhortatory. Indeed, there remain many legal difficulties which effectively prohibit the development of the principle. Furthermore, if the experience of the ICCBD and the FAO are representative in this regard, then it will be some time before these problems are resolved and States are legally obliged to take measures which implement internalisation. In the international context this is significant because internalisation is likely to be resisted by those who will have to pay for the benefits, primarily developed countries. Thus, it must be considered unlikely that these countries will start to fulfil their moral duty until it becomes a legal one, or in other words that the compensation and motivation for conservation and sustainable use starts to make an impact upon the rate of biodiversity loss.

International and State Practice

Countries are beginning to unilaterally declaring sovereignty over the components of biodiversity within their jurisdiction, placing controls on access and starting to demand a share of the commercial benefits which result from their use, citing the CBD as providing them the legal authority to carry out such measures.[73] Although these unilateral actions illustrate support for the concept of full cost resource pricing, in order to fully realise the potential of internalisation these measure must be developed on a multilateral basis and not unilaterally. This is even more imperative than usual, due to the potential for conflict with mainstream trade issues which measures to implement full cost resource pricing have. Thus, in order to avoid conflict with the World Trade Organisation, GATT, the World Intellectual Property Organisation and the international intellectual property regime, a conflict which considerations motivated primarily by environmental concerns will inevitably lose, despite the provisions of Article 22 of the CBD, countries must act in a co-ordinated and multilateral manner to have any chance of developing systems which internalise the international benefits of biodiversity.

Furthermore, despite the growing enthusiasm for internalisation, many governments still pursue policies which are either contrary to internalisation or have

---

[72] A/CONF.164/13/22.

[73] For instance, Costa Rica declared all wild flora and fauna to be part of the "national patrimony" in its Wildlife Protection Act October 1992 which implements the CBD. This has also happened in Western Australia under a 1994 amendment to the Conservation and Land Management Act and is being contemplated by the Federal Government of Australia (see the proposed Biological Diversity Act or the Australian Federal Government's in *A National Strategy for the Conservation of Australia's Biological Diversity* (The Biological Diversity Advisory Committee, 1993)). Many other countries have started increasing their control over the transfer of genetic material out of their country insisting that they retain "ownership" of any specimens which are taken from their country (personal communication with Roger Smith, Kew Royal Botanic Gardens, September 1993), or declaring that they are in the process of developing new rules governing such matters (e.g. see the Indian Government's declarations reported on Greenett, 29 August 1994).

unintended effects which work against it. For example, governments continue to intervene in agricultural commodity markets through price support schemes and other macroeconomic interventions. One example of such price distortions are the low economic rents earned from natural resource extraction and harvesting by private concessionaries. The often cited illustration of this is the low or reduced rental rates on timber. Distortions also continue to be encouraged by deliberate policy choices, such as the use of economic incentives for land conversion and clearance, which often lead to the loss of forest and wetlands. Elimination of these policies present as many difficult problems as developing property rights over the components of biodiversity.

Many public international organisations also still pursue policies which ignore externalities. For instance, although the World Bank and the IMF have made an effort to incorporate full cost resource pricing in sectors such as energy and food, complete internalisation for all products, whether by regulation or economic instruments is not yet a feature of structural adjustment plans. Since these programmes are likely to define governmental policies for the foreseeable future in a large number of developing countries, this failure to push strongly for full internalisation must be considered an important defect.

Conclusion

Despite the enigmatic nature of the term sustainable use, it appears that some basic features of the term are increasingly being accepted. There is growing acceptance that it now requires more than simply quotas, that it requires preservation; management on the basis of biological unity; a holistic ecosystem approach to management; rehabilitation of denuded aspects of biodiversity; integrated approach; intergenerational equity; research efforts; monitoring the effects of use; establishment of flexible management systems; and the precautionary approach are all arguably corollary duties implied by the concept and, to the extent that this is so, represent manifestations of the development of the concept from a guiding principle to a binding commitment. Importantly, UNCED and the CBD have also started to develop, in a legal sense, the economic understanding of the term. This is significant, not only because of the poor record of regulatory approaches in actually achieving sustainable use, but also due to the fact that economic instruments have a greater potential to address the fundamental causes of biodiversity loss. Although great progress has been made in international law, these developments still do not require or oblige specific actions; they remain in the realm of guidance as opposed to legal obligations. Many difficult issues need to be resolved before sustainable use will be a binding commitment and before there is even partial internalisation at the international level. Nevertheless, the process has begun and the will to address these problems and to implement the necessary changes will be a true test of the commitment to ensuring the conservation of biodiversity.*

* The author would like to thank Mr Lyle Glowka of the Interim Secretariat for his helpful comments on an earlier draft.

4

The Contribution of Existing Agreements for the Conservation of Terrestrial Species and Habitats to the Maintenance of Biodiversity

Robin Churchill

Introduction

For many years before the Convention on Biological Diversity was signed, there had been in existence a considerable number of international agreements concerned with the conservation of fauna and flora and the protection of habitats and ecosystems. Although none of these agreements refers to the concept of biological diversity (indeed some of the agreements predate the introduction of this term by scientists and environmentalists), they clearly contribute to the maintenance of biodiversity. The Biodiversity Convention indeed recognises the value of these existing agreements. Its preamble refers to the desire of its parties to "enhance and complement existing international arrangements for the conservation of biological diversity", whilst Article 22 provides that the Convention "shall not affect the rights and obligations of any Contracting Party deriving from any existing international agreement, except where the exercise of those rights and obligations would cause a serious damage or threat to biological diversity".[1] Existing species and habitat agreements also represent examples of *in situ* conservation for which the Biodiversity Convention calls in Article 8, and many of the conservation techniques listed in that article are found in existing agreements.

The aim of this chapter is to try to assess the contribution these existing agreements for the conservation of terrestrial species and habitats have made to the maintenance of biodiversity. The next two sections of the chapter will deal with the main agreements for the conservation of species and for the protection of habitats and ecosystems. In each section the format will be the same. First, a brief outline of the relevant agreements will be given. This will be followed by an analysis of the techniques of conservation utilised by the agreements, and an evaluation of their

[1] For an exploration of the implications of this provision, see the chapter by Alan Boyle in this volume.

International Law and the Conservation of Biological Diversity (C. Redgwell and M. Bowman, eds.: 90 411 0863 7: © Kluwer Law International: pub. Kluwer Law International, 1995: printed in Great Britain), pp. 71–89

contribution to the maintenance of biodiversity. These two sections will be followed by two sections looking at agreements which contribute indirectly to the conservation of species and habitats. Finally, some conclusions will be drawn. It must be stressed at the outset that because of constraints of space and the number of agreements to be surveyed, much of the discussion must necessarily be very brief and condensed, and perhaps somewhat impressionistic.

Before turning to examine the agreements, it may be desirable to pause a moment and ask why States have thought it necessary to enter into international agreements to conserve species and habitats. Why are not measures at the national level sufficient? There seem to be two main reasons why international action is thought necessary. First, many mammals, birds and insects migrate, often great distances, during their life-cycle. Measures to conserve a particular species by a State in which that species spent part of its lifecycle would clearly be undermined if another State where the species spent some other part of its lifecycle allowed its indiscriminate slaughter. Second, some States are slow or reluctant to take conservation measures unless pressurised into action by other States by means of international agreement or unless other States agree to take similar measures. Often in such cases economic and political factors are the reason for tardy action. A State may be reluctant to set aside land as a nature reserve and thus prevent its development for industry, agriculture or transport links unless other States are prepared to do the same. Developing States are often reluctant to take conservation measures unless there is some financial inducement for them to do so.

It is also worth pausing another moment to spell out the assumption made above that existing conservation agreements contribute to the conservation of biodiversity. In its definition of biological diversity, the Convention includes "diversity within species, between species, and of ecosystems" (Article 1). Clearly, agreements for the conservation of species, especially of rare and endangered species, help to promote "diversity ... between species". Equally, agreements that protect habitats from destruction or degradation (as a result of agricultural, industrial, urban or other development) help to promote a "diversity ... of ecosystems". Such agreements also help to conserve species, as without sufficient suitable habitat species will not survive. What existing agreements do not do, however, is to promote the third form of biodiversity, "diversity within species" (i.e. genetic diversity within a species), except insofar as greater numbers of a species and distribution over different geographical areas will encourage greater genetic diversity. Although the existing international agreements that will be examined in this paper help to promote diversity between species and diversity of ecosystems, their contribution to the maintenance of biodiversity, although representing the conventional approach to preserving biodiversity, should not be exaggerated. In particular, these agreements do not address the root causes of loss of biodiversity. These root causes have been identified by the Global Biodiversity Strategy (produced by UNEP, the World Conservation Union and World Resources Institute) as population growth, increasing consumption of resources, ignorance about species and ecosystems, poorly conceived government policies, and economic causes such as the effects of global trading systems, inequity in resource distribution and the failure of economic systems to account for the value of biological resources.[2] Thus, agreements to

---

[2] World Resources Institute, *World Resources 1992–93* (1992), pp. 134–136. For a similar view see also J.A. NcNeely, "The Loss of Biological Diversity", in P. Brackley (ed.), *World Guide to Environmental Issues and Organisations* (1990), p. 79 at 79–81.

conserve species and habitats should perhaps be seen as making a secondary, rather than a primary, contribution to the preservation of biodiversity.

Existing Agreements for the Conservation of Terrestrial Species

This section will begin by outlining briefly the provisions of the main multilateral agreements concerned directly with the conservation of terrestrial species of fauna and flora. It will then go on to analyse the conservation techniques employed by these agreements and assess the effectiveness of their contribution to the maintenance of biodiversity.

Existing Agreements in Outline

There are about a dozen agreements which require consideration.[3]

Convention on the Conservation of Migratory Species of Wild Animals, 1979 (hereafter referred to as the Bonn Convention)[4]

The Convention requires its forty or so parties that are range States of the endangered migratory species listed in Appendix I to prohibit the killing and hunting of such species, to remove obstacles to the migration of such species, and to prevent the further endangering of such species. The Convention also requires parties that are range States of those species with an "unfavourable conservation status" listed in Appendix II to try to conclude agreements to provide for the conservation and management of such species so as to restore them to a favourable conservation status. Although the Bonn Convention has been in force since 1983, only one such agreement for terrestrial species has so far been concluded – the Agreement on the Conservation of Bats in Europe, 1991[5] – although a number of other agreements are under negotiation.

Convention on Nature Protection and Wildlife Preservation in the Western Hemisphere, 1940 (hereafter the Western Hemisphere Convention)[6]

The Convention requires its eighteen parties, whose combined territories account for about 25% of all species on earth and the largest intact tropical forests,[7] to prohibit the hunting and killing of the fauna, and the taking of the flora, found in their national parks except where authorised; to adopt suitable laws and regulations for the

---

[3] The following agreements will not be considered here: the International Convention for the Protection of Birds, 1950, because it is moribund; the EC Directives on the Conservation of Wild Birds (1979) and on the Conservation of Natural Habitats and of Wild Fauna and Flora (1992), which are dealt with in the chapter by Patricia Birnie; the Agreed Measures for the Conservation of Antarctic Fauna and Flora, 1964 and the Protocol to the Antarctic Treaty on Environmental Protection, 1991, which are discussed in the chapter by Catherine Redgwell; the International Convention for the Protection of Plants, 1951, which is dealt with in the chapter by Greg Rose; and, for reasons of space, all bilateral agreements. For a brief survey of the latter, see P.H. Sand (ed.), *The Effectiveness of International Environmental Agreements* (1992), pp. 468–470.

[4] (1980) 19 ILM 15. For full discussion of the Convention, see S. Lyster, *International Wildlife Law* (1985), pp. 278–298; P.W. Birnie and A.E. Boyle, *International Law and the Environment* (1992), pp. 470–475 and literature cited there.

[5] UKTS 9 (1994), Cm 2472. The Agreement came into force in January 1994.

[6] 161 UNTS 193. The Convention came into force in 1942. Further on the Convention, see Lyster, *op. cit.*, n. 4, pp. 97–111.

[7] Sand, *op. cit.*, n. 3, at p. 60.

protection and preservation of flora and fauna not included in national parks or nature reserves; to adopt appropriate measures for the protection of migratory birds of "economic and aesthetic value"; to protect as completely as possible those species listed in the annex to the Convention; and to control trade in protected fauna and flora. For many years the Convention has been largely moribund, with no periodic meetings of its parties, no updating of its annex and a lack of any administrative structure, but in the past few years there have been signs of attempts to revive the Convention.[8]

African Convention on the Conservation of Nature and Natural Resources, 1968 (hereafter the African Convention)[9]

In relation to flora, the Convention requires its thirty or so parties to take all necessary protective measures and to ensure its best utilisation and development, *inter alia* through conservation and management plans for forests and rangeland: in addition, parties are to conserve plant species or communities which are threatened or of special scientific or aesthetic value by ensuring that they are included in conservation areas. In relation to fauna, the parties are to manage wildlife populations inside designated areas in accordance with the objectives of such areas and manage exploitable wildlife populations outside such areas to an optimum sustained yield. Certain methods of hunting and killing animals are also prohibited. The killing or taking of those animal and plant species threatened with extinction and listed in the Annex to the Convention is normally prohibited. Trade in such species and other animals is to be regulated. The level of activity under the Convention is very low: meetings of the parties are rare (the last being in 1985) and no attempt is made to monitor compliance with the Convention.

Convention on Conservation of Nature in the South Pacific, 1976 (hereafter the Apia Convention)[10]

The Convention requires its parties to prohibit the taking of flora and the hunting and killing of fauna in their national parks, except where duly authorised, and to maintain their national reserves "inviolate".[11] In addition to the protection given to fauna and flora in national parks and reserves, the parties are to protect indigenous fauna and flora from "unwise exploitation "and other threats that may lead to their extinction: species that are threatened with extinction are to be listed by the parties and protected as completely as possible. The Convention lacks any machinery for its review or for monitoring its compliance, and only about a quarter of South Pacific States have become parties to it.

Convention on the Conservation of European Wildlife and Natural Habitats, 1979 (hereafter the Berne Convention)[12]

As the Convention was concluded under the auspices of the Council of Europe, its twenty-five parties are mainly from Western Europe, but there are a number

---

[8] *Ibid.*, pp. 60–62.

[9] 1001 UNTS 3. The Convention came into force in 1969. Further on the Convention, see Lyster, *op. cit.*, n. 4, pp. 112–128.

[10] Text in B. Simma and B. Ruster (eds.), *International Protection of the Environment* (1975 onwards), Vol. 20, pp. 10, 359. The Convention came into force in 1990.

[11] For the meaning of the terms "national park" and "national reserve", see the discussion of the Convention in section on the protection of terrestrial habitats, below.

[12] UKTS 56 (1982), Cmnd 8738. The Convention came into force in 1982. Further on the Convention, see Lyster, *op. cit.*, n. 4, pp. 129–155.

(Bulgaria, Cyprus, Greece, Hungary, Turkey, Senegal and Burkina Faso) from outside this area. The Convention imposes a general obligation on its parties to maintain the population of fauna and flora at, or adapt it to, "a level which corresponds in particular to ecological, scientific and cultural requirements, whilst taking account of economic and recreational requirements and the needs of sub-species, varieties or forms at risk locally" (Article 2). More specifically, the Convention requires its parties to prohibit the taking of those species of plants listed in Appendix I; to prohibit the capture and killing of those species of fauna listed in Appendix II; and to regulate the exploitation of those species of fauna listed in Appendix III in order to keep their populations out of danger. The appendices have been amended and extended from time to time, and it is noteworthy that whereas Appendices II and III in their original form included only species of mammals, birds, reptiles and amphibians, they have subsequently been extended to include species of fish, arthropods (insects and spiders) and molluscs.

Protocol to the Convention for the Protection of the Mediterranean Sea against Pollution concerning Mediterranean Specially Protected Areas, 1982 (hereafter the Mediterranean Protocol)[13]

Although this Protocol is primarily concerned with the marine environment, its area of application also includes "wetlands or coastal areas designated by each" of its nineteen parties (Article 2). Within the protected areas established by the parties (on which see further the discussion on protection of habitats below), the parties are "progressively [to] take the measures required, which may include" the regulation of the taking of fauna and flora, a prohibition on such taking, and the regulation of any activity likely to harm or disturb such fauna and flora. In addition, the parties must not allow any activity within protected areas which would cause the extinction or substantial reduction of any species within the protected ecosystems, or any ecologically connected species, particularly migratory, rare, endangered or endemic species.

ASEAN Agreement on the Conservation of Nature and Natural Resources, 1985 (hereafter the ASEAN Agreement)[14]

The Agreement was drawn up and signed by the six ASEAN States (Brunei, Indonesia, Malaysia, Philippines, Singapore and Thailand). The Agreement requires its parties to maintain maximum genetic diversity by taking action aimed at ensuring the survival and promoting the conservation of all species under their jurisdiction and control. To this end the parties are to ensure the sustainable use of harvested species (*inter alia* by maintaining or restoring harvested populations at or to levels which ensure their stable recruitment and the stable recruitment of dependent or related species); protect endangered species (listed in an appendix to the Agreement to be adopted), *inter alia* by prohibiting the taking of such species and adopting all other measures to restore their populations to the highest possible level; conserve endemic species by taking all necessary measures to maintain populations at the highest possible level; and take measures to prevent the extinction of any species. The ASEAN Agreement is perhaps the most advanced and sophisticated of the species conservation agreements, not only in its conservation provisions but also for its institutional machinery for review of its implementation: this is no doubt due to

---

[13] (1982) OJ, C278/5. The Protocol came into force in 1986.
[14] I. Rummel-Bulska and S. Osafo (eds.), *Selected Multilateral Treaties in the Field of the Environment* (1991), Vol. 2, p. 343.

the fact that the Agreement was drafted with help from environmental organisations. These factors may explain why the Agreement has not yet come into force.

Protocol to the Convention for the Protection, Management and Development of the Marine and Coastal Environment of the Eastern African Region concerning Protected Areas and Wild Fauna and Flora in the Eastern African Region, 1985 (hereafter the East African Protocol)[15]

Although this Protocol is primarily concerned with the marine environment, it also applies to the "coastal areas" (a term not further defined) of its parties. The latter are required to prohibit the taking of those species of flora listed in Annex I and the capture, killing or disturbance of those endangered species of fauna listed in Annex II (which include a number of terrestrial species), and to regulate the exploitation of the depleted or threatened species of fauna listed in Annex III (which amongst its terrestrial species includes the elephant and warthog) in order to restore and maintain the populations of such species at "optimum levels". In addition, the parties are to coordinate their efforts for the protection of the migratory species listed in Annex IV: the only species with any terrestrial connection included in the latter are various species of turtle. Within the protected areas to be established by the parties, the latter are to take the same protective measures as those prescribed in the Mediterranean Protocol: these obviously apply primarily to species not listed in the Annexes.

Protocol to the Convention for the Protection and Development of the Marine Environment of the Wider Caribbean Region concerning Specially Protected Areas and Wildlife, 1990 (hereafter the Caribbean Protocol)[16]

Although the Protocol is again primarily concerned with the marine environment, its area of application includes waters on the landward side of the baseline and "such related terrestrial areas (including watersheds) as may be designated" by the parties. Within protected areas established by the parties, the latter are to regulate or prohibit the hunting or taking of endangered or threatened species of fauna and flora and any activity likely to harm or disturb such species. Like the East African Protocol, the Caribbean Protocol does not confine its provisions on protection of species to those in protected areas. Each party must prohibit the disturbance or taking of endangered and threatened species of fauna and flora listed in Annexes I and II, and regulate or prohibit the disturbance or taking of such species found in its territory but not so listed. In relation to species listed in Annex III, the parties shall adopt appropriate measures to ensure their protection and recovery. Finally, there is a general obligation on the parties, "to the extent possible, consistent with each Party's legal system, [to] manage species of fauna and flora with the objective of preventing species from becoming endangered or threatened" (Article 3(3)).

Draft Convention on the Conservation of Arctic Fauna and Flora

As part of the so-called Rovaniemi process set in train by the eight Arctic States in 1991 to tackle the growing environmental problems of the Arctic, there are plans to draw up a convention on the conservation of Arctic fauna and flora. At the time of writing, it was unclear what form such a convention would take or when its negotiation might be finalised.

---

[15] (1986) OJ, C253/18. The Protocol is not yet in force.

[16] (1990) 5 *International Journal of Estuarine and Coastal Law* 369. The Protocol is not yet in force. Further on the Protocol, see D. Freestone in *ibid.*, at pp. 362–368 and (1991) 22 *Marine Pollution Bulletin* 579–81.

Agreement on the Conservation of Polar Bears, 1973[17]

The Convention obliges its parties, the five Arctic rim States, to manage polar bear populations in accordance with sound conservation practices, and in particular to prohibit the hunting, killing or capturing of polar bears, subject to certain limited exceptions. The Agreement appears to have had the effect of reversing the previous decline in the populations of polar bears and allowing them to reach a stable and acceptable level.

Convention for the Conservation and Management of the Vicuna, 1979[18]

This Convention, which replaces an earlier convention of 1969, provides that its four parties (Bolivia, Chile, Ecuador and Peru) are to conserve and manage the vicuna so that its population is increased until the grazing capacity of a specific area has been reached, and thereafter to maintain a balance between population and grazing capacity by means such as translocation or culling. The hunting of vicuna is prohibited, and trade in vicuna and its products prohibited or very strictly controlled. Probably as a result of the Convention the population of vicuna increased from 59,000 in 1978 to 160,000 in 1990.[19]

The above agreements are of a number of types. The Polar Bear and Vicuna Conventions are single species agreements, covering the entire world populations of those species. The other agreements each apply to a wide variety of fauna and, in most cases, flora. The Bonn Convention is global in its coverage, the remaining agreements regional.

Techniques of Conservation

The agreements outlined above demonstrate quite a wide variety of techniques for seeking to conserve terrestrial species of fauna and flora. First, there are two basic techniques of conservation which are used: to protect all or certain species found in nature reserves or other specially protected areas, and to protect certain listed species wherever found. The former technique is employed by the Mediterranean Protocol, whilst the Bonn, Berne, ASEAN, Polar Bear and Vicuna agreements rely primarily on the list system. The remaining agreements utilise both techniques of conservation.

Most agreements distinguish between species according to their conservation status. The taking or killing of what are variously described as "endangered" (Bonn, Berne, ASEAN and East African), or "threatened with extinction" (African, Apia) or "endangered and threatened" species (Caribbean) is normally prohibited. Most of the conventions also address less vulnerable species. Thus the Bonn Convention identifies species of "unfavourable conservation status" (which require conservation and management in accordance with certain criteria); the Apia Convention indigenous species (which must be protected from "unwise exploitation" and other threats which might lead to their extinction); the ASEAN Agreement endemic species (whose populations must be maintained at the highest levels); and the East African

---

[17] (1974) 13 ILM 13. The Convention came into force in 1976. Further on the Convention, see Lyster, *op. cit.*, n. 4, pp. 55–61 and A. Fikkan, "Polar Bears: Hot Topic in Cold Climate", (1990) 10 *International Challenges* 32.

[18] Text in Rummel-Bulska and Osafo, *op. cit.*, n. 14, Vol. 2, at p.74. The Convention came into force in 1982. Further on the Convention, see Lyster, *op. cit.*, n. 4, pp. 88–94.

[19] Sand, *op. cit.*, n. 3, at p. 103.

Protocol depleted and threatened species (which must be so regulated that their populations are at optimum levels). Several conventions have provisions concerning species generally, requiring that they be protected and preserved (Western Hemisphere), managed to optimum sustained yields (African), maintained at levels which correspond in particular to ecological, scientific and cultural requirements (Berne) or so managed that they are not endangered or threatened (Caribbean Protocol), or that their survival be ensured and their conservation promoted (ASEAN). In nearly all cases the obligations relating to non-threatened and non-endangered species are qualified in various ways.

The techniques of conservation just outlined are supplemented in a number of ways. Many of the agreements prohibit the introduction of exotic or non-native species; the reason for this is that such species often have a deleterious effect on native species. The Berne Convention, as well as prohibiting the introduction of exotic and non-native species, goes a step further by requiring its parties to encourage the reintroduction of native species where this would contribute to the conservation of endangered species. Several of the agreements (e.g. African, Berne, Caribbean, Polar Bear) prohibit the use of certain methods of killing: the point of this provision is to make it more difficult to kill species of fauna (where this is not prohibited) and to reduce possible disturbance to those fauna that are not killed. The prohibition or regulation of the taking of flora and fauna is reinforced in the case of most of the agreements by a prohibition or strict control of trade in protected species. Finally, virtually all the agreements attempt to protect species by also having provisions to protect habitat: in many ways these provisions (which are discussed in the next section) are the most important way in which species are protected, and it is perhaps somewhat artificial to separate (as this chapter does) the protection of species from the protection of habitats.

An Evaluation of the Species Agreements

As mentioned in the Introduction to this chapter, the species agreements do not address the root causes of loss of biodiversity. The one exception is those agreements (mainly the more recent ones) which attempt to overcome ignorance about species and ecosystems by providing for the promotion of, and co-operation over, scientific research. But if they do not address the primary causes of loss of biodiversity, how do the agreements stand as a secondary contribution to combating such loss?

While most of the agreements, as we have seen, look good on paper, there are many practical weaknesses about them which limit their effectiveness. A first weakness is that most of the agreements suffer from a relatively low number of ratifications. Indeed, three of the agreements (the ASEAN Agreement and East African and Caribbean Protocols) have not yet received sufficient ratifications to enter into force. The Bonn Convention presents a particular problem in this respect. As regards the endangered migratory species listed in Annex I, it is only in a very few cases that most or all of the range States are parties to the Convention: unless all the range States of a particular species are parties, the Convention will clearly be of very limited value. Related to the lack of ratifications is the patchy geographical coverage of the agreements. Apart from the Bonn Convention, the agreements are regional in their application. There is, however, no agreement applicable to most of Eastern Europe; no agreement in force for any part of Asia and most of the Pacific; and moribund or semi-moribund agreements applicable to the Americas and Africa. Only in Western Europe (and Antarctica) are there agreements which are in force and functioning in a reasonably active way. Thus, there are no effective agreements applicable to tropical rain forests, which are thought to contain between 50 and 90% of the earth's species, or to most Mediterranean-climate regions, which also contain large numbers of species and have high levels of endemism, or to most islands, which also have high levels of endemism (for example, over 10% of the world's species of birds are each confined to a single island).[20]

A second group of weaknesses concerns the list system on which most of the agreements rely wholly or substantially. The lists have tended to concentrate on fauna, and, perhaps not surprisingly, on higher life forms, especially mammals and birds. Large, visible species of mammals, bird and plants, however, make up less than 5% of the world's species.[21] Thus, the lists provide little protection for the vast numbers of species of lower life forms, and obviously can provide no protection at all for species that have not yet been identified (and fewer than 1.4 million of the world's 5 to 30 million species have so far been named).[22] A further weakness with a list system is that in some agreements the procedures for amending and adding to lists to take account of new knowledge and the changed conservation status of particular flora and fauna are slow and cumbersome: for example, the list in the Western Hemisphere Convention has not been amended since 1967.[23] On the positive side the lists do include many threatened and endangered species of the higher life forms, and are by no means confined to species that are migratory or shared between two or more States.

A third group of weaknesses relates to implementation and compliance. Many of the agreements lack proper machinery, such as regular meetings of the parties and an effectively funded secretariat, for monitoring the implementation of, and compliance with, the agreement by States parties. This is often compounded (for example, in the case of the Bonn and African Conventions) by parties failing to report on implementation and compliance when required by the agreement to do so. To some extent these deficiencies have been remedied by unofficial reports by non-governmental environmental organisations. Where reporting and monitoring systems have worked reasonably well, they have shown that most parties have been defective in implementing and complying with the agreement concerned.[24] None of the agreements has any mechanism for dealing with non-compliance.

The various agreements surveyed above also suffer from a number of other weaknesses. Some of the obligations, particularly in relation to non-threatened and non-endangered species, are too vague and/or qualified by other factors (such as nutritional needs and the desire for economic development) to be very effective. None of the agreements containing such obligations addresses the issue of predator–prey relationships, which in the case of non-vegetarian species is crucial to the question of population levels. There has sometimes been a poor response where further action is required for a treaty to be effective: for example, the very slow rate at which Agreements have been developed under the Bonn Convention, or the failure (mentioned above) to update lists timeously. There is also some overlap

---

[20] McNeely, *op. cit.*, n. 2, at p. 81.
[21] World Resources Institute, *op. cit.*, n. 2, at p. 128.
[22] *Ibid.*, p. 134.
[23] Sand, *op. cit.*, p. 62.
[24] See, for example, International Union for Conservation of Nature and Natural Resources, *Implementation of the Berne Convention* (1986); Swiss Federal Office of the Environment, *Results of the Questionnaire regarding Seven Environmental Conventions: Participation and Implementation* (1993), pp. 5–7 (on the Berne and Bonn Conventions).

between agreements, with the same species sometimes being protected by two or more instruments. Some of the agreements, particularly the older ones, fail to provide for the parties to co-operate over scientific research and exchange of information. Given the lack of knowledge of so many species, it is highly desirable that there should be such obligations.

The two most successful species agreements, perhaps the only successful such agreements, are the Polar Bear and Vicuna Agreements. The reason for their success is probably because they are each limited to a single species, that species is a large, visible mammal, and there is small number of parties which had at the outset a high degree of consensus as to what action needed to be taken. Thus, these two single species agreements can hardly serve as a role model for species agreements in general (except perhaps for certain agreements under the Bonn Convention): apart from anything else, given the number of species requiring protection, it would obviously be impractical to have hundreds, and possibly thousands, of single species agreements.

Existing Agreements for the Protection of Terrestrial Habitats

This section will follow the same format as the preceding section. Thus, it will begin by outlining briefly the provisions of the main multilateral agreements concerned directly with the protection of terrestrial habitats.

Existing Agreements in Outline

Nearly all the species agreements surveyed in the preceding section also contain provisions relating to the protection of habitats. These will be outlined first, in the same order as before. This will then be followed by a discussion of four agreements which are concerned with habitat only and not species.

Bonn Convention

In the case of the endangered migratory species listed in Annex I the range State parties are to "endeavour to conserve and, where feasible and appropriate, restore those habitats of the species which are of importance in removing the species from danger of extinction" (Article III(4)). In the case of the species listed in Annex II, the Convention lays down a number of guidelines concerning habitat protection for the Agreements relating to such species: these include the conservation, and where required and feasible, the restoration of the habitats of importance in maintaining a favourable conservation status, and the protection of such habitats from disturbance. The one Agreement so far concluded relating to terrestrial species, the European Bats Agreement, provides that its parties are to protect sites which are important for bats from damage or disturbance, taking into account as necessary economic and social considerations. The parties are also to try to reduce the effect of pesticides and timber treatment chemicals on bats.

Western Hemisphere Convention

The Convention has a single, and rather weak, provision on habitat protection (Article II) which stipulates that the parties are to explore the possibility of establishing in their territories national parks (defined as areas established for the protection and preservation of flora and fauna of national significance); national reserves (regions established for the conservation and utilisation of natural resources under government control); nature monuments (which include species of flora and fauna to which strict protection is given); and strict wilderness reserves, where no commercial developments shall be permitted and which shall be preserved inviolate.

African Convention

The Convention provides that its parties are to maintain existing conservation areas (such as game reserves, nature reserves and national parks – all terms extensively defined in Article III – where interference with habitats is normally prohibited or strictly regulated), and to assess the necessity of establishing additional conservation areas in order to protect those ecosystems which are peculiar to, or most representative of, their territory and to ensure the conservation of species. More particularly, in relation to flora, there is a series of obligations for protecting habitats by requiring the parties to control bush fires, forest exploitation, land clearing and grazing; to set aside areas for forest reserves; and to establish botanical gardens. There are also provisions requiring the parties to combat soil erosion and to conserve water and prevent its pollution.

Apia Convention

The Convention requires its parties to encourage the creation of national parks (defined as an area established for the protection and conservation of ecosystems containing animals, plant species and their habitats of special scientific interest) and national reserves (areas established for the protection and conservation of nature), which, together with existing parks and reserves, will safeguard representative samples of natural ecosystems, particular attention being given to endangered species. The killing or taking of fauna and flora in national parks is normally to be prohibited, whilst national reserves shall be maintained inviolate as far as practicable.

Berne Convention

As originally drafted, the provisions of the Convention concerning habitat protection were, to a considerable extent, unclear and unworkable. Although these provisions have not been formally amended, they must now be read in the light of Resolution No. 1 of the Standing Committee, adopted in 1989. Thus read, these provisions involve the following obligations. First, under Article 4 the parties are required to take appropriate and necessary legislative and administrative measures to ensure the conservation of those habitats which have been determined by the Committee as requiring specific habitat conservation measures. Second, under Article 6 the parties are required to conserve breeding or resting sites of those species and those sites which the Committee has determined require protection. In pursuance of these provisions the Committee has addressed a considerable number of recommendations to the parties.

Mediterranean Protocol

Article 3 of the Protocol requires its parties to establish, "to the extent possible", protected areas in order to safeguard in particular "sites of biological and ecological value; the genetic diversity, as well as satisfactory population levels, of species, and their breeding grounds and habitats; representative types of ecosystems, as well as ecological processes". The parties are to "endeavour to undertake the action necessary to protect" such areas, which may include introducing a planning and management system for the area, prohibiting pollution, regulating any activity involving the exploitation of the subsoil of the area, and taking any other measure aimed at safe-guarding ecological and biological processes. In taking such measures the parties must take into account the traditional activities of their local populations, but this must not be done in such a way as to endanger the maintenance of ecosystems protected under the Protocol. The parties are also to co-operate to try to create a network of protected areas in the Mediterranean. Finally, the parties may establish buffer areas around pro-tected areas in order to strengthen the latter, in which activities are less severely restricted whilst remaining compatible with the purposes of the protected area.

ASEAN Agreement

This Agreement has the most far-reaching provisions on habitat protection of the agreements being surveyed. For reasons of space and because the Agreement is not in force, only the briefest indications of its provisions can be given here. The Agreement imposes a general obligation on its parties to take the measures "neces-sary to maintain essential ecological process and life-support systems" (Article 1(1)). More specifically, the parties are required to conserve the habitats of animal and plant species in order to maintain maximum genetic diversity, and to this end to create and maintain protected areas; to prevent changes in the ecosystems of har-vested species which are not reversible over a reasonable period of time; and to protect the habitats of endangered species by ensuring that sufficient portions are included in protected areas. The protected areas referred to, which should include representative samples of all types of ecosystems found in the Agreement area, are to be regulated and managed in such a way as to further the objectives for which they have been created. This is to include preparing a management plan for the area, establishing buffer zones around the area, prohibiting pollution of the area, and con-trolling activities outside the area which are likely to cause disturbance or damage to the ecosystems the area is designed to protect. The Agreement also contains provi-sions aimed at ensuring the conservation of vegetation and forest cover; the preven-tion of soil erosion and degradation; the conservation of water resources; the preservation of air quality; the prevention of other forms of environmental degrada-tion, including pollution; the integration of natural resource conservation into land use planning; and the use of environmental impact assessments before permitting any activity which may significantly affect the natural environment.

East African Protocol

The Protocol has provisions on the establishment of protected areas very similar to those of the Mediterranean Protocol. In addition, the parties to the Protocol are to endeavour to protect and preserve rare or fragile ecosystems and the habitats of rare, depleted, threatened or endangered species of fauna and flora; prohibit activities having adverse effects on the habitats of the flora listed in Annex I and the en-dangered species of fauna listed in Annex II; and safeguard the critical habitats of the depleted or threatened species of fauna listed in Annex III in protected areas.

Caribbean Protocol

The Protocol has provisions on the establishment of protected areas similar to those of the Mediterranean and East African Protocols. In addition, the parties to the Protocol are to regulate and prohibit activities having adverse effects on the habitats and ecosystems of endangered or threatened species of flora and fauna. Finally, environmental impact assessments are to be carried out in relation to proposed projects that would have a negative environmental impact and significantly affect areas afforded special protection under the Protocol.

Polar Bear Agreement

Article II of the Agreement provides that the parties are to take "appropriate action to protect the ecosystems of which polar bears are a part, with special attention to habitat components such as denning and feeding sites and migration patterns". In pursuance of this obligation each party has taken at least some measures to protect important polar bear habitats.[25]

Vicuna Convention

Article 5 of the Convention provides that the parties are to "maintain and develop national parks and reserves and other protected areas containing vicuna populations, and to extend the areas of repopulation managed as wildland areas". In pursuance of this provision at least six protected areas for vicuna have been established.[26]

Convention on Wetlands of International Importance especially as Waterfowl Habitat, 1971 (hereafter the Ramsar Convention)[27]

This is the first of the agreements in this section concerned only with habitat, rather than both species and habitat, protection. The Convention requires its eighty or so parties to designate at least one suitable wetland in its territory for inclusion in a List of Wetlands of International Importance. Wetlands should be selected for inclusion on account of "their international significance in terms of ecology, botany, zoology, limnology or hydrology" (Article 2): in the first instance these should include wetlands of importance to waterfowl at any season. Parties are required to "formulate and implement their planning so as to promote the conservation" of wetlands on the List, and as far as possible the wise use of wetlands (in accordance with guidelines elaborated in 1987) in their territory (Article 3). Finally, parties are to promote the conservation of wetlands and waterfowl by establishing nature reserves on wetlands, whether included in the List or not, and endeavour through management to increase waterfowl populations on appropriate wetlands. By April 1992, 549 wetlands in sixty-five States parties had been placed on the Convention List.[28]

Convention for the Protection of the World Cultural and Natural Heritage, 1972 (hereafter the World Heritage Convention)[29]

The Convention places a general duty on its 130 or so parties to conserve and protect the natural heritage, the definition of which includes the habitats of "threatened species of animals and plants of outstanding universal value from the point of view of science or conservation" (Articles 2 and 4). To that end the parties are to integrate protection of the natural heritage into comprehensive planning programmes, work out such operating methods as are necessary to counteract threats to that heritage, and take the necessary measures to protect and conserve such heritage.

---

[25] For examples, see Lyster, *op. cit.*, n. 4, p. 60.

[26] *Ibid.*, p. 90.

[27] 996 UNTS 245. The Convention came into force in 1975. Further on the Convention, see Lyster, *op. cit.*, n. 4, pp. 183–207 and Birnie and Boyle, *op. cit.*, n. 4, pp. 465–468.

[28] Birnie and Boyle, *op. cit.*, n. 4, p. 466.

[29] 1037 UNTS 151. The Convention came into force in 1975. Further on the Convention, see Lyster, *op. cit.*, n. 4, pp. 208–238 and Birnie and Boyle, *op. cit.*, n. 4, pp. 468–470.

These rather vague obligations have been spelt out in more detail in Operational Guidelines for the Implementation of the Convention.[30] Each party is to submit to the World Heritage Committee an inventory of property forming part of the natural heritage suitable for inclusion in the World Heritage List. The Committee is to draw up a List of World Heritage in Danger, for the conservation of which major operations are necessary. The Committee can also provide financial assistance from the World Heritage Fund for sites on this List. As of 1991, ninety-four natural heritage sites in forty-three States had been placed on the World Heritage List.[31] According to the Operational Guidelines a site will only be placed on the List if it contains "the most important and significant natural habitats where threatened species of animals or plants of outstanding universal value from the point of science or conservation still survive".

Convention on the Protection of the Alps, 1991 (hereafter the Alps Convention)[32]

This Convention, which is a framework one, is only partly concerned with nature conservation. It is a response to the increasing threat to the Alps and their ecology from growing human exploitation. In the preamble the parties recognise that the Alps are "an indispensable habitat and refuge for numerous endangered plants and animals". Article 2 provides that the parties are to maintain a comprehensive policy of protection and preservation of the Alps. To attain this objective they are to take appropriate measures relating *inter alia* to the protection of nature with a view to "assuring the protection, management, and if necessary, the restoration of nature and the countryside in such a way as to guarantee the lasting functioning of ecosystems, the preservation of flora and fauna and their habitats". Measures are also to be taken concerning air quality, soil preservation, water quality and the management of forests. The parties are to agree on protocols to give further effect to the Convention. No such protocols have yet been concluded.

Convention for Co-operation in the Protection and Development of the Marine and Coastal Environment of the West and Central African Region, 1981 (hereafter the West African Convention)[33]

Like the Mediterranean, East African and Caribbean Protocols, the West African Convention is primarily concerned with the marine environment, but it does also apply to "coastal zones and related inland waters" (Article 1). It has one provision relating to habitat protection. Article II requires the parties to "take all appropriate measures to protect and preserve rare or fragile ecosystems as well as the habitat of depleted, threatened or endangered species." To this end the parties are to endeavour to establish protected areas, and to prohibit or control any activity likely to have adverse effects on the species, ecosystems or biological processes in such areas. What effect this provision has had in practice it has not proved possible to ascertain.

Of the fifteen agreements surveyed in this section, eleven are concerned with the conservation of both species and habitats, and four with habitat only. Three agreements (the Bonn, Ramsar and World Heritage Conventions) are global in scope, the

---

[30] Doc. WHC/2 Revised (27 March 1992) quoted in M.C. Maffei, "Evolving Trends in the International Protection of Species" (1993) 36 *German Yearbook of International Law* 131, at p. 141.

[31] T. Scovazzi and T. Treves (eds.), *World Treaties for the Protection of the Environment* (1992), p. 379.

[32] (1992) 31 ILM 767.

[33] (1981) 20 ILM 746. The Convention came into force in 1984.

remainder are regional. The agreements fall into two broad types, though a few display features of both types. The first type of agreement is that which is designed to protect a particular kind of habitat or ecosystem – for example the Polar Bear Agreement (the ecosystem of which polar bears are a part); the Ramsar Convention (wetlands); and the Alps Convention – or representative types of habitats or ecosystems to be selected by the parties (African, Apia, Mediterranean, ASEAN, East African, Caribbean, World Heritage and West African agreements). The second type of agreement is aimed at protecting the habitat of particular species, which are either identified precisely (as in the Bonn, Bats, Berne, East African, Polar Bear and Vicuna agreements) or generically, eg threatened or endangered species (as in the Western Hemisphere, ASEAN, East African, Caribbean, World Heritage and West African agreements). The first type of agreement obviously contributes more directly to the protection of ecosystems and promotion of their diversity.

Techniques of Protection

Both types of agreement employ a number of techniques for protecting habitats. First, most of the agreements (Western Hemisphere, African, Apia, Mediterranean, Asean, East African, Caribbean, Vicuna, Ramsar, World Heritage and West African) require their parties to establish special areas (variously referred to as protected areas, reserves, parks, etc.), in which activities which would harm habitats or ecosystems are to be prohibited or strictly regulated. By contrast, some agreements do not make use of special areas but instead require their parties not to damage or disturb the habitats of particular species (the Bats and Polar Bear Agreements; the East African and Caribbean Protocols also make use of this technique in addition to protected areas) or simply to conserve the habitats of particular species (the Bonn, Berne, World Heritage and Alps Conventions).

A number of the agreements using the technique of special areas (Mediterranean, ASEAN, East African, Caribbean) supplement this by encouraging their parties to establish buffer zones around such areas. A number of agreements, particularly more recent ones, have wide-ranging provisions concerned with such matters as soil erosion, water and air quality, other forms of environmental degradation and pollution, all of which help to protect habitats. The most far-reaching agreements in this regard are the African, ASEAN and Alps agreements, whilst the Bats, Mediterranean, East African and Caribbean agreements are more limited, being chiefly concerned with pollution. What is noteworthy about all these agreements is that they integrate two matters, conservation and pollution, which in the past have tended to be considered and dealt with separately.

Two of the agreements (the ASEAN and Caribbean) have specific provisions requiring environmental impact assessment, which should help to protect habitats. Going in the same direction, but rather less far, are three agreements (Ramsar, World Heritage and Berne) which require their parties to formulate planning and development policies so as to promote the protection of habitats.

Finally, some of the agreements (principally Mediterranean, ASEAN, East African and Caribbean) require their parties to safeguard or maintain ecological processes. Although this term is nowhere defined, it presumably refers to maintaining predator–prey relationships, food supplies and so on in order to allow an ecosystem to function in such a way that the species inhabiting that ecosystem are not threatened with extinction or declining populations. This technique therefore contributes more to the maintenance of ecosystems than habitat as such.

An Evaluation of the Habitat Agreements

The technique found in most of the agreements of requiring the parties to designate special areas is in principle a very effective tool for conserving diversity of habitats and ecosystems. However, there are weaknesses in the way in which this technique is employed in the agreements surveyed. The principal one is the wide degree of discretion generally given to parties in designating special areas, seen at its widest in the Western Hemisphere Convention which simply requires its parties to explore the possibility of establishing special areas. The effectiveness of such agreements thus depends entirely on how States parties exercise their discretion. Little attempt is made by the agreements' institutional machinery to oversee how the parties exercise their discretion, except in the case of the Ramsar and World Heritage Conventions.

In practice so far, protected areas have been established, especially in developing countries, to preserve the so-called charismatic species (e.g. elephants, tigers, bears) or spectacular vistas, rather than expressly to conserve biodiversity.[34] Often such areas tend to be in relatively isolated pockets. This is not very satisfactory, particularly for migratory species, unless such pockets happen to coincide with their need for resting places along a migration route, and especially if the region surrounding the protected areas is heavily exploited. The management of such regions therefore needs to be integrated into conservation programmes: as has been seen, only a few agreements have begun to address this issue. Some of the more recent agreements have begun to make use of buffer zones, but none has yet begun to use the more modern technique of bio-regional management.[35]

A number of other general weaknesses are common to the special area approach, whether implemented as part of an international agreement or as a purely national measure. Area boundaries tend to follow political, rather than ecological, lines; many areas are too small to be effective; areas are often imposed on the local population giving rise to conflict; and areas suffer from ineffective and underfunded management.[36]

Returning to the agreements themselves, the obligation to conserve habitats is seldom an absolute one and is often subject to economic and development considerations, as seen particularly in the Bats, Mediterranean, East African and Caribbean agreements. These agreements give no indication of how economic and habitat conservation interests are to be weighed one against the other. Those agreements that require the conservation of the habitats of certain listed species are obviously much more limited than agreements basing themselves on special areas because they cover only listed species, but it will usually be the case that in conserving the habitats of listed species the habitats of many non-listed species will receive indirect protection – indeed, if the food supplies of non-vegetarian listed species are to be maintained, it will be essential to maintain the habitats of their prey. A broad view of the obligation to conserve the habitats of listed species would thus include conserving the habitats of sufficient of their prey. The weaknesses just mentioned point to a defect common to virtually all the agreements, which is that so many of their provisions are formulated in a vague and highly generalised way.

In the section concerning the evaluation of species agreements above a number of comments were made about weaknesses in such agreements concerning the number of

---

[34] World Resources Institute, *op. cit.*, n. 2, at p. 136.
[35] As to which, see *ibid.*, p. 137.
[36] *Ibid.*, p. 136.

ratifications, geographical coverage, implementation and compliance. How far are these comments also applicable to the agreements surveyed in this section concerned only with habitat protection? As far as concerns ratifications, the two global agreements, the Ramsar and World Heritage Conventions, have a relatively high number of parties, including many developing States. One reason for this is no doubt because both agreements, unlike any of the species agreements, are able to offer financial assistance to developing States to help them carry out their obligations.[37] The two regional agreements, those for the Mediterranean and Alps, have been ratified by most of their potential parties. As regards geographical coverage, whilst the high number of ratifications of the Ramsar and World Heritage Conventions means that in principle there is greater geographical coverage, it must be noted that the Ramsar Convention applies only to one type of habitat – wetlands – which are not in general characterised by high levels of species diversity and endemism,[38] whilst the World Heritage Convention applies only to a limited number of, although of course very important, sites. There thus remain many important habitat areas, especially in developing tropical States, to which no international agreement currently applies.

Turning now to implementation and compliance, the Ramsar and Alps Conventions both have rather more effective machinery than most of the species agreements for trying to oversee parties' compliance with their obligations. As regards this issue, the picture appears to be somewhat mixed. For example, in the case of the Ramsar Convention it has been reported that over 10% of 61 important international wetlands face ecological degradation,[39] whilst in the case of the World Heritage Convention many World Heritage sites are under threat from development, inadequate management, and other factors.[40]

Agreements Contributing Indirectly to the Conservation of Species

Apart from the agreements surveyed above which provide directly for the conservation of species and habitats, there is a number of international agreements which contribute indirectly to such conservation, which will be examined in this and the following section. For reasons of space such an examination must necessarily be very brief.

The principal agreement contributing indirectly to the conservation of species is the Convention on International Trade in Endangered Species of Wild Fauna and Flora, 1973.[41] This Convention prohibits or strictly regulates trade in listed endangered or potentially endangered species (amounting now to some 24,000 species). Although the Convention does not require its 120 or so parties to prohibit the killing or taking of such species, by strictly controlling trade in such species and products of such species the Convention removes some of the incentive and reason for such killing or taking. As with the other agreements surveyed, experience of its practical operation has been mixed, but the Convention has certainly achieved a number of successes.[42]

---

[37] For details, see *ibid.*, p. 140.
[38] McNeely, *op. cit.*, n. 2, at p. 91.
[39] *International Environmental Reporter*, 30 June 1993, at p. 479.
[40] World Resources Institute, *op. cit.*, n. 2, at p. 297.
[41] 983 UNTS 243. The Convention came into force in 1975. Further on the Convention, see Lyster, *op. cit.*, n. 4, pp. 239–277; Birnie and Boyle, *op. cit.*, pp. 475–480; and Sand, *op. cit.*, pp. 79–86.
[42] For examples, see Birnie and Boyle, *op. cit.*, n. 4, pp. 478–480 and World Resources Institute, *op. cit.*, n. 4, p. 133.

Agreements Contributing Indirectly to the Conservation of Habitats

There is quite a variety of agreements which contribute indirectly to the conservation of habitats. First, there are agreements concerned with pollution, which can degrade habitats and render them unable to sustain certain species. These agreements include the UN Economic Commission for Europe's Convention on Long-Range Transboundary Air Pollution (1979)[43] together with its protocols on sulphur (1985 and 1994)[44] and nitrogen oxides (1988):[45] these are aimed at curbing acid rain, which is thought to be responsible for damaging trees and causing acidification of freshwater streams and lakes, leading to the death of fish and many other aquatic species. There are also agreements concerned with reducing pollution in rivers and lakes: these include both agreements concerned with individual rivers, such as the Rhine[46] and Elbe,[47] and the ECE's general framework Convention on the Protection and Use of Transboundary Water-courses and International Lakes, 1992.[48]

A second group of agreements is concerned with forests. The Treaty for Amazonian Co-Operation (1978)[49] and the International Tropical Timber Agreement (1983)[50] both have the potential to reduce the rate of exploitation and destruction of tropical rain forests, but in practice neither appears to have done so to any significant degree.[51] More may be expected of the 1994 International Tropical Timber Agreement,[52] the objectives of which now specifically recognise the need for tropical timber to be produced from forests managed on a sustainable basis. The 1994 Agreement was inspired by the Statement of Principles for a Global Consensus of the Management, Conservation and Sustainable Development of All Types of Forests, adopted at the 1992 Rio Conference on Environment and Development.[53]

Finally, there are a number of miscellaneous agreements. These include the ILO's Indigenous and Tribal Peoples Convention 1989,[54] which requires its parties to safeguard the environment of indigenous peoples. Brief mention may also be made of the framework convention to combat the spread of deserts which is currently being negotiated under the auspices of the UN Commission on Sustainable Development.[55]

---

[43] 1302 UNTS 217.

[44] 1985 Protocol on the Reduction of Sulphur Emissions or Their Transboundary Fluxes by at least 30 Per Cent, (1988) 27 ILM 707; 1994 Protocol on Further Reduction of Sulphur Emissions, UN Doc.ECE/EB.AIR/R.84.

[45] 1988 Protocol concerning the Control of Emissions of Nitrogen Oxides or Their Transboundary Fluxes, UKTS 1 (1992), Cm 1787.

[46] 1976 Convention on the Protection of the Rhine against Chemical Pollution, (1986) 90 RGDIP 297; 1976 Convention concerning the Protection of the Rhine against Pollution by Chlorides, (1986) 90 RGDIP 290.

[47] 1990 Convention on the International Commission for the Protection of the Elbe (1990) OJ, L321/25.

[48] Text in Scovazzi and Treves, *op. cit.*, n. 31, at p. 132.

[49] (1978) 17 ILM 1045.

[50] Cmnd 9240.

[51] See further on this question J. Woodliffe, "Tropical Forests", in R.R. Churchill and D.A.C. Freestone (eds.), *International Law and Global Climate Change* (1991), pp. 57–74 and World Conservation Monitoring Centre, *Conservation Status of Tropical Timbers in Trade* (1991).

[52] (1994) 33 ILM 1014.

[53] (1992) 31 ILM 881.

[54] (1989) 28 ILM 1382.

[55] *International Environment Reporter*, 30 June 1993, p. 478. See also (1994) 33 ILM 1328.

Conclusions

It was suggested at the beginning of this chapter that agreements for the conservation of species and habitats do not address the fundamental causes of loss of biodiversity. Their contribution to combating such loss is essentially secondary. Within these terms, how successful have such agreements been? Certainly the agreements have made some contribution, by providing protection for a number of rare and endangered species, either directly or indirectly through conserving their habitat. However, the agreements suffer from a number of major weaknesses which substantially limit their ability to contribute to the preservation of biodiversity. First, few geographical areas of the world are covered by effectively functioning agreements, and such areas are not amongst those which are richest in biodiversity. Second, the record of States parties in implementing and complying with the agreements is at best patchy: this is partly caused by the fact that most agreements lack effective mechanisms for supervising parties' compliance. Third, agreements for the protection of species are inherently limited: protection of fauna and flora through measures other than habitat protection (such as prohibiting their killing and taking) is possible only in the case of higher species.

What then can be done to improve the contribution of species and habitat agreements to maintaining biodiversity? First, it is necessary to get more States (especially developing States) to ratify existing agreements; to bring into force those agreements which are not yet in force; and to revive those agreements (such as the Western Hemisphere Convention) which are moribund. All these things are more likely to happen if financial aid were made available to developing States parties to help them implement their obligations under the agreements (as is the case with the Ramsar and World Heritage Conventions), and if more effort were made, especially by developed States, to address the fundamental causes of loss of biodiversity. Even if existing agreements were more widely ratified and moribund agreements revived, this would still leave significant gaps in the species and types of habitat covered. One possibility might be the adoption of further regional treaties, especially for those areas rich in biodiversity, and further treaties for particular species and habitats. The negotiation of such treaties and their bringing into force would inevitably take some time, however, and, especially in the case of species treaties, risk diluting existing conservation efforts. Before following this course, the prospects for addressing the fundamental causes of biodiversity loss should therefore be considered to see whether this might not be more cost-effective in time and effort than concluding new treaties.

One issue which can and should be addressed in the case of existing agreements concerns their implementation and impact. This could be improved, first, by offering financial aid to developing States to help them implement the agreements (as already mentioned); and secondly, by amending existing agreements in order to introduce more effective mechanisms to supervise and monitor States parties' implementation of, and compliance with, the obligations they impose. Examples from other international environmental agreements, particularly in the areas of air and marine pollution, show both that this is possible and how it can be done.

5

The Conservation of Marine Ecosystems under International Law

David Freestone

The problems of addressing the conservation of marine ecosystems and the maintenance of biodiversity in the oceans are qualitatively different from those of terrestrial
systems. Because mankind is a terrestrial creature there is, perhaps inevitably, a terrestrial bias in our understanding of species and of ecosystems as well as in the means
which have been developed for their protection. This bias is reflected in the
Convention on Biological Diversity itself.[1] Article 2 of the Convention defines "biological diversity" to include the "variability amongst living organisms from all sources
including ... marine and other aquatic ecosystems and the ecological complexes of
which they are a part", however it goes on to specify that "this includes diversity
within species, between species and of ecosystems". Nowhere else in the Convention
is specific reference made to the protection of marine biodiversity[2] although Article
22(2) does specifically provide that contracting States "shall implement the
Convention with respect to the marine environment consistently with the rights and
obligations of States under the law of the sea".[3] In fact the whole approach of the

---

[1] United Nations Environment Programme Convention on Biological Diversity, signed in Rio de
Janeiro 5 June 1992. Text in (1992) 31 ILM 841 and S. Johnson, *The Earth Summit* (1993), at
pp. 82–102, also reproduced in an appendix to this volume.
[2] This was one of the reasons why the USA initially refused to become a party, see Tucker Scully, "The
Protection of the Marine Environment and the UN Conference on Environment and Development",
in *The Law of the Sea: New Worlds, New Discoveries*, Proceedings of the 26th Annual Conference of the
Law of the Sea Institute, Genoa, 22–25 June 1992. He suggests that in relation to marine biodiversity
the Convention is "poorly drafted and a weak instrument" and that "one could read its obligations as a
set back": *ibid.*, p. 148.
[3] On this issue see M. Chandler "The Biodiversity Conventions: Selected Issues of Interest to the
International Lawyer" (1993) 4 *Colorado Journal of International Environmental Law and Policy* 141–175.
Although it is apparent from this paper by a US State Department legal adviser that US concerns in
introducing Article 22(2) were primarily related to issues such as freedom of navigation, it can be
argued that this provision incorporates the environmental principles of customary international law of
the sea (codified in the 1982 Convention) into the 1992 Treaty. See further below, n. 83.

International Law and the Conservation of Biological Diversity (C. Redgwell and M. Bowman, eds.:
90 411 0863 7: © Kluwer Law International: pub. Kluwer Law International, 1995: printed in Great
Britain), pp. 91–107

Convention – directed as it is to finance and biotechnology issues and, arguably, to a concept of national ownership of biological resources based on assumptions about endemic species[4] – bypasses some of the key issues of marine biodiversity conservation.

This chapter will seek to identify the particular problems of the conservation of marine ecosystems and marine biodiversity and the threats with which it is confronted. It will then assess the various means by which current international law could be said to address the need for conservation of marine ecosystems. Awareness of the importance of ecosystems or of ecosystem conservation and management is relatively new in the international arena. Few international instruments actually use this precise terminology, so this survey will take an historical approach, examining the ways in which treaty regimes have sought to address the issue of the conservation of various components of marine biodiversity, by the protection of key species and areas. It will then examine those few conventions which do expressly espouse an ecosystem approach, and set these various components into the wider context of the 1982 Law of the Sea Convention and emerging norms of customary marine environmental law. Finally it will address the issue of whether there can be said to have arisen, through these mechanisms, a general obligation on all States to conserve marine ecosystems.

Marine Biodiversity: Why it is Important and How it is Threatened

The oceans cover some 70% of the planet yet far less is known about the marine environment than the terrestrial; 80% of all known species are terrestrial;[5] only sixteen of the 6,691 species officially classified as endangered are marine and fourteen of these are mammals and turtles – creatures which have some affinity with terrestrial creatures.[6] Because of the fluid nature of the marine environment, scientists suggest that there has been less opportunity or need for speciation in marine organisms, as there has been in land organisms in which species and subspecies have developed as they have become separated from each other by physical factors such as mountains, deserts, ocean or inland waters or simply by distance. This does not mean, however, that the oceans are a single amorphous system. Apart from the obvious variations in the oceans at different latitudes or depths, the existence of closed and semi-enclosed seas and of major currents, confluences and gyres in the open ocean means that there is a wide variety of different ecosystems within the marine environment. It is equally clear however that these bear little relation to the various legal jurisdictional zones established by customary international law and now to be found codified in the 1982 Law of the Sea Convention (LOSC).[7]

As seen above, the definition of biodiversity in the 1992 Convention is centred upon variation within, and between, species. This reflects the common practice of

---

[4] See Scully, *op. cit.*, n. 2.

[5] G.C. Ray, "Ecological Diversity in Coastal Zones and Oceans", in E.O. Wilson (ed.), *Biodiversity* (1988), at pp. 36–50.

[6] B. Thorne-Miller and J. Catena, *The Living Ocean: Understanding and Protecting Marine Biodiversity* (1991).

[7] The United Nations Convention on the Law of the Sea, signed in Montego Bay, Jamaica on December 1982, came into force on 16 November 1994. It recognises different regimes for internal waters, archipelagic waters, territorial waters, contiguous zones, exclusive economic zones and continental shelves of coastal states as well as for the high seas. See further R.R. Churchill and A.V. Lowe, *The Law of the Sea* (2nd edn., 1989).

terrestrial biologists of assessing biological diversity and richness in terms of the numbers of species and subspecies in a particular ecosystem, especially the numbers of those which are unique or endemic. However, this approach does load the dice against marine diversity where speciation is low and endemism is uncommon, but where richness should properly be assessed in different terms. In the oceans there is far greater variety of organisms amongst the higher taxonomic orders than species or subspecies – family, genera, orders or phylum. Indeed one third of all phyla (the highest taxonomic order) are found in the oceans, and the majority of the known animals in the sea are benthic.[8] In the last few years entirely new life forms which thrive in the boiling waters around deep ocean thermal vents have been discovered which offer exciting opportunities for development of medical and industrial processes.[9]

In fact, the health of marine ecosystems and marine life forms is at least as important to life on the planet as that of terrestrial systems. Marine and coastal systems provide important food sources, and marine creatures offer a multitude of different substances which may be of significance to the medical and chemical industry.[10] It is well established that the oceans play a key role as sinks for greenhouse gases,[11] but also, and perhaps most significantly, there is increasing evidence that marine biota play an important role in global chemical processes which may affect climate change.[12] Thorne Miller and Catena suggest that the concentration on genetic, species and ecological diversity reflected in the work of terrestrial biologists (and strongly represented in the 1992 Convention) overshadows what has been termed functional diversity – which reflects the biological complexity of an ecosystem. In their words:

> In the face of environmental change, the loss of genetic diversity weakens a population's ability to adapt; the loss of species diversity weakens a community's ability to adapt; the loss of functional diversity weakens an ecosystem's ability to adapt; and the loss of ecological diversity weakens the whole biosphere's ability to adapt.[13]

The evidence suggests that marine ecosystems are rich in functional diversity, and that there are therefore dangers in transferring to the marine environment concerns about lower order diversity and about the protection of rarity which have been developed in a terrestrial context. Angel has pointed out that 90% of pelagic species are consistently rare and that little is understood about the ecology of rarity in the oceans. However, enough evidence does exist for him to conclude that "the small scale management approach developed for terrestrial ecosystems will be totally ineffective if applied uncritically to the oceans".[14]

---

[8] M.V. Angel, "Biodiversity in the Pelagic Ocean" (1993) 7/4 *Conservation Biology* 760, at p. 762. He reports that the oceans have twenty-eight phyla of animals, thirteen of which are endemic. In contrast terrestrial ecosystems have just eleven phyla, none endemic. Just as only eleven phyla have made the transition from aquatic to terrestrial habitats, only eleven have succeeded in becoming permanent inhabitants of the pelagic realm; *ibid.*

[9] See e.g. W.T. Burke, "State Practice, New Ocean Uses and Ocean Governance under UNCLOS", a paper presented to the 28th Annual Conference of the Law of the Sea Institute, Honolulu, Hawaii, 11–14 July 1994.

[10] *Ibid.*; see also e.g. the extraction of chemicals from sponges which are being used in the development of anti cancer drugs, *The Guardian*, 18 February 1995.

[11] This has been consistently highlighted by e.g. the work of the Intergovernmental Panel on Climate Change (IPCC): see J.T. Houghton, G.J. Jenkins and J.J. Ephraums (eds.), *Climate Change: the IPCC Scientific Assessment* (1990).

[12] *Op. cit.*, n. 6, at pp. 10–14; E.A. Norse (ed.), *Global Marine Biological Diversity* (1993), at pp. 27–29.

[13] *Op. cit.*, n. 6, at p. 10.

[14] *Op. cit.*, n. 8, at p. 769.

The majority of writers seem agreed that the most critical dangers for marine bio-diversity are posed by systemic threats to the very maintenance of marine ecosystems themselves.[15] These threaten the ecological balance of the oceans. The most significant threats are posed by marine pollution from a variety of sources, but princi-pally from land-based sources and activities,[16] from over-exploitation or indiscrim-inate exploitation of marine species,[17] as well as from the destruction of coastal habitats. A large proportion of sea creatures depend on inshore or coastal areas for an important part of their breeding or lifecycles. The destruction or degradation of coastal habitats or the degradation of coastal water quality therefore has a major impact on a wide spread of marine life. This does suggest that the protection of rare and endangered species and of key and representative ecosystems (which is the primary approach of terrestrial biodiversity conservation) may also be appropriate to certain aspects of marine biodiversity conservation. It does not however mean that marine biodiversity can be adequately protected simply by extending the network of protected areas into the sea. These protected areas cannot in themselves provide pro-tection from marine pollution, which may emanate from vessel sources which may be on the high seas or from land-based sources which are in areas under the jurisdic-tion of other States. It must be said also that despite the fact that the 1982 Law of the Sea Convention specifically recognises that "the problems of ocean space are closely interrelated and need to be considered as a whole"[18] the maritime jurisdictional zones recognised by the LOSC, which inevitably make arbitrary divisions in ocean ecosys-tems, do not assist an holistic approach to management of these issues.[19]

The holistic approach to both marine and terrestrial ecosystems is firmly endorsed by Agenda 21, approved at the 1992 UNCED. It is significant to note that from a marine biodiversity point of view much of the key wording is to be found in the Protection of the Oceans Chapter 17[20] rather than in Chapter 15 on the Conservation of Biological Diversity.[21] This perhaps provides more evidence, should it be needed, that the guiding agenda of the biodiversity debate was the issue of ownership and exploitation of biotechnology rather than conservation.

---

[15] See *op. cit.*, n. 6, at pp. 87–154 (this book is the compilation of the work of nearly 500 authors and reviewers); also Thorne-Miller and Catena, *op. cit.*, n. 6, at pp. 7–22.

[16] Also significant is marine debris and atmospheric deposition: *ibid*. There is growing evidence that marine pollution is affecting migratory species like whales. For a recent example see reports of the examination of four whales found dead on the Belgian coast conducted by Professor Claud Joiris of Brussels University in which high levels of toxic materials, including cadmium, mercury, DDT and PCBs were found in their blubber: *The Guardian*, 14 January 1995.

[17] See generally D. Freestone, "The Effective Conservation and Management of High Seas Living Resources: Towards a New Regime?" (1995) *Canterbury Law Review* 341–362.

[18] Preambular para. 3.

[19] The difficulties that jurisdictional zones pose to holistic management are highlighted in the debates over straddling stocks and highly migratory fish stocks. See generally Freestone, *op. cit.*, n. 17; B. Kwiatkowska, "The High Seas Fisheries Regime: At a Point of No Return?" (1993) 8 IJMCL 327–358; M. Hayashi, "The Management of Transboundary Fish Stocks under the LOSC Convention" (1993) 8 IJMCL 245–262.

[20] Chapter 17.1 starts with the statement: "The marine environment – including the oceans and all seas and adjacent coastlines – forms an integrated whole that is an essential component of the global life-support system and a positive asset that presents opportunities for sustainable development."

[21] The equivalent introductory statement in Chapter 15 stresses utilisation: "Our planet's essential goods and services depend on the variety and variability of genes, species, populations and ecosystems. Biological resources feed and clothe us and provide housing, medicines and spiritual nourishment. The natural ecosystems of forests, savannahs, pastures and range lands, deserts, tundras, rivers, lakes and seas contain much of the Earth's biodiversity."

The recognition by international environmental law of the importance of ecosystem management is of relatively recent origin. The earliest "environmental" treaties relate simply to species protection – in the first instance in order to ensure their continued use by man.[22] In the development of a typology it is possible for the purposes of analysis to group general classes of relevant international legal obligations: first, those that address specific systemic threats to the marine environment and therefore to the marine ecosystem – this network of obligations relating to matters such as the control of pollution and of fisheries is well documented[23] and will not be addressed in detail here; second, those obligations that address the conservation of what might be called ecosystem components – marine species and specific marine areas; and finally those obligations that require conservation of marine ecosystems *per se*. Naturally, not all these classes are 'watertight' in the sense that many instruments will be found to contain more than one of these elements. Nevertheless such a classification may serve to identify the strengths as well as defects and *lacunae* in the current legal regimes.

Regimes for the Conservation of Marine Ecosystem Components

Historically, the two main techniques which have been utilised by international conventions for the conservation of marine species are derived from those taken for terrestrial species, namely the regulation or prohibition of the taking of designated species and the protection of habitat by the designation of protected areas – whether standing alone or as part of a system representing migration routes. To these two techniques has more recently been added the regulation of international trade in endangered species and their products under the 1973 Washington Convention, which has increasing significance for marine systems with the growth of the aquarium and coral and shell curio trade.

Protection of Species

The protection of designated species has habitually been addressed by the imposition of restrictions or prohibitions on the harvesting, taking or killing of target species. This approach was taken by the 1946 Whaling Convention,[24] by the various seal hunting regulatory agreements[25] and by the 1973 Polar Bears

---

[22] For example the 1902 Convention for the Protection of Birds *useful to* Agriculture [emphasis added], 102 BFSP 969, cited in A. Kiss and D. Shelton, *International Environmental Law* (1991), at p. 33.

[23] There is a voluminous literature on marine pollution law, see e.g. reading cited in Churchill and Lowe, *op. cit.*, n. 7; on fisheries see W.T. Burke, *The New International Law of Fisheries* (1994).

[24] 1946 International Convention for Regulation of Whaling (Washington) 161 UNTS 143; UKTS 5 (1949) Cmd 7604; amended 1956 338 UNTS 366. See further P.W. Birnie, *The International Regulation of Whaling* (1985), 2 vols.

[25] On the international regulation of seal and polar bear hunting see generally S. Lyster, *International Wildlife Law* (1984), at pp. 39–61. Notable multilateral agreements are the 1911 Treaty for the Preservation and Protection of Fur Seals, 104 BFSP 175; the 1957 Interim Convention on the Conservation of North Pacific Fur Seals, 314 UNTS 105, amended by Protocol in 1963, 1969, 1976, 1980; and the 1972 Convention for the Conservation of Antarctic Seals (which establishes a closed season 1 March to 31 August), 11 ILM 251. Bilateral agreements include the 1957 Norway/USSR Agreement on measures to regulate Sealing and to Protect Seal Stocks in the North-eastern part of the Atlantic Ocean, 309 UNTS 269; and the 1971 Canada/Norway Agreement on Sealing and the Conservation of Seals Stocks in the North-west Atlantic. See also D. McGillivray, "Seal Conservation in the UK: past, present and future" (1994) 10 IJMCL 19–51.

Agreement.[26] Such a strategy is still maintained as part of the approach adopted by more modern generic or regional protected species treaties such as the global 1979 Bonn Convention on the Conservation of Migratory Species of Wild Animals,[27] and the regional treaties concluded under the UNEP Regional Seas Programme, namely the protocols concluded under the Nairobi[28] and the Cartagena[29] Conventions as well as legislation produced under the auspices of the European Community.[30] These latter treaties give equal attention to the issue of habitat or ecosystem loss as a cause of species decline.[31] In the marine context, some other treaties also impose restrictions and/or prohibitions on the incidental affects of non-selective catching techniques which impact on species other than the target species.[32] Modern wildlife treaties also endorse a more proactive approach to species management, requiring rehabilitation, rescue and recovery planning. The Caribbean SPAW Protocol goes further than most in adopting a precautionary approach requiring that States "shall manage species of flora and fauna with the objective of preventing species becoming threatened or endangered".[33]

---

[26] 1973 Agreement on the Conservation of Polar Bears, 13 ILM 13, which contains in Article II a specific injunction to maintain polar bear ecosystems:

> Contracting parties shall take appropriate action to protect the ecosystems of which polar bears are a part, with special attention to habitat components such as denning and feeding sites and migration patterns.

[27] Article 6 of the 1979 Bonn Convention, (1980) 19 ILM 15. It recognises that "wild animals in their innumerable forms are an irreplaceable part of the earth's natural systems" (which must be conserved for the good of mankind) (Preamble, para. 1). A key concept in its effectiveness is the concept of "conservation status" of a migratory species, defined as "the sum of influences acting on the migratory species that may affect its long term distribution and abundance" (Article I.1.b). This status is only taken to be favourable when the following conditions, which include important habitat considerations, are satisfied:

(a) population dynamics data indicate that the migratory species is maintaining itself on a long term basis as a viable component of its ecosystems;

(b) the range of the migratory species is neither currently being reduced, nor is it likely to be reduced on a long-term basis;

(c) there is, and will be in the foreseeable future, sufficient habitat to maintain the population of the migratory species on a long-term basis; and

(d) the distribution and abundance of the migratory species approach historic coverage and levels to the extent that potentially suitable ecosystems exist and to the extent consistent with wise wild life management.

[28] Protocol concerning Protected Areas and Wild Fauna and Flora in the Eastern African Region, concluded at the same time as the 1985 framework Convention for the Protection, Management and Development of the Marine and Coastal Environment of the East African Region; text in P.H. Sand, *Marine Environmental Law* (1988), at p. 156.

[29] 1990 Kingston Protocol on Specially Protected Areas and Wildlife (SPAW). Annexes 1, 2 and 3 list, respectively, flora and fauna which are completely protected and those which may be harvested on a "rational and sustainable basis". Text reproduced in D. Freestone, "Protected Areas and Wildlife in the Wider Caribbean" (1990) 5 *International Journal of Estuarine and Coastal Law* 562. The text of the 1983 Cartagena Convention for the Protection and Development of the Marine Environment of the Wider Caribbean Region, is in Sand, *ibid.*, n. 28, at p. 135.

[30] Directive 79/409 on the Conservation of Wild Birds, OJ L103/1, 24.4.1979.

[31] See, for example, the 1973 Polar Bear Agreement, Article II, *op. cit.*, n. 26.

[32] See, for example, the precautionary quotas on krill fishing in Antarctica *infra*, n. 62; prohibitions on driftnetting *infra* n. 90; regarding other non selective methods of killing/capturing wildlife see the Berne Convention, Appendix, and EC Directive 79/405, Article 8; see also the SPAW Protocol, Article 11(1)(c)(i)(a).

[33] SPAW, *op. cit.*, n. 29, Article 3(3).

Another significant feature of the SPAW Protocol is the way that the species protection provisions have been interpreted by the parties in such a way as to allow a marine ecosystem approach. This has been done by the simple but innovative expedient of designating corals, mangroves and sea grasses as protected species under Annex III of the Protocol that "may only be utilised on a rational and sustainable basis".[34] This concept was not included in the draft of the Protocol itself in 1990 but was introduced at the stage at which the lists of species for the Annexes were being drawn up. The relevant paragraph of the interpretative statement made by the Parties at the Plenipotentiary Meeting in June 1991 is worth reproducing in full:

> in the case of species essential to the maintenance of fragile and vulnerable ecosystems (such as mangrove forests and coral reefs), the listing of such species was felt to be an "appropriate measure to ensure protection and recovery" of the ecosystem which they constitute and hence to fulfil the requirements for listing under Article 11(c) of the Protocol; because these systems as a whole are subject to anthropogenic changes, as well as large scale natural disturbances (such as the consequences of sea-level and temperature rise induced by global warming), appropriate protection should be focused on the system as a whole, rather than on individual specimens; this approach was thought to be appropriate to foster comprehensive national and regional policies for managing these fragile and threatened ecosystems.

Protected Areas

The second key technique, often used in combination with protection of species (above) and in modern treaties increasingly merging with it,[35] is the establishment of protected areas either to protect the habitat of specific species or as representative examples of ecosystems or habitat. These may be important isolated areas or take their place within a systematic network permitting for example transnational migration.[36] In the terrestrial environment between 5–8% of the total world land mass now lies in protected areas. Despite the fact that the sea covers more than two and a half times the land area, marine protected areas cover an area less than a half that of terrestrial protected areas.[37] Of these the overwhelming proportion are coastal and wetland sites in areas under national jurisdiction. A major impetus for the designation of marine protected areas under national law has been the development of the programme developed by the International Union for the Conservation of Nature and Natural Resources (IUCN) Commission of National

---

[34] See Article 1(l), SPAW Protocol, *op. cit.*, n. 29, and further D. Freestone, "Protection of Wildlife and Ecosystems in the Wider Caribbean" (1991) 23 *Marine Pollution Bulletin* 578–581.

[35] Notable examples would be the 1990 Kingston SPAW Protocol, *op. cit.*, n. 29, and the 1990 Agreement on the Conservation of Seals in the Wadden Sea, which combines specimen protection (Article VI) with management (Article IV) and an obligation to create a "network of protected areas" (Article VII) as well as to reduce pollution (Article VIII). For the of text of the Wadden Sea Agreement, see D. Freestone and T. IJlstra, *The North Sea: Perspectives on Regional Environmental Co-operation* (1991), at pp. 280–287.

[36] This is the objective of the 1979 Berne Convention and of the 1992 EC Habitats Directive, *infra*, n. 44.

[37] R. Kenchington, C. Bleakey and S. Wells, *A Global Representative System of Marine Protected Areas*, Draft Report, Great Barrier Reef Marine Park Authority IUCN, 1995.

Parks and Protected Areas (CNPPA).[38] A web of treaties provides international recognition for such national sites, including the 1971 Ramsar Convention[39] and the 1972 UNESCO World Heritage Convention.[40] Regional treaties include the 1940 Washington Convention on Nature Protection and Wildlife Preservation in the Western Hemisphere[41] which was, in its time, a revolutionary treaty which introduced many of the early techniques of modern ecosystem protection. There are now four UNEP Regional Seas Protocols on Protected Areas in the Mediterranean, East Africa, the Wider Caribbean and South-east Pacific. Other regional treaties include the 1976 Apia Convention on the Conservation of Nature in the South Pacific,[42] the Berne Convention,[43] as well as the EC Wild Birds Directive which implements it and the 1992 Habitats Directive.[44]

[38] The goals of this programme were established by Resolution 17.38 of the IUCN General Assembly in 1986:

> To provide for the protection, restoration, wise use, understanding and enjoyment of the marine heritage of the world in perpetuity through the creation of a global, representative system of marine protected areas and through the management, in accordance with the principles of the World Conservation Strategy, of human activities that use or affect the marine environment.

Ibid., p. 5.

[39] The 1971 Ramsar Convention on Wetlands of International Importance, Especially as Waterfowl Habitat. The increase in knowledge of the significance of wetlands since 1971 has meant that designations are increasingly made in recognition of the importance of wetlands as support systems for marine life. The Cagliari Criteria–developed in 1980 at the Cagliari conference of the parties – recognise this change of emphasis. In the first category of sites are those of significance for waterfowl, but in the second category are included any wetland which: (a) supports an appreciable number of rare, vulnerable or endangered species or subspecies of plant or animal; or (b) is of special value for maintaining the genetic and ecological diversity of a region because of the quality and peculiarities of its flora and fauna; or (c) is of special value as the habitat of plants or animals at a critical stage of their biological cycles; or (d) is of special value for its endemic plant or animal species or communities. In the third category are sites which are "particularly good examples of a specific type of wetland characteristic of its region". Although subject to further amendment, the 1990 Montreal COP recommended that further amendment of the criteria be avoided as far as possible to assist uniform application of the Convention. See M.J. Bowman, "The Ramsar Convention Comes of Age" (1995) XLII *NILR* 1–51, p. 21.

[40] 1972 UNESCO Convention for the Protection of the World Cultural and Natural Heritage, (1972) 11 ILM 1358; UKTS 2 (1985), Cmnd 9424. These sites may be on land or at sea. Sites which have a marine or coastal component include The Great Barrier Reef, the Lord Howe Island Group, (Australia) the Galapagos Islands and Sangay National Park (Ecuador). The "Operational Guidelines" adopted in 1979 (revised 1984) by the Intergovernmental World Heritage Committee (established under Article 8) envisage that sites will only qualify for designation if, *inter alia*, they:

> (ii) [are] outstanding examples representing significant ongoing geological processes, biological evolution and man's interaction with his natural environment; as distinct from periods of the earth's development, this focuses upon ongoing processes in the development of communities of plants and animals, land forms and marine areas and freshwater bodies; or …
> (iv) contain the most important and significant natural habitats where threatened species of animals or plants of outstanding universal value form the point of view of science or conservation still survive.

[41] 12 October 1940, 161 UNTS 193; USTS 981. It envisaged the establishment of protected areas (with different categories of protected area in Article 1) and called on all state parties to protect and preserve all flora and fauna within their territories whether or not in protected areas. It also envisaged international co-operation and special measures for migratory species (Article VII – listed in Annex I).

[42] The 1976 Apia Convention, which only came into force in June 1990, aims to encourage the establishment of a network of protected areas, with the objective, *inter alia*, of safeguarding representative samples of natural ecosystems. It has no independent institutional framework nor machinery for monitoring or enforcement. For text see W. Burhenne (ed.), *International Environmental Law: Multilateral Treaties* (1974–), at p. 976.

[43] UKTS 56 (1982) Cmnd. 8738, ETS 104.

[44] Directive 92/43 on the conservation of natural habitats and of wild flora and fauna, OJ L206/7, 22.7.1992.

These measures of course relate to the international recognition of areas designated under national law. At a different level, a number of instruments do actually provide international recognition for areas outside national jurisdiction, albeit for specific purposes. Inevitably wider ranging areas are involved. Protection from whaling has for example been conferred by the parties to the International Convention for the Regulation of Whaling on the Indian Ocean Whale Sanctuary, and on the newly agreed Antarctic Whale Sanctuary. For shipping purposes, under the 1973/78 Convention for the Prevention of Pollution from Ships (MARPOL),[45] extensive ocean areas, such as the Wider Caribbean and the North Sea, have been designated "Special Areas" for the various purposes of the different annexes to that convention. In each of these special areas, which is designated "for recognised technical reasons in relation to its oceanographical and ecological condition and to the particular characteristics of its traffic",[46] more rigorous requirements apply to discharges from vessels than in the wider oceans. The International Maritime Organisation also takes environmental factors into account in authorising other restrictions on shipping and may also identify Particularly Sensitive Sea Areas (PSSAs) at risk from shipping to assist States in developing further protective measures.[47]

A more ambitious proposal was formulated within the US during the preparations for the 1992 Rio Summit for the designation of "Wild Ocean Reserves" – protected areas of high seas or deep sea bed outside national jurisdiction for reasons of cultural as well as biological sensitivity.[48] Protecting areas of the global commons involves not only a considerable number of legal but also political obstacles (relating for example to freedom of fishing and transit) and the full proposal did not survive the work of the UNCED PrepCom. All that remained of the concept in the final text of Agenda 21 is a reference to the need "to preserve high seas marine habitats and other ecologically sensitive areas".[49]

Regulation of Trade in Wild Species

The 1973 Washington Convention on International Trade in Endangered Species of Wild Fauna and Flora (CITES)[50] provides a key component in this strategy, aimed at taking away the economic incentive for depredation of endangered species or species important for habitat, namely prohibitions or restrictions on trade. It is clear that one of the main threats to marine species such as sea turtles is the commercial trade in products such as turtle shell. Other well known threats to marine ecosystems are posed by over-exploitation of shells and corals for the tourist souvenir trade and of reef fish for the aquarium trade. CITES provides the main regulation of such trade, although a number of treaties provide independent proscriptions or regulation of such trade.[51]

---

[45] (1978) 17 ILM 546. For designation as a special area, an area has to be listed in an Annex by decision of the IMO under the requirements laid down in Article 16. See further Gjerde and Freestone, *infra*, n. 47.

[46] MARPOL Convention 1973/78, *ibid.*

[47] Under the "Guidelines for the Designation of Special Areas and the Identification of Particularly Sensitive Sea Areas" adopted by IMO Assembly Resolution A.720(17) of 6 November 1991. For wider discussion of this issue see K. Gjerde and D. Freestone, *Particularly Sensitive Sea Areas: An Important Environmental Concept at a Turning Point* (1994), a special Issue of the IJMCL published by Graham & Trotman.

[48] See US National Ocean Service, *Report on the International Meeting on Wild Ocean Reserves, Honolulu, October 1991* (Washington, DC, 1992).

[49] Agenda 21, para. 17.46; see also para. 17.7.

[50] (1973) 12 ILM 1085.

[51] See, for example, the SPAW Protocol, *op. cit.*, n. 29, Article 11, and note Article 25 which specifically relates to the relationship with CITES.

Conservation of Marine Ecosystems

It was commented above that the recognition of the importance of management of ecosystems themselves, rather than simply those of their components which may be of immediate significance to mankind, is a relatively recent phenomenon. Crucial steps in this development were the 1972 Stockholm Declaration[52] and the 1980 IUCN World Conservation Strategy which formed the basis for the 1982 UN General Assembly World Charter for Nature,[53] and which popularised the concept of, as well as the term, "life support systems" and which stressed the interrelationship of these with other ecological processes and genetic diversity.[54]

Against this background and time scale must be viewed the important regional regime created by the 1980 Canberra Convention on the Conservation of Antarctic Marine Living Resources[55] which is arguably the first to be centred on an ecosystem approach to conservation and has been described as "a model of the ecological approach".[56] Even the geographical scope of this treaty itself is unique, in that it is designed around the Antarctic ecosystem. Under Article 1 the Convention applies: "to the Antarctic marine living resources of the area south of 60 degrees South latitude and to the Antarctic marine living resources of the area between that latitude and the Antarctic Convergence which form part of the Antarctic marine ecosystem". The Antarctic marine ecosystem is then defined by Article 1(3) as "the complex of relationships of Antarctic marine living resources with each other and with their physical environment".

The sole objective of the Convention is declared to be "the conservation of Antarctic marine living resources";[57] conservation is however defined to include "rational use".[58] To achieve this end any harvesting or associated activities has to be conducted in accordance with declared principles:

(a) prevention of decrease in the size of any harvested population to levels below those which ensure its stable recruitment. For this purpose its size should not be allowed to fall below a level close to that which ensures the greatest net annual increment;

---

[52] Principle 4 of that Declaration provides that:

> Man has a special responsibility to safeguard and wisely manage the heritage of wildlife and habitat which are [sic] now gravely imperilled by a combination of adverse factors. Nature conservation, including wildlife, must therefore receive importance in economic planning.

[53] UN GA Res. 37/7, 28 October 1982, reproduced in (1983) 23 ILM 455–460.

[54] The IUCN 1980 World Conservation Strategy specifically required:

> – maintenance of essential ecological processes and life support systems, soil regeneration and protection, recycling of nutrients, and water purification;
> – preservation of the genetic diversity on which depends the functioning of most processes and life support systems, together with breeding programmes necessary to the protection and improvement of cultivated plants, domestic animals and micro-organisms;
> – sustainable utilisation of species and ecosystems (fish and other wildlife, forests and grazing lands) by humans in both industrial zones and the countryside.

[55] (1982) 19 ILM 841; UKTS 48 (1982), Cmd 8714. See also Redgwell, this volume.

[56] Kiss and Shelton, *op. cit.*, n. 22, at p. 254.

[57] Article II.1.

[58] Article II.2.

(b) maintenance of the ecological relationships between harvested, dependent and related populations of Antarctic marine living resources and the restoration of depleted populations to the levels defined in subparagraph (a) above; and

(c) prevention of changes or minimisation of changes in the marine ecosystem which are not potentially reversible over two or three decades, taking into account the state of available knowledge of the direct and indirect impact of harvesting, the effect of the introduction of alien species, the effects of associated activities on the marine ecosystem and of the effects of environmental changes, with the aim of making possible the sustained conservation of Antarctic marine living resources.[59]

These objectives, which clearly relate to the maintenance of the ecosystem rather than its exploitation, are implemented by the Commission for the Conservation of Antarctic Marine Living Resources (CCAMLR) which co-ordinates research on Antarctic marine living resources and adopts appropriate conservation and management measures.[60] After a slow start,[61] the Commission has been developing "precautionary" catch limits on fish stocks within its jurisdiction to ensure that over fishing does not damage the food chain. In 1991 for example it established a precautionary catch limitation on *Euphausia superba*[62] and a working group of the Scientific Committee was established to develop "precautionary measures on krill fishing" in order to prevent the "unregulated expansion of the fishery at a time when the information available for predicting potential yield [was] very limited".[63] A further formal step in the protection of the Antarctic Ecosystem was taken with the conclusion of the 1991 Madrid Protocol to the Antarctic Treaty on Environmental Protection.[64]

Another treaty of major potential significance but, unfortunately, still not in force after a decade is the 1985 ASEAN Convention on the Conservation of Nature and Natural Resources.[65] This Convention reflects in its wording the concepts contained in the 1980 IUCN World Conservation Strategy, embracing a clear ecosystem approach to conservation. Kiss and Shelton describe this treaty as "the most comprehensive approach to viewing conservation problems that exists today".[66]

The Convention recognises "the interdependence of living resources, between them and with other natural resources, within ecosystems of which they are a part". It is divided into eight chapters. Chapter I is devoted to Conservation and Development. The parties commit themselves in Article 1 to the "Fundamental Principle" that they will:

undertake to adopt singly, or where necessary and appropriate through concerted action, the measures necessary to maintain essential ecological processes and life support systems,

---

[59] Article II.3.
[60] See also R.R. Churchill, *EEC Fisheries Law* (1987), at p. 188. Its powers include the establishment of quantities to be harvested, designation of protected species and closed seasons as well as regulation of gear.
[61] See M. Howard, "The Convention on the Conservation of Antarctic Marine Living Resources: A Five-Year Review" (1989) 38 ICLQ 104.
[62] CCAMLR Conservation Measure 32/X on Precautionary Catch Limitations on *Euphasia superba* in Statistical Area 48.
[63] Working Group Report, para. 6.34. The impetus for that work was the statement at the Ninth meeting of the Commission by the USSR, Japan and Korea that they were not in principle opposed to the idea of a precautionary limit of krill fishing, but that "the quantitative basis for such a precautionary limit on fishing should have scientific justification based on assessments performed by the Scientific Committee" (CCAMLR-IX, para. 8.7).
[64] (1991) 30 ILM 1461.
[65] Kuala Lumpur, 9 July 1985, reproduced in Kiss and Shelton, *op. cit.*, n. 22, at p. 493.
[66] *Op. cit.*, n. 54, at p. 279.

to preserve genetic diversity, and to ensure the sustainable utilisation of harvested natural resources under their jurisdiction in accordance with scientific principle and with a view to attaining the goal of sustainable development.

Chapter II is specifically devoted to Conservation of Species and Ecosystems. Article 3 obliges the parties to maintain, wherever possible, maximum genetic diversity and take appropriate measures to conserve "animal and plant species whether terrestrial, marine and[sic] freshwater" and *inter alia* to conserve "natural terrestrial, freshwater and coastal and marine habitats". The measures envisaged include regulation of harvested species, protection of endangered and endemic species, protection of soil to protect both land-based and coastal and marine habitats such as coral reefs.[67] Chapter III is devoted to Conservation of Ecological Processes,[68] whilst Chapter IV covers Environmental Planning Measures including the designation of protected areas.

It is therefore tempting to observe that the major treaties calling for marine ecosystem conservation considered thus far are either limited in geographic scope, or not yet in force,[69] or both.

The Law of the Sea and Customary International Law

In order to seek an answer to the wider question of whether there is a general obligation on all States to conserve marine ecosystems it is necessary to look beyond specific treaty obligations at customary international law. The starting point for an assessment of this is the 1982 Law of the Sea Convention (LOSC) which only came into force in November 1994[70] but which is widely regarded as reflective of customary law. Customary law recognises the division of the ocean into a series of juridical regimes which reflect criteria related to coastal States' sovereignty and resource exploitation rather than considerations of ecosystem integrity. The nature of the obligations which customary international law, and now the 1982 Convention, imposes on States in relation to the marine environment does to a large extent depend upon the juridical nature of the particular waters under consideration, consequently these jurisdictional divisions can constitute a major obstacle to the rational management of ecosystems or species which cross or straddle more than one such zone. Broadly, the oceans are divided into the following maritime zones: internal waters – behind the coastal State baseline; a belt of territorial waters up to 12 nautical miles (nm) in breadth, a 24 nm contiguous zone with restricted

---

[67] Article 7.2.b.

[68] By, for example, the reduction and control of environmental degradation (Article 10) and pollution (Article 11). These objectives would be attained by, *inter alia*, the promotion of sound agricultural practices, promotion of pollution control and environmentally sound industrial processes, together with economic and fiscal incentives. Other factors to be taken in account include the cumulative effect of new activities or new discharges of contaminants with existing activities and discharges as well as possible interactions and regulation of activities or pollutants which may adversely affect "ecologically essential" processes.

[69] To the ASEAN Convention could perhaps be added the African Convention on the Conservation of Nature and Natural Resources, Algiers, 15 September 1968, 1001 UNTS 3. The treaty makes no explicit reference to the marine or coastal environments of contracting states, and none of the species listed for protection appear to be marine species. Nevertheless, there appears to be nothing in the treaty to exclude marine/coastal ecosystems from its ambit. It too is not yet in force.

[70] For an analysis of the circumstances of the coming into force of the Convention and the UN General Assembly Resolution A/48/263 of 28 July 1994 adopting the Agreement Relating to the Implementation of Part IX of the United Nations Convention on the Law of the Sea of 10 December 1982, see D. Freestone and G.J. Mangone, *The Law of the Sea Convention: Unfinished Agendas and Future Challenges* (1995), a special issue of the IJMCL.

enforcement jurisdiction, a 200 nm exclusive economic or fishing zone, and the high sea areas beyond those limits. Within each of these zones the Convention envisages a different balances of rights and duties between coastal and other States.

The 1982 Convention contains a number of provisions of general significance for the protection of marine ecosystems. Nevertheless, it would probably be a mistake to think this was a conscious drafting objective *per se*. It is certainly possible to read into the provisions of Part XII of the Convention (on Protection and Preservation of the Marine Environment) endorsement for a marine ecosystem approach to marine conservation,[71] although these obligations are even less precise than those relating to pollution control. Article 192 LOSC recognises a general obligation to "protect and preserve the marine environment". In so far as this goes beyond simple protection, it can be interpreted as being an obligation to behave in a precautionary way.[72] Article 194(5) specifically requires that "measures taken in accordance with this Part [i.e. Part XII] shall include those necessary to protect and preserve rare or fragile ecosystems as well as the habitat of depleted, threatened or endangered species and other forms of marine life". As this provision is located within the general provisions of Part XII this requires all States to protect these special ecosystems and habitats from the effects of pollution originating from all sources in addition to other more general conservation measures.[73] In addition Article 196 requires States to take all measures to prevent, reduce and control pollution from "the use of technologies" under either their jurisdiction or control. This could be read to mean biotechnology or any polluting technology. The rest of the paragraph requires States to prevent, reduce and control the "intentional or accidental introduction of species, alien or new, to a particular part of the marine environment which may cause significant and harmful changes thereto". However, the definition of "pollution" adopted by the LOSC[74] does not make explicit reference to impacts on marine ecosystems – a defect which has been remedied in some regional conventions.[75]

In the Exclusive Economic Zone (Part V) coastal States are obliged to ensure "through proper conservation and management measures that the maintenance of the living resources [in the EEZ] is not endangered by over-exploitation" (Article 61(2)), taking into consideration the effects on species associated with or dependent upon harvested species with a view to maintaining or restoring populations of such associated or dependent species above levels at which their reproduction may become seriously threatened" (Article 61(4)). Similar provisions apply to such

---

[71] M.H. Belsky, "Management of Large Marine Ecosystems: Developing a New Rule of Customary International Law" (1985) 22 *San Diego Law Review* 733–763.

[72] The view that the LOSC intends a proactive, rather than simply passive, role is supported by the Commentary produced by the University of Virginia, Centre for Oceans Law and Policy, see M.H. Nordquist (ed.), *United Nations Convention on the Law of the Sea: A Commentary*, Volume IV (edited by S. Rosenne and A. Yankov), at pp. 35ff. On the development of the precautionary principle in international environmental law see D. Freestone, "The Precautionary Principle" in R. Churchill and D. Freestone (eds.), *International Law and Global Climate Change* (1991), at pp. 21–40. This is developed further in D. Freestone, "The Road from Rio: International Environmental Law after the Earth Summit" (1994) 6 *Journal of Environmental Law* 193–218.

[73] See also K. Gjerde, "The Law of the Sea", in J. Broadus and R. Vartanov (eds.), *The Oceans and Environmental Security: Shared US/USSR Perspectives* (1994), at p. 233.

[74] Article 1.1.

[75] E.g. 1992 Paris Convention for the Protection of the Marine Environment of the North East Atlantic, Article 1(d); 1992 Helsinki Convention on the Protection of the Marine Environment of the Baltic Sea Area, Article 2(1).

species in High Sea fisheries.[76] But as Gjerde has pointed out these provisions "only aim to maintain the viability of such species, and ... [not] to protect their role within the food web or the functioning of the marine ecosystem as a whole".[77]

The requirements which the Convention imposes in relation to monitoring the risks or effects of pollution can also be taken to support the ability of the Convention to protect ecosystems. The Convention requires States to monitor their ongoing activities and to assess the potential effects of their planned activities to determine whether they are likely to pollute the marine environment or cause "significant and harmful changes".[78]

Under general customary international law there is also a general obligation first promulgated in Principle 21 of the Stockholm Declaration and now to be found in Principle 2 of the 1992 Rio Declaration to ensure that "activities within their juris-diction or control do not cause damage to the environment of other States or *of areas beyond the limits of national jurisdiction*".[79] Minimal and general though this obligation is in relation to global commons areas such as the high seas, the wording is clear to the extent that this obligation not to cause damage extends not simply to activities physically located on State territory but also to the activities under State jurisdiction (including State registered vessels) and arguably, to the extent that they are subject to national jurisdiction, even to nationals.

Although this obligation not to cause transboundary environmental damage has primarily been discussed to date in the context of damage to global commons – notably the atmosphere or the ozone layer – it can be argued that customary law would impose on States responsibility under this principle for a range of activities which impact on marine ecosystems whether such activities take place within areas of national jurisdiction (and therefore coming perhaps within the first leg of Principle 21, above) or outside or straddling such areas (and therefore within the second leg).[80] Activities which could be argued to cause damage to marine ecosys-tems and to fall foul of this principle would include marine pollution – particularly that emanating from land-based sources and activities;[81] it could also be extended to fishing and related activities which impact upon rare and endangered marine species or their habitat. Certainly as our understanding of Large Marine Ecosystems develops it becomes apparent that fisheries management must fit into an holistic

---

[76] See further, D. Freestone, "The Requirements of Proof for Conservation in High Seas Fisheries", UN FAO Legal Office (forthcoming).

[77] Gjerde, *op. cit.*, n. 73, at p. 234.

[78] Articles 204–206.

[79] Johnson, *op. cit.*, n. 1, at p. 118 (emphasis added).

[80] Report of Experts Group on Environmental Law of the World Commission on Environment and Development, published as *Environmental Protection and Sustainable Development: Legal Principles and Recommendations* (1987), J.G. Lammers (ed.), at p. 80. In their "Legal Principles for Environmental Protection and Sustainable Development" the Experts Group suggest, in relation to a general obliga-tion to prevent and abate transboundary environmental interference, that:

> the obligation ... exists only to the extent that it is reasonably foreseeable that *substantial* harm [original italics] is caused, or that there is *significant risk* [emphasis added] that such harm will be caused...
> ... the nature and the extent of the measures to be taken, of course, depends on the nature and the extent of the extra-territorial harm which must be prevented or abated.

[81] The international community already accepts the principle of flag state responsibility for vessel source pollution, see *op. cit.*, n. 72. This principle is also utilised for high seas fishing in the recent UN FAO Agreement to Promote Compliance with International Conservation and Management Measures by Fishing Vessels on the High Seas, 1994.

view of the protection of the marine environment.[82] Such a view appears to be endorsed also by the Convention on Biological Diversity itself, which in its jurisdictional scope (under Article 4(b)) applies its obligations on States to "processes and activities, regardless of where their effects occur, carried out under [a contracting State's] jurisdiction or control, within the area of its national jurisdiction or beyond the limits of national jurisdiction".[83]

It can also be argued that the precautionary principle[84] is emerging as a discrete principle of general international environmental law and thus must be incorporated into the general environmental obligations of the LOSC. While arguments may persist amongst international lawyers on this point,[85] there can be little doubt that its acceptance at a political level in both the Rio Declaration and Agenda 21 represents a major change in environmental policy. It requires a shift in decision making in favour of a bias towards safety and caution.[86] Its wide acceptance in regional and sectoral treaty obligations relating to the marine environment would strongly support such an argument.[87] So too would its endorsement in the preamble of the Biodiversity Convention which says that "where there is a threat of a significant reduction or loss of biodiversity, lack of full scientific certainty should not be used as a reason for postponing measures to avoid or minimise such a threat".

The significance of the endorsement of the precautionary approach or principle in relation to the general obligation imposed by Article 192 LOSC to protect the marine environment is twofold. The first aspect relates to the relevant sphere of activity. A precautionary approach requires a broader and longer range perspective on the potential impacts of particular activities on the marine environment. If the

---

[82] In 1984 the American Association for the Advancement of Science introduced the concept of the Large Marine Ecosystem (LME) which can cover whole ocean systems on the basis that "local processes are always embedded in a matrix of larger scale processes". See K. Sherman, L. Alexander and B. Gold (eds.), *Large Marine Ecosystems: Patterns, Processes and Yields* (1990), at p. 167; see also Freestone, *op. cit.*, n. 76.

[83] And, as suggested above, Article 22 could be interpreted as an incorporation of the environmental provisions of the LOSC into the 1992 Convention.

[84] The precautionary principle, accepted by virtually every intergovernmental environmental forum since 1990 and now widely included in regional and sectoral treaty obligations, was endorsed by Principle 15 of the 1992 Rio Declaration in the following terms:

> In order to protect the environment, the precautionary approach shall be widely applied by States according to their capabilities. Where there are threats of serious or irreversible damage, lack of full scientific certainty shall not be used as a reason for postponing cost-effective measures to prevent environmental degradation.

There is now a large literature on this, see e.g. *op. cit.*, n. 72 and sources discussed there, such as: the 1992 UN Framework Convention on Climate Change, Article 3(3), and the 1992 Convention on Biological Diversity, preambular para. 9, both reproduced in (1993) 31 ILM at p. 848 and p. 818 respectively and in Johnson, *op. cit.*, n. 1; the 1992 Helsinki Convention on the Protection of the Marine Environment of the Baltic Sea Area, Article 3(2), reproduced in (1993) 8 IJMCL 215; 1992 Helsinki Convention on the Protection and Use of Transboundary Watercourses and Lakes, Article 2(5)(a), reproduced in (1993) 31 ILM 1312; the 1992 Maastricht Treaty on European Union, Article 130R (2), reproduced in (1993) 32 ILM 1693; and the 1992 Paris Convention for the Protection of the Marine Environment of the North East Atlantic, Article 2(2)(a), reproduced in (1993) 8 IJMCL 50.

[85] Cf. Freestone (1994), *op. cit.*, n. 72, and P. Birnie and A. Boyle, *International Law and the Environment* (1992), at p. 98.

[86] Indeed it can be seen as the crystallisation of the general obligation which Article 192 LOSC imposes to take proactive action in relation to protection of the marine environment, including the protection of marine ecosystems.

[87] See *op. cit.*, n. 84.

goal is to preserve marine biodiversity, it requires an holistic/comprehensive approach which can address the whole range of threats to marine species or ecosystems, including ecological processes and functions: in other words, an ecosystem approach. The regimes developed for the Antarctic or the ASEAN region are possible models for such an approach. It requires, *inter alia*, the development of principles and procedures to overcome the arbitrary juridical zones established by the LOSC where this is necessary to achieve these objectives.[88]

The second issue relates to the burden of proof and the timing of action. In situations where specified activities constitute a serious threat to marine ecosystems, a precautionary approach would shift the traditional burden of proof from those who seek to argue that harm has or will be caused onto the shoulders of those advocating a particular activity to show that serious harm will not be caused. This has been controversial in relation to a number of issues which might impact marine ecosystems such as the management of straddling fish stocks or highly migratory fish stocks,[89] or the use of pelagic drift nets,[90] although there is increasing evidence of a move towards precautionary approaches in these areas.[91] There is also evidence of its adoption *de facto* in a number of new treaty regimes, including the treaty designed to address the problems of fishing in the now famous Bering Sea high seas enclave or "donut hole".[92]

Conclusions

Despite the fact that marine biodiversity is included within the ambit of the 1992 Convention on Biological Diversity and that the jurisdictional scope of the Convention makes is clear that it covers the high seas as well as national territory, the particular problems of the conservation of marine ecosystems and biodiversity have been largely overlooked by the Convention. It is paradoxical that although the particular problems of conservation of many marine creatures, particularly pelagic

---

[88] Note for example the "compatibility" approach being developed at the UN Straddling Stocks Conference, which would require compatibility in the conservation and management of stocks inside and outside 200 limits. See Draft Agreement for the Implementation of the Provisions of the United Nations Convention on the Law of the Sea 1982 relating to the Conservation and Management of Straddling Fish Stocks and Highly Migratory Fish Stocks, prepared by the Chairman of the Conference, A/CONF.164/22, 23 August 1994, particularly Article 7.

[89] See e.g. Draft Agreement, *ibid.*, Article 6 and Annex 2.

[90] The practice of driftnetting has resulted in a series of UN General Assembly Resolutions: 44/225 of 22 December 1989; 45/197 of 21 December 1990; 46/215 of 20 December 1991 and 48/445 of 21 December 1993. Regional action has also been taken in the form of the 1989 Wellington Convention for the Prohibition of Fishing with Long Driftnets in the South Pacific and EC Regulation 345/92, both of which set an upper limit of 2.5 km on such lengths (with some exceptions). These actions are still highly controversial. See, for example, the arguments adduced by Professor Kazua Sumi, "International Legal Issues Concerning the Use of Driftnets with Special Emphasis on Japanese Practices and Responses" in *The Regulation Of Drift Net Fishing On The High Seas: Legal Issues* (1991), FAO Legislative Study No. 47, at pp. 45–70; see also W.M. Burke, *op. cit.*, n. 9, at pp. 13–32 and now *The New International Law of Fisheries, op. cit.*, n. 23.

[91] The various actions on large scale pelagic driftnetting, *ibid.*, are clearly precautionary. The precautionary principle is being elaborated by the current UN Conference On Straddling Fish Stocks and Highly Migratory Fish Stocks, *op. cit.*, n. 88, as well as within the preparatory meetings for the 1995 Washington Conference on Land Based Sources Of Marine Pollution, see e.g. UNEP/ICL/IG/1/2.

[92] The treaty provides a precautionary approach in that no fishing in the area will take place unless the Aleutian Basin Pollock biomass is scientifically calculated to exceed 1.68 million metric tons. See Annex to 1994 Convention on the Conservation and Management of Pollock Resources in the Central Bering Sea, reproduced in (1995) 10 IJMCL 127–135; for commentary see Dunlap, *ibid.*, at pp. 114–126.

creatures, make them particularly suitable to regulation at an international level under a treaty on biological diversity, in fact the most important discussions concerning conservation of marine biological diversity are currently taking place in the context of other forums – those relating to land-based sources, straddling fish stocks or at a regional or sectoral level.

As this chapter has sought to show, there is in existence a substantial body of treaty law which seeks to address one or more aspects of marine ecosystem conservation. Although few treaties actually commit themselves to this, it is clear that a larger number of treaty regimes are developing an ecosystem approach through their parties interpretation of their existing treaty obligations.[93] At the level of customary international law the coming into force of the 1982 Law of the Sea Convention must be seen as a most positive force in the crystallisation of the general obligations of States to protect the marine environment. Nevertheless, important though the obligations of Part XII are in this respect, they too require further substantial elaboration and implementation.[94]

The recognition by the 1992 Convention of the issue of the "conservation of biological diversity as a matter of common concern of humankind"[95] implies that all States have a legal interest in the issue as well as a positive responsibility to safeguard it. However, this "common concern" still requires a more obvious focus than national actions or diverse regional or sectoral actions, for, as this chapter has sought to suggest, much of the action needed has to be taken in international waters as well in coastal waters or in ways that will reflect natural ecosystem boundaries rather than national maritime jurisdictional boundaries. The 1992 Convention envisages the elaboration of protocols.[96] A protocol on the conservation marine biodiversity in the context of the protection of marine ecosystems would be an obvious way of seeking to remedy the *lacunae* of the existing regimes and refocusing attention on this crucial, but somewhat neglected, aspect of the biodiversity debate.*

---

[93] E.g. the SPAW Protocol, the Ramsar and the UNESCO Conventions.

[94] See e.g. J.I. Charney, "The Marine Environment and the 1982 United Nations Convention on the Law of the Sea" (1994) *International Lawyer* 879, at p. 882, who suggests that the LOSC regime "represents an effort to establish optimal progressive norms along with a framework for the evolution of new norms as developments require".

[95] Preambular para. 3, 1992 Convention. See comments on this concept in the context of climate change by A.E. Boyle, "International Law and the Protection of the Global Atmosphere: Concepts, Categories and Principles" in Churchill and Freestone, *op. cit.*, n. 72, at p. 19.

[96] Article 28.

* I am grateful to Kristina Gjerde for reading earlier drafts and making a number of important suggestions for improvement.

6

The Protection of the Antarctic Environment and the Ecosystem Approach

Catherine Redgwell

Introduction

The Antarctic environment and its associated and dependent ecosystems form a distinctive component of planetary biological diversity. Although there is some dispute as to the number of plant and animal species endemic to the Antarctic region, hundreds of native species of land flora and fauna have been identified.[1] There is a stark contrast between land and sea, with an immense richness of marine resources and a paucity of terrestrial life, tempting comparisons between the Antarctic continent and the moon or Mars.[2] Indeed, for more than half of each year most of the continent of Antarctica is virtually lifeless.[3] The contrasting richness of the marine environment is deceptive, however, since this relates to high individual species populations rather than to species diversity, which is relatively low when compared with low and mid-latitude areas.[4]

The Antarctic ecosystem is characterised by three broad types of population at different trophic levels. The higher organisms of Antarctica, particularly those at the top trophic level (whales, seals, birds), apply a "K-strategy" in the exploitation of environmental resource: pluri-annual development dominating the mature stages of

---

[1] See Tables 2 and 3 in the IUCN's *A Strategy for Antarctic Conservation* (1991), at pp. 13–14.

[2] J.A. Gulland, "The Management Regime for Living Resources", in C.C. Joyner and S.K. Chopra (eds.), *The Antarctic Legal Regime* (1988), at p. 219.

[3] G.A. Llano, "The Ecology of the Southern Ocean Region" (1978) 33 *University of Miami Law Review* 357, at p. 362. There are no trees on the Antarctic continent, no indigenous land invertebrates, and only two flowering plant species: C.C. Joyner, *Antarctica and the Law of the Sea* (1992), at pp. 13–14.

[4] R. Tucker Scully, "The Marine Living Resources of the Southern Ocean" (1978) 33 *University of Miami Law Review* 341, at p. 344. It has been asserted that productivity in the Southern Ocean is 400% higher than the average for primary productivity in the world's oceans: Llano, *op. cit.*, p. 359, citing El-Sayed, "Biology of the Southern Ocean" (1975) 18 *Oceanus* 40, at p. 41. However, more recent scientific studies have indicated that this figure may be exaggerated and that, in any event, it fails to take into account considerable variability within the region: Joyner, *op. cit.*, p. 22.

International Law and the Conservation of Biological Diversity (C. Redgwell and M. Bowman, eds.: 90 411 0863 7: © Kluwer Law International: pub. Kluwer Law International, 1995: printed in Great Britain), pp. 109–128.

the ecological system with a low rate of renewal.[5] This explains the apparent richness, yet fragility, of the Antarctic ecoystem.[6] With only three trophic levels, the Antarctic food chain is very short: for example, the link between plant plankton, krill and baleen whales. This has particular consequences for the implementation of the ecosystem approach under the Convention on the Regulation of Antarctic Marine Living Resources, discussed further below.

There are linkages between land and marine ecosystems,[7] which together comprise a wide variety including:

1. Dry valleys and other oases in the barren ice deserts of the continent;
2. Ice pools at the surface of ice shelves…;
3. Shore areas of the continental coasts;
4. The multitude of ice-free areas on the great variety of Antarctic and Subantarctic Islands;
5. Littoral and sublittoral zones;
6. Shelf zones of the Southern Ocean …; [and]
7. The deep Southern Ocean, which is subdivided by local current systems and by the extent of the seasonal sea ice cover.[8]

Geographically these ecosystems encompass both the land continent of Antarctica, its ice shelves, a number of sub-Antarctic islands, and the ocean up to the Antarctic Convergence. Situated at approximately 50° South latitude, the Antarctic Convergence is a biological boundary which marks a change in the temperature and salinity of the surface waters of the Southern Ocean where warmer northern surface waters meet colder southern waters.[9]

In addition to the unique flora and fauna found in Antarctica, the IUCN's *A Strategy for Antarctic Conservation* notes that it also "affords unique opportunities as a reference against which to assess the impacts of pollutants on global ecosystems and processes in the atmosphere, on land and at sea elsewhere in the world. These linkages provide added reasons why the conservation of Antarctica's natural environment and dependent ecosystems is so important and why a comprehensive conservation strategy is required."[10] The Second World Conservation Strategy Project, *Caring for the Earth: A Strategy for Sustainable Living* (1991), similarly calls for the development of "a comprehensive and integrated conservation regime for Antarctica and the Southern Ocean" which "should provide for adequate scientific information, monitoring, open reporting, and effective compliance procedures". Both the 1980 and 1991 World Conservation Strategies[11] work within the context of the existing

---

[5] G. Billen and C. Lancelot, "The Functioning of the Antarctic Marine Ecosystem: A Fragile Equilibrium" in J. Verhoeven, P. Sands and M. Bruce (eds.), *The Antarctic Environment and International Law* (1992), at p. 49.

[6] *Ibid.*, p. 50.

[7] *Op. cit.*, n. 1, at p. 15, para. 36.

[8] G. Hempel *et al.*, "Antarctic Ecosystems: Change and Conservation. Review of the Fifth Symposium on Antarctic Biology", in K.R. Kerry and G. Hempel (eds.), *Antarctic Ecosystems: Ecological Change and Conservation* (1990), at pp. 412–413.

[9] See further Joyner, *op. cit.*, n. 3, Chapters 1 and 5. He points out a number of deficiencies with the Convergence as a boundary, including seasonal fluctuations and the breadth of the convergence zone: *ibid.*, p. 18.

[10] *Op. cit.*, n. 1, at p. 21, para. 55.

[11] The Conservation Strategy put forward by the IUCN in 1991 is based on three fundamental principles of conservation elaborated in the 1980 IUCN/WWF/UNEP *World Conservation Strategy* namely: (a) that essential ecological processes and life-support systems must be maintained; (b) that genetic diversity must be preserved; and (c) that any use of species and ecosystems must be sustainable.

Antarctic Treaty System, urging the parties to fill the significant gaps in protection perceived within the System as well as the strengthening of existing measures.

Though clearly shaped by external pressures, measures for Antarctic conservation have emerged from within the Antarctic Treaty System itself and it is these measures upon which this chapter will focus. There are two sources of law within the Antarctic Treaty System of relevance to environmental protection. First, there is the umbrella 1959 Antarctic Treaty (AT) and measures in effect thereunder. Article IX of the AT provides for measures[12] to be recommended for adoption by Governments by the Antarctic Treaty consultative parties (ATCPs) at consultative meetings. Consultative parties are entitled to participate at the biennial (which, from 1991, became annual)[13] Antarctic Treaty consultative meetings (ATCMs) at which, since 1983, non-consultative parties (NCPs) have been observers. ATCPs comprise the original signatories to the Antarctic Treaty and those parties to the AT which have demonstrated their "interest in Antarctica by conducting substantial research activity there, such as the establishment of a scientific station or the despatch of a scientific expedition".[14] Presently twenty-six of the forty-two States party to the AT have consultative status.[15]

Second,[16] there are separate conventions interlinked with the Antarctic Treaty and constituting part of the Antarctic Treaty System (ATS), namely, the Convention on the Conservation of Antarctic Seals (CCAS), in force since 1978,

---

[12] The legal status of recommendations "is not free from difficulty"; Watts refers to measures adopted at Consultative Meetings as coming "close to being made in exercise of a quasi-legislative function": Sir Arthur Watts, *International Law and the Antarctic Treaty System* (1992), at p. 24. Effectiveness depends upon approval by "all Consultative Parties whose representatives were entitled to participate in the meetings held to consider those measures" (Article IX.4, AT). Whether "effective" means "legally binding" depends upon the nature of the recommendation; an exhortatory recommendation will remain an exhortation, whereas a recommendation drafted in obligatory terms and intended to create legally binding obligations "will become "effective" in that sense when it has received the necessary approvals, and a legal obligation in the terms of the measure will thereupon arise" (*ibid.*).

[13] There was a two-year gap between the XVIIth ATCM in Venice in 1992 and the XVIIIth ATCM in Kyoto in 1994. The XIXth ATCM will take place in Seoul, Republic of Korea, in May 1995.

[14] Article IX.2 AT. Once the Protocol enters into force an additional requirement for consultative status will be ratification, acceptance, approval or accession to the Protocol: Article 22 EP.

[15] At the XVIIIth Antarctic Treaty Consultative Meeting held in Kyoto, Japan, from 11–22 April 1994, the succession to Czechoslovakia of the Czech Republic and Slovakia was reported, bringing the total number of States party to the Antarctic Treaty to forty-two: *Draft Final Report of the XVIIIth Antarctic Treaty Meeting*, XVIII ATCM/WP 37, 22 April 1994, para. 20.

[16] There are two further sources which are beyond the scope of the present chapter but which have effect within Antarctica, namely, general principles of international environmental law and environmental treaties. See further Joyner, *op. cit.*, n. 3, Chapter 5 (Protection of the Antarctic Marine Environment) and L. Pineschi, "The Antarctic Treaty System and General Rules of International Environmental Law", in F. Francioni and T. Scovazzi (eds.), *International Law for Antarctica* (1987). In connection with other environmental treaties, at the XVIIIth ATCM the Chilean delegation submitted a Working Paper (ATCM XVIII/WP 31) on the relationship of the Environmental Protocol to other global ageeements, including: the 1989 Convention on the Control of Transboundary Movement of Hazardous Wastes and their Disposal; the 1992 Convention on Biodiversity; the 1992 United Nations Framework Convention on Climate Change; the 1985 Vienna Convention for the Protection of the Ozone Layer and 1987 Protocol on Substances that Deplete the Ozone Layer; the 1972 Convention on the Prevention of Marine Pollution by Dumping of Wastes and other Matters; and the International Convention for the Prevention of Pollution from Ships and 1978 Protocol (MARPOL 73/78): *Draft Final Report of the XVIIIth Antarctic Treaty Meeting*, XVIII ATCM/WP 37, 22 April 1994, paras. 51–55. The Meeting agreed that primary responsibility for co-ordination between these treaty regimes and the Antarctic Treaty lay with the Parties to the Antarctic Treaty that were Parties to the other agreements: *ibid.*, para. 55.

and the Convention on the Conservation of Antarctic Marine Living Resources (CCAMLR), in force since 1985. The Convention on the Regulation of Antarctic Mineral Resource Activities (CRAMRA), concluded in 1988, has not yet entered into force and is unlikely to do so.[17] A direct consequence of opposition to CRAMRA and the desire for enhanced environmental protection of Antarctica is the most recent addition to the Antarctic Treaty System, the Protocol on Environmental Protection to the Antarctic Treaty (EP), concluded at Madrid on 4 October 1991, but which has not yet entered into force.[18]

The Evolution of Environmental Protection in Antarctica

The 1959 Antarctic Treaty

The 1959 Antarctic Treaty (AT) applies to the area south of 60° South latitude. It reserves the continent of Antarctica for peaceful purposes, preserves freedom of scientific research and declares Antarctica a nuclear-free zone.[19] Claims of territorial sovereignty in Antarctica are put on hold, or "frozen", by operation of Article IV of the Treaty. Although there is no express mention of the Antarctic environment nor of the Antarctic ecosystem,[20] the Treaty does contain provisions of obvious relevance to environmental protection[21] of the Antarctic Treaty Area.[22] Article V prohibits any nuclear explosions in Antarctica and the disposal there of radioactive waste material, and Article IX provides that the consultative parties may make recommendations including measures regarding "preservation and conservation of living resources in Antarctica".

In practice the ATCPs have assumed that they have competence to regulate the Antarctic environment and that protection of the Antarctic environment is one of the objectives of the Antarctic Treaty.[23] In 1970 the ATCPs explicitly recognised that they "should assume responsibility for the protection of the environment and the wise use of the Treaty Area" in the light of the particular vulnerability of the ecosystem of the Antarctic Treaty Area to human interference.[24] This was also the

---

[17] For background see this author, "Environmental Protection in Antarctica: The 1991 Protocol" (1994) 43 ICLQ 599.

[18] According to Article 23(1) thereof, the Protocol will enter into force on the thirtieth day following the date of ratification, acceptance, approval or accession by all States which were Antarctic Treaty Consultative Parties at the date on which the Protocol was adopted. Of these twenty-six Consultative Parties, 9 have become Parties to the Protocol (Ecuador, France, Peru, Norway, Argentina, Australia, Sweden, the Netherlands, and Spain).

[19] Article V AT.

[20] Environmental protection was not wholly outwith the delegates' consideration in 1959. See Oscar Pinochet de la Barra, "Reminiscences of the 1959 Antarctic Treaty Conference" (1992) XXVII: 2 *Antarctic Journal of the United States* 9, at p. 10.

[21] Though, as Blay observes, this is a far cry from comprehensive environmental protection: S.K.N. Blay, "New Trends in the Protection of the Antarctic Environment: The 1991 Madrid Protocol" (1992) 86 AJIL 377, at p. 379.

[22] Defined as the area south of 60° South latitude, including all ice shelves: Article VI AT. This is expressly without prejudice to the rights of any State under international law in respect of the high seas within this area.

[23] Cf. B.A. Bozcek, "Specially Protected Areas as an Instrument for the Conservation of the Antarctic Nature" in Wolfrum (ed.), *Antarctic Challenges III*, at p. 74; and R. Bilder, "The Present Legal and Political Situation in Antarctica", in J. Charney (ed.) *The New Nationalism and the Use of Common Spaces: Issues in Marine Pollution and the Exploitation of Antarctica* (1982), at p. 175.

[24] Recommendation VI-4: Man's Impact on the Antarctic Environment, reproduced in J. Heap, *Handbook of the Antarctic Treaty System* (7th edn., October 1990), Part 3, at p. 2101.

first acknowledgement of an Antarctic "ecosystem", which is the focus of the innovative approach to conservation found in the 1980 Convention on the Conservation of Marine Living Resources and in the 1991 Protocol to the Antarctic Treaty, both discussed further *infra*. Indeed, this recommendation embodies what might be viewed as the first set of environmental principles for Antarctica, providing that the Scientific Committee on Antarctic Research (SCAR)[25] be invited to identify the types and impact of human interference in the Treaty Area and to propose measures to minimise harmful interference; that research on the impact of man on the Antarctic ecosystem be encouraged; and that interim measures (unspecified) to reduce known causes of harmful interference (unidentified) be taken.

In 1975 the ATCPs acknowledged in Recommendation VIII-13 that the results of human interference could lead to major alterations in the Antarctic environment with consequences of global significance.[26] This is the only recommendation specifically titled "The Antarctic environment". It was a response to post-Stockholm developments, specifically an overture from the Executive Director of the newly formed United Nations Environment Programme offering co-operation in questions of scientific and technical interest of relevance to the Antarctic environment. The recommendation, in effect, puts down a marker indicating "that prime responsibility for Antarctic matters, including protection of the Antarctic environment, lies with the States active in the area which are parties to the Antarctic Treaty".[27] The Environmental Protocol may be viewed in a similar light, serving as a marker indicating that primary responsibility for protection of the Antarctic environment rests with the Antarctic Treaty parties. This has been successful, for the legitimacy of the Antarctic Treaty System is implicitly acknowledged in paragraph 17.105 of Agenda 21.[28]

Evidence of the Antarctic Treaty parties' concern with protection of the Antarctic environment is found in the fact that the topic "Man's impact on the Antarctic environment" was on the agenda of every Antarctic Treaty consultative meeting (ATCM) between 1970, when the item first appeared, and 1989, when comprehensive measures for the protection of the Antarctic environment were adopted. It is also reflected in the output of the ATCMs, which has always included one or more recommendations regarding human impact on the Antarctic environment (except for the XIth ATCM, which adopted only three recommendations). There are now a total of 141 recommendations which have been adopted since the Antarctic Treaty came into force in 1961 whose primary purpose is related to environmental management or protection, or 69% of the 204 recommendations made up to and including the XVIIIth ATCM in 1994.[29] Recommendations concerned with environmental management or protection cover

---

[25] SCAR was established by the International Council of Scientific Unions as an umbrella for scientific research in Antarctica. It is an observer at ATCMs and on the Commission for the Conservation of Antarctic Marine Living Resources. SCAR will have the right to participate in the meetings of the Committee on Environmental Protection (CEP) under the Protocol when this body is established following the entry into force of the Protocol.

[26] Recommendation VIII-13: The Antarctic Environment, reproduced in Heap, *op. cit.*, n. 24, at p. 2102.

[27] Recommendation VIII-13 (1975), first indent.

[28] *Report of the United Nations Conference on Environment and Development* (1992), A/CONF.151/26/ Rev. 1 (Vol. I).

[29] C. Harris and J. Meadows, "Environmental Management in Antarctica: Instruments and Institutions" (1992) 25: 9–12 *Marine Pollution Bulletin* 239, at p. 240 (with figures updated to take into account the XVIIth ATCM (1992) in Venice and the XVIIIth ATCM (1994) in Kyoto); see, generally, Sir Gareth Evans, "Protection of the Environment: An Ongoing Effort", in J. Handmer and M. Wilder (eds.), *Towards A Conservation Strategy for the Australian Antarctic Territory* (1993).

four broad categories: conservation of flora and fauna; creation of protected areas; human impacts on the environment; and information exchange.[30]

One of the problems with recommendations is that, unless expressed in mandatory language, itself a rare phenomenon, they are not strictly speaking legally binding.[31] Of course, in practice parties may well implement recommendations in their domestic law and/or through administrative procedures which apply to the bulk of activities in Antarctica under State control. Full compliance is left to individual national interpretation; on-site observation of compliance with AT and other international obligations is costly and difficult. There are no well-established procedures for fact finding in connection with compliance, though the exchange of information requirements under the Protocol may go some way to improving the situation.[32] Even where non-compliance is detected, pursuing the defaulting individual or State is difficult because of the underlying disputes about sovereignty in Antarctica, reflected in the fact that there is no compulsory dispute settlement under the Antarctic Treaty in respect of failure to comply with measures. Thus compliance has traditionally been sought through persuasion rather than confrontation.[33]

The 1964 Agreed Measures for the Conservation of Antarctic Fauna and Flora

One of the most significant early measures under the AT is the 1964 Agreed Measures for the Conservation of Antarctic Fauna and Flora (Recommendation III-VIII). Their purpose is "to promote and achieve the objectives of protection, scientific study, and rational use of [Antarctic] fauna and flora". Although no reference is made to "ecosystem" as such, the 1964 Agreed Measures refer to maintenance of the balance of ecological systems and the variety of species in designated specially protected areas. The concept of a specially protected area was introduced in the 1964 Agreed Measures in order to preserve the unique natural ecological system of areas of outstanding scientific interest. In 1972 the ATCPs responded to a SCAR report which indicated that existing SPAs "are not fully representative of the major Antarctic land and freshwater ecological systems" and recommended that the SPAs designated under the Agreed Measures should include:

(a) representative examples of the major Antarctic land and freshwater ecological systems;
(b) areas with unique complexes of species;
(c) areas which are the type locality or only known habitat of any plant or invertebrate species;
(d) areas which contain specially interesting breeding colonies of birds or mammals;
(e) areas which should be kept inviolate so that in future they may be used for purposes of comparison with localities that have been disturbed by man.[34]

---

[30] Harris and Meadows, *ibid*. The most recent (and only) recommendation adopted by the ATCPs at the XVIIIth ATCM in Kyoto falls within the third category of human impacts on the environment: Recommendation XVIII-1: Tourism and non-Governmental Activities, reproduced in (1994) 30(174) *Polar Record* 236. This sets forth guidance for visitors to Antarctica explaining the application of the Environmental Protocol to tourists and non-governmental activities. It also serves as an excellent example of the ATCPs' provisional ratification of the Protocol pending its entry into force.

[31] See n. 12, *supra*.

[32] See, for example, Articles 6, 13, 14 and 17 EP.

[33] *Op. cit.*, n. 1, at p. 28, para. 78.

[34] Recommendation VII-2 (1972).

More particular and restrictive conditions attach to the permitting of activities within specially protected areas, with entry into an SPA prohibited without a permit. A permit is effective provided that (a) it is issued for a compelling scientific purpose which cannot be served elsewhere and (b) the actions permitted thereunder will not jeopardise the natural ecological system existing in that area (Article VIII.4 AM). Evaluation of both these criteria is left to the government concerned, with an obligation only to "take appropriate action" to carry out the Agreed Measures and to exchange information regarding such implementation.

In 1985 a comprehensive review of protected area status was published by SCAR with the consequence that in 1989 recommendations were adopted by the Antarctic Treaty parties establishing two new categories of protected area: multiple-use planning areas and specially reserved areas. This was explicit recognition of the limitations of the existing protection system which was based on the evaluation of undefined criteria with insufficient description of "the precise ecological system or the components of the system that protected area designation is intended to preserve".[35] Moreover only 0.15% of the ice-free portions of the Antarctic continent had been designated protected areas under the Agreed Measures by the mid-1980s, though it is in these areas that most human activity occurs. This suggests the possible prioritisation of human over non-human use where these conflict.[36]

The ATCPs went on to adopt two recommendations which provide management plans for specially protected areas. The first amends Article VIII of the Agreed Measures to include a further condition to the efficacy of permits in SPAs, namely, that "the actions permitted thereunder are in accordance with any Management Plan accompanying the description of a Specially Protected Area" (Recommendation XV-8). The second stipulates the information to be contained in a Management Plan which is designed to remedy two perceived defects in the Agreed Measures: (a) there is now a requirement for a description of the natural ecological systems and components thereof in the SPA which designation as such is designed to preserve; and (b) SPA descriptions are now to contain an indication of "the types of activities that could or could not be carried out without harming or damaging any of the components of the natural ecological systems that the areas are intended to preserve".[37] Provisional area management plans are to be developed for existing SPAs with the criteria for future proposals for SPA designation more closely defined (including "descriptions of the measures necessary to ensure preservation of the area's unique or representative natural ecological systems"). At the XVIIth ATCM in Venice in 1992, revised descriptions and proposed management plans were recommended in respect of four specially protected areas following the pattern set forth in the above recommendations and in Article 5 of Annex V of the Protocol.[38]

As a mechanism for controlling human activity in Antarctica, the success of protected areas has been mixed. By 1990 the sixteen SPAs designated remained essentially unchanged from their initial designation in 1966.[39] There have been flagrant

[35] Report of the XVth ATCM, Paris, 1989, Para. 127.
[36] J. Handmer, M. Wilder and S. Dovers, "Australian Approaches to Antarctic Conservation", in Handmer and Wilder, op. cit., n. 29, at p. 27.
[37] Recommendation XV-9 (1989).
[38] Recommendation XVII-2: Revised Descriptions and Proposed Management Plans for Specially Protected Areas, reproduced in SCAR Bulletin No.111, (1993) 29 (171) Polar Record 353.
[39] Mannheim, HR 3977 statement, p. 115.

national examples of breaches. One notorious example involved a major SPA on King George Island, ultimately dedesignated because of the siting of a major scientific station there without prior review of the status of the area.[40] This was only done *ex post facto*. However, the regime has been streamlined and improved through the elevation of area protection to treaty status in Annex V to the 1991 Protocol, discussed further below.

The 1972 Convention for the Conservation of Antarctic Seals

Of the six species of seal found in Antarctica, only two have been subjected to significant exploitation by man.[41] The Seals Convention was in fact negotiated when no industry exploiting Antarctic seals was in existence,[42] but with seals representing "by far the world's largest unexploited mammal stock".[43] In force since 1978, the Seals Convention applies to seas south of 60° South latitude. Contracting parties also report on seal catches made on ice floes north of 60° South latitude by ships flying their flag.[44]

Unlike the later CCAMLR, the Seals Convention does not adopt an ecosystem approach, although the preamble does refer to contracting parties' desire "to promote and achieve the objectives of protection, scientific study and rational use of Antarctic seals, and to maintaining a satisfactory balance within the ecological system." Instead it adopts what has been termed a "traffic lights" approach to resource regulation.[45] Exploitation of Antarctic pelagic seals is permitted in accordance with the measures in the Annex[46] to the Convention – the green light. Any contracting party may request a meeting if the Scientific Committee on Antarctic Research (SCAR) reports that the harvest of any species of Antarctic seal regulated by the Convention is having a significant harmful effect on the total stocks or the ecological system in any particular locality – the amber light.[47] The red light is signalled where permissible catches are likely to be exceeded, with contracting parties obliged to prevent the taking of that species until the contracting parties decide otherwise. This could put the burden of

---

[40] *Op. cit.*, n. 1, at p. 37, para. 112.

[41] S. Lyster, *International Wildlife Law* (1985), at p. 48.

[42] A fact referred to in the statement made by the Representative of the United States on signing the Final Act of the Conference, reproduced in Heap, *op. cit.*, n. 24, at p. 4104.

[43] Brian Roberts, reporting on the outcome of the intergovernmental conference to negotiate the Seals Convention, as quoted in J. Heap, "Has CCAMLR Worked? Management Policies and Ecological Needs", in A. Jorgensen-Dahl and W. Ostreng (eds.), *The Antarctic Treaty System in World Politics* (1991), at p. 45.

[44] Article 5(7). Jurisdiction is nationality based under the Seals Convention, with each Contracting Party obliged to adopt for its nationals and vessels "such laws, regulations and other measures, including a permit system as appropriate, as may be necessary to implement the Convention" (Article 2(2)). Apart from those States which participated at the Conference (Argentina, Australia, Belgium, Chile, France, Japan, New Zealand, Norway, South Africa, the USSR, UK and USA), the Convention is open for accession by any State which is invited to accede with the consent of the Contracting Parties (Article 12). All participating States save for New Zealand are parties to the Convention; in addition, consent was obtained for the accession of Poland (1980), Germany (1987), Canada (1990), Brazil (1991) and Italy (1992).

[45] Heap, *op. cit.*, n. 43, at p. 46.

[46] The quotas set for the more abundant seal species are extremely conservative: D. Overholt, "Antarctic Protection in Antarctica: Past, Present, and Future" (1990) 28 CYIL 227, at p. 238.

[47] Article 6(3); there is also provision in Article 6 for a meeting of the Contracting Parties to consider, inter alia, "the provision of further regulatory measures, including moratoria".

proof on harvesting States to demonstrate that sealing may be resumed, one attractive feature of the "traffic lights" approach.[48]

In practice commercial sealing has not been resumed and the safeguards within the Seals Convention have not been employed. The provisions in Article 6 which envisage the establishment of a Commission and a Scientific Advisory Committee have not been invoked, nor have regular five-yearly reviews of the operation of the Convention after its entry into force in 1972, as envisaged by Article 7, taken place.[49] Despite this, the Seals Convention undoubtedly contributed to the innovative approach of CCAMLR. Heap attributes the Antarctic Treaty consultative parties' willingness to extend the ambit of CCAMLR beyond the biologically irrelevant 60° South latitude to the Seals Convention, which "so accustomed the consultative parties to circumscribing the exercise of their rights on the high seas within the Antarctic Treaty Area (ATA), that they felt able to spread their wings and, in addition, circumscribe those rights in waters adjacent to the ATA".[50]

The 1980 Convention on the Conservation of Antarctic Marine Living Resources (CCAMLR)

Negotiations to conclude CCAMLR began in the 1970s as a response to Japanese and Soviet fishing for krill, which is the foundation of the Antarctic ecosystem. The Convention was concluded in 1980 and came into force in 1982,[51] before a major Antarctic krill fishery had emerged; thus, like the Seals Convention, CCAMLR "is one of the few international treaties concerned with wildlife conservation to be concluded prior to heavy commercial pressure on the species it was designed to protect".[52] The Convention encompasses all marine living resources except seals and whales, which are covered separately in the Seals Convention (part of the ATS) and the International Convention for the Regulation of Whaling (not part of the ATS).[53] Geographically its coverage extends beyond the Antarctic Treaty Area to the Antarctic Convergence,[54] thus encompassing the entire Antarctic ecosystem

---

[48] Heap, *op. cit.*, n. 43, at p. 47.

[49] Watts, *op.cit.*, n. 12, at p. 40. A review was held in 1988: for the text of the Report of the Meeting see Heap, *op. cit.*, n. 24, at pp. 4112–4116.

[50] *Op. cit.*, n. 43, at p. 46. It should be noted that both the Seals Convention and CCAMLR are free standing and thus open to non-Antarctic Treaty Parties.

[51] There are twenty-nine parties to CCAMLR, with a 21-member Commission. All States fishing in the Southern Ocean are members of the Commission save for the Ukraine and Bulgaria. The latter has acceded to the Convention and is likely to be invited to join the Commission shortly, having satisfied the criteria in Article VII; the Ukraine has been invited to attend CCAMLR meetings as an observer: K.-H. Kock, "Fishing and Conservation in Southern Waters" (1994) 30 (172) *Polar Record* 3, at pp. 9–10.

[52] Lyster, *op. cit.*, n. 41, at p. 157. However, large-scale exploitation of many Antarctic fish stocks had preceded the conclusion of CCAMLR, with many overexploited at the time CCAMLR came into force: Kock, *ibid.*; see also the tables in Billen and Lancelot, *op. cit.*, n. 5. The weakness of CCAMLR is not, then, that it has presided over the decimation of Southern Ocean fish stocks, but rather that it has failed to ensure their recovery: Heap, *op. cit.*, n. 43, at p. 49.

[53] Some commentators have suggested that the Seals Convention could be merged within CCAMLR as part of a common management approach to marine living resources in the Southern Ocean: see Kock, *op. cit.*, n. 43, at p. 16. Heap notes proposals of this kind, but observes that the absence of a red light mechanism in CCAMLR would require some harmonisation of the two regimes: *ibid.* On cooperation between CCAS, IWC and CCAMLR, see Kock, *ibid.*

[54] Unlike the Seals Convention and the Agreed Measures which apply the same boundary as the Antarctic Treaty, viz. 60° South latitude, CCAMLR applies to the marine living resources of the area south of 60° South latitude *and* the area between that latitude and the Antarctic Convergence. The Convergence is defined by a set of co-ordinates in Article I(4) CCAMLR because of its variability.

rather than following an artificial geographic boundary.[55] Furthermore, the presence of migratory or straddling stocks is anticipated in Article XI CCAMLR which provides for co-operation with contracting parties exercising jurisdiction in respect of marine areas beyond CCAMLR's area of application.

CCAMLR defines the Antarctic marine ecosystem as "the complex of relationships of Antarctic marine living resources with each other and with their physical environment".[56] The purpose of CCAMLR is the conservation of Antarctic marine living resources, with conservation defined to include "rational use" in deference to harvesting States. Without this explicit recognition of rational use[57] there is some doubt that CCAMLR could have been concluded. None the less it sets up a tension within the Convention between conservation and exploitation which is not entirely resolved by Article II. This article is the heart of CCAMLR. It explicitly adopts an ecosystem approach, thus distinguishing it from other marine living resource management regimes which characteristically take a single species approach to conservation.[58] This multi-species or ecosystem approach has been described as enabling CCAMLR to be many things to all parties, harvesting and non-harvesting States alike, and thus to operate with residual political efficacy.[59] It is also explicit recognition of the impracticality of applying traditional resource management techniques to the Southern Ocean because of the interdependence of species and their low diversification.[60] In negotiating CCAMLR, concern was particularly felt in relation to the exploition of krill, a crucial species in the Antarctic food chain. To set catch quotas for krill without regard to the impact on predator species could have catastrophic effects on the Antarctic ecosystem. This was confirmed in reports by SCAR and the United Nations Food and Agriculture Organisation in the 1970s which "highlighted the importance of krill in the Antarctic marine ecosystem; irrational, large scale exploitation of krill could have severe repercussions on the birds, seals and whales of the Antarctic which depend on krill for their food".[61]

The conservation principles adopted in Article II represent a compromise between the maximum sustainable yield approach sought by harvesting States and the conservation standard sought by non-harvesting States. The result is an article for which "there is no unambiguous interpretation".[62] Three principles of conservation are set forth in Article II(3):

(a) prevention of decrease in size of any harvested population to levels below those which ensure stable recruitment. For this purpose, its size should not be allowed to fall below a level close to that which ensures the greatest net annual increment;

(b) maintenance of the ecological relationships between harvested, dependent and related populations of Antarctic marine living resources and the restoration of depleted populations to the levels defined in sub-paragraph (a) above; and

---

[55] Joyner, op. cit., n. 3, at pp. 226, 241.
[56] Article I(3) CCAMLR.
[57] Itself undefined: see further the discussion at note 64 below.
[58] Whether CCAMLR has successfully maintained this distinction is evaluated below.
[59] Joyner, op. cit., n. 3, at p. 230.
[60] Overholt, op. cit., n. 46, at p. 239.
[61] Heap, op. cit., n. 24, at p. 4201.
[62] M. Basson and J. Beddington, "CCAMLR: The Practical Implications of an Eco-System Approach", in Jorgenson-Dahl and Ostreng (eds.), op. cit., n. 43, at p. 61; Joyner titles his discussion "The Paradoxes of Article II", op. cit., n. 3, at p. 242.

(c) prevention of changes or minimization of the risk of changes in the marine ecosystem which are not potentially reversible over two or three decades, taking into account the state of available knowledge of the direct and indirect impact of harvesting, the effect of the introduction of alien species, the effects of associated activities on the marine ecosystem and of the effects of environmental changes, with the aim of making possible the sustained conservation of Antarctic marine living resources.

This may be distinguished from the "traffic lights" approach of the Seals Convention, which arguably places the burden of proof on harvesting States to show if sealing may be resumed. One criticism of CCAMLR's ecosystem approach is that the burden of proof rests with conservationists; one suggestion for reform would be to reverse the basic presumption and prohibit fishing until the objectives of "rational use" can be met.[63] Rational use is undefined;[64] other suggestions for reform of CCAMLR include "establishing a set of operational definitions of the relevant terms and objectives" contained in Article II.[65]

Under CCAMLR the implementation of Article II is left to the Commission for the Conservation of Antarctic Marine Living Resources, which is charged with giving effect to the general principles and objectives of CCAMLR through regular annual meetings. It is not a plenary body; membership is restricted to the original contracting parties to the Convention and to States[66] which have "engaged in research or harvesting activities in relation to the marine living resources" to which the Convention applies.[67] Based in Hobart, Tasmania, the Commission was the first permanent body to be established under the Antarctic Treaty System. The Commission's functions are defined in Article IX and include identification of conservation needs and the efficacy of conservation measures, formulating and adopting such measures on basis of the best scientific evidence available, and implementing the system of observation and inspection established under CCAMLR. The Commission is also charged with drawing the attention of non-CCAMLR States to activity affecting the implementation of CCAMLR (Article XI). In keeping with the obligation to adopt measures "on the basis of the best scientific evidence available", the Commission is to take full account of the recommendations and advice of the Scientific Committee, also established under CCAMLR. The latter is thus not a decision-making body but "a consultative body to the Commission"[68] with no independent power of action. Advice is also provided by SCAR. Measures are adopted by the Commission by consensus and, in addition, are subject to an

---

[63] Heap, *op. cit.*, n. 43, at p. 52; and Joyner, *ibid.*, p. 243 ("'traffic light' approach akin to that in the Seals Convention would probably be more effective").

[64] Heap, *ibid.*, would employ the Soviet definition put forward at CCAMLR-V as implying "inter alia, obtaining maximum output of the highest quality with the minimum amount of effort during the course of an indefinitely long period of time".

[65] Basson and Beddington, *op. cit.*, n. 62, at p. 67.

[66] Unlike under the Antarctic Treaty, the European Community is able to become a party to CCAMLR pursuant to Article XXIX and became so with effect from 21 May 1982. It is also a member of the Commission.

[67] Article VII CCAMLR. A number of Contracting Parties do not meet this "relevant activities" criterion, even though they are Contracting Parties to the Convention, e.g. Uruguay, Greece, Finland, Peru, Canada and the Netherlands: Watts, *op. cit.*, n. 12, at pp. 43–44.

[68] Article XIV(1) CCAMLR. Article XV further defines the functions of the Scientific Committee as, *inter alia*, "a forum for consultation and co-operation". The early days of CCAMLR were marked by an uneasy relationship between the two bodies.

objection or opt out procedure, a common and unfortunate feature of fishery conservation agreements.[69] This double veto is the result of the issue of unresolved sovereign claims putting pressure on the negotiating States to adopt consensus for decision making within the Commission, the decision-making procedure used in respect of measures adopted under the Antarctic Treaty – the first veto – whilst more traditional conservation-harvesting issues led to the adoption of the objection procedure – the second veto.[70]

Quite apart from the procedural difficulties in achieving consensus regarding a measure, which blocked the adoption of more stringent conservation measures certainly in the first five years of CCAMLR's operation,[71] the ecosystem approach under CCAMLR is not one easily implemented from a scientific viewpoint. It requires complex scientific modelling and independent data in addition to the species specific information collected incidentally in commercial fishing operations. One of the innovatory features of Article II CCAMLR is its emphasis upon the use of applied research as a basis for the assessment of the impacts of resource use and for directing the course of national research to answer resource management questions.[72] A particularly thorny issue has been the predator–prey relationship and the development of common parameters for monitoring. The further development of the ecosystem approach is actively under dicussion in a CCAMLR Working Group responsible for ecosystem monitoring (CEMP). This marks the first comprehensive approach to monitoring in Antarctica which seeks to distinguish changes in the ecosystem brought about by neutral causes or commercial fishing.[73]

As for environmental monitoring, this is vital not only to conservation purposes but also as a vital input into basic research.[74] But attracting funding and personnel for monitoring activities is difficult even where a standardised methodology has been agreed.[75] Nor is it practically or economically feasible to monitor everything. Indeed, CCAMLR is based on the assumption that it is not; monitoring programmes developed under CCAMLR have thus "focused on the elements of and species in the Antarctica marine ecosystems where change is most likely to be perceived".[76] An additional problem in adopting an ecosystem approach is the lack of homogeneity in the Antarctic ecosystem. Indeed, given the variability of species in space and time within the Southern Ocean it is inaccurate to speak of "an Antarctic ecosystem". Figures based on ocean-wide distributions, for example of krill, mislead in the absence of "[i]nformation on systems and processes at scales of kilometres and

---

[69] Article IX(6) CCAMLR.

[70] Orrego Vicuna, "The Effectiveness of the Decision-Making Machinery of CCAMLR: An Assessment", in Jorgensen-Dahl and Ostreng (eds.), *op. cit.*, n. 43, at p. 28; see also R.F. Frank, "The Convention on the Conservation of Antarctic Marine Living Resources" (1983) 13:3 *Ocean Development and International Law* 291, at p. 310 (reference to "triple veto" for France in respect of measures affecting her undisputed sovereignty over the Kerguelen and Crozet Islands).

[71] Vicuna, *ibid.*; for a fuller account of the first five years of operation of CCAMLR, see M. Howard, "The Convention on the Conservation of Antarctic Marine Living Resources: A Five-Year Review" (1989) 38 ICLQ 104.

[72] *Op. cit.*, n. 1, at p. 38, para. 118.

[73] L. Kimball (1989) 24 *Antarctic Journal of the United States* 9.

[74] S.B. Abbott and W.S. Benninghoff, "Orientation of Environmental Change Studies to the Conservation of Antarctic Ecosystems", in Kerry and Hempel (eds.), *op. cit.*, n. 8, at p. 402.

[75] G. Hempel *et al.*, *op. cit.*, n. 8 at p. 412.

[76] "Experts gather to discuss long-term environmental monitoring" (1989) XXIV *Antarctic Journal of the United States* 4, at p. 5 (comments of Dr. Draggan).

days".[77] What is required is information on local concentrations, their accessibility, and reliability/predictability.[78] To an extent this variability is catered for in the measures adopted by the Commission, which relate to species in specific statistical and geographic areas rather than applying to the Southern Ocean as a whole.

There is thus little doubt that uncertainty regarding the working of Antarctic ecosystems makes practical management difficult, not least because in the early days of CCAMLR harvesting nations blocked more rigorous conservation measures on the basis of inadequate scientific evidence, an argument familiar in other fisheries commissions. But the undeniable national interest at stake in objecting on such grounds should not obscure the very real difficulties in gathering such evidence. Indeed, some scientists consider that "[i]t is unlikely that a well established functional relationship could be developed between rate of exploitation of krill and the effect on reproductive capacity of top-level predators in the foreseeable future",[79] a calculation at the heart of the application of CCAMLR's ecosystem approach to this fishery. These difficulties have led the Commission to take a reactive case-by-case approach focusing on single species, rather than ecosystem management, and it has attracted criticism for consequently failing to take a precautionary approach.[80] Lack of knowledge about the Southern Ocean renders a precautionary approach to conservation absolutely vital. It will be recalled that one of the functions of the Commission is to "formulate, adopt and revise conservation measures on the basis of the best scientific evidence *available*".[81] However, it was only at its third meeting in 1985 that any conservation measures were adopted by the CCAMLR Commission in the face of further decline in already depleted fish stocks in the Southern Ocean – the prohibition of directed fishery on *Notothenia rossii* (a species of cod) around South Georgia.[82] It was not until the sixth meeting of the Commission that it was able to move from reactive to proactive management through the setting of total allowable catches; this followed upon the adoption in 1987 of a better harvesting strategy, with maximum sustainable yield replaced by a fishing mortality value of FO.1.[83] Indeed, 1987 was something of a watershed for CCAMLR, when "the nature of the conservation measures adopted ... changed dramatically".[84] Precautionary total allowable catches, one mechanism for ensuring ecologically sustainable fishing effort particularly where stock assessments are tentative or non-existent, have recently been set for crab near South Georgia (1992) and on Patagonian toothfish around the South Sandwich Islands (1992) and, finally, for krill (1991).[85] In response to continuing difficulties in modelling the predator–prey relationship in respect of krill exploitation, CCAMLR is also considering an interim

---

[77] Hempel *et al.*, *op. cit.*, at p. 410.

[78] *Ibid.*, at p. 411.

[79] Kock, *op. cit.*, n. 51, at pp. 16–17.

[80] Sir A. Parsons, *Antarctica: The Next Decade* (1987), at p. 127.

[81] Article IX(1)(f) CCAMLR, emphasis added.

[82] Conservation Measure 3/IV, reproduced in Heap, *op. cit.*, n. 24, at p. 4220. The Commission had regard to a report from the Scientific Committee indicating concern regarding this stock. According to Kock there is little evidence that this species, a primary target species in the early 1970s, has recovered: *ibid.*, at p. 12; see also Basson and Beddington, *op. cit.*, n. 62, at p. 61.

[83] Kock, *ibid.*

[84] Vicuna, *op. cit.*, n. 70, at p. 28. At the Sixth Meeting of CCAMLR, the Scientific Committee requested the Commission to provide guidance regarding management policy, to which it responded by requesting the Committee to assess alternatives for achieving management objectives in relation to fish populations: *ibid.*, p. 32.

[85] Kock, *op. cit.*, n. 51, at p. 18.

measure "to restrict the catch of krill within the foraging range of land-based preda-tors – that is, within 100 km from the shoreline during the critical period of the year – to minimize potential adverse effects of the krill fishery to predators".[86]

Whilst there are signs of a growing awareness within CCAMLR of the need for precautionary management techniques in the face of biological uncertainties, there is little doubt that "precise implementation of the CCAMLR objectives still lies in the future".[87] Even with rapid advances in the scientific information available in respect of dynamic processes in the Southern Ocean, there remains the difficulty of enforc-ing measures once adopted. Primary responsibility rests with contracting parties under Article XXI to ensure that appropriate measures within their competence[88] are taken to ensure compliance with the Convention. A "tattletale provision"[89] requires contracting parties "to exert appropriate efforts" to ensure that no one engages in activities contrary to the Convention, and to report any such activities to the Commission.

But what of institutional enforcement measures? These are still embryonic within CCAMLR. Article XXIV calls upon the contracting parties to establish a system of observation and inspection in order to promote, and ensure the observance of, the Convention.[90] Implementation of such a system is one of the tasks of the Commission. However, a system of inspection was not established until the 1989/90 season. Inspectors have authority to inspect the fishing gear, including nets, the catch and general harvesting activities and spend only a few hours aboard the vessel. They are appointed by national authorities and, given the logistics involved, tend to be of the same nationality as the vessel inspected. Kock reports that for the 1991/92 season, sixteen of the eighteen vessels inspected fell within this category.[91] Violations must be reported to the Commission, as must the steps taken by the flag State to sanction such violations, though here again there is some evidence of such offences going unreported and unpunished.[92] As for an observation system, this was not implemented until the 1992/93 season amidst fears that observers would become *de facto* inspectors. Even then it has only been implemented on a bilateral basis with observers nationals of the State which nominates them, and thus falls far short of an independent scheme of scientific observation operated by the Commission itself.[93] Information gathering on fishing activities, particularly biological data on the catch and rates of incidental mortality, is the purpose behind the observation system. The routine collection of such information would enable much more accurate conserva-tion measures to be adopted, and would complete the information loop with vital feedback where measures have been adopted.

The scientific verdict on the ambitious ecosystem approach adopted in CCAMLR is that, whilst very far from being fully developed, it may be worthy of emulation in other fora, particularly where the management of straddling and highly

---

[86] *Ibid.*, p. 17, citing the Reports of the eleventh and twelfth meetings of the Scientific Committee in 1992 and 1993.
[87] Vicuna, *op. cit.*, n. 70;; see also National Science Foundation (US), *Science & Stewardship in Antarctica* (1993), at p. 49.
[88] Wording designed to maintain the bifocal approach of Article IV CCAMLR by finessing the sover-eignty issue: Joyner, *op. cit.*, n. 3, at p. 245.
[89] Joyner's characterisation of Article XXII: *ibid.*
[90] Article XXIV CCAMLR.
[91] *Op. cit.*, n. 51, at p. 14.
[92] *Ibid.*
[93] See, generally, Kock, *ibid.*, p. 15.

migratory stocks is at issue.[94] In response to the query "Has CCAMLR worked?", Heap concludes that it is working and its effectiveness is improving, but that it could work better – a qualified yes rather than a definitive no.[95] It is economic considerations and market demands which have had the greatest impact on the exploitation of Antarctic marine living resources, with CCAMLR designed to break the "boom and bust" cycle of over-exploitation which has previously typified human endeavour in the region. Ironically it is the reduction in economic susbsidies for fishing activity in the Southern Ocean which has recently had a significant beneficial impact, with a dramatic decline in krill catches in the 1990s as compared with the previous two decades. Though this decline is unlikely to be an irreversible one in terms of global food demand, it may provide a welcome window of opportunity to further develop conservation measures within CCAMLR without the intrusion of overwhelming national interests.

The 1991 Protocol on Environmental Protection to the Antarctic Treaty

It will be recalled that the Protocol was negotiated to provide for comprehensive environmental protection in Antarctica, and to serve as a complement to the existing Antarctic Treaty System. This is reflected in Article 4 of the Protocol, which expressly states that the Protocol supplements the Antarctic Treaty and neither modifies nor amends it.[96] The "integrity and continuing efficacy" of the Antarctic Treaty system is thus preserved.[97] For the first time in the Antarctic Treaty system, the Protocol extends on a treaty basis environmental protection measures to the whole of the Antarctic Treaty area,[98] and elevates to treaty status the conservation of the Antarctic environment.[99] Under Article 2 the parties "commit themselves to the comprehensive protection of the Antarctic environment and dependent and associated ecosystems and hereby designate Antarctica as a natural reserve, devoted to peace and science".[100] Thus the Protocol adopts an explicitly ecosystem approach. In the planning and conduct of all activities in the Antarctic Treaty area, Article 3 stipulates the following as "fundamental considerations": the protection of the Antarctic environment and dependent and associated ecosystems; the intrinsic value[101] of Antarctica, including its

---

[94] Kock, *ibid.*, p. 19; see also Basson and Beddington, *op. cit.*, n. 62.

[95] Heap, *op. cit.*, n. 43, at p. 51.

[96] Article 4(2) further provides that "Nothing in the Protocol derogates from the rights and obligations of the Parties to this Protocol under the other international instruments in force within the Antarctic Treaty system." One criticism of the Protocol is that effective environmental protection in fact requires that the other components of the Antarctic Treaty system should be subordinated to the Protocol.

[97] F. Francioni, "The Madrid Protocol on the Protection of the Antarctic Environment" (1993) 28: 1 *Texas Journal of International Law* 47, at p. 55. See also Article 5, which requires the Parties to consult and cooperate to achieve consistency between the components of the ATS.

[98] That is, south of 60° South latitude. The geographical scope of the Protocol is thus the same as that of the Antarctic Treaty and Seals Convention. Unlike CCAMLR, the Protocol does not extend to the Antarctic Convergence.

[99] The IUCN's Strategy for Antarctic Conservation recommended the elevation of the objective of "conservation of the Antarctic environment, with its outstanding wilderness qualities and unique dependent and associated ecosystems" to the same level of emphasis as the use of Antarctic only for peaceful purposes, and the prohibition on nuclear explosions and the deposit of radioactive waste, in the Antarctic Treaty: *op. cit.*, n. 1, at p. 25, paras. 66–67.

[100] Article 2 EP. The 1964 Agreed Measures declared Antarctica a "Special Conservation Area".

[101] For general discussion of intrinsic value, see further Bowman, this volume.

wilderness and aesthetic values; and its value as an area for the conduct of scientific research, in particular research essential to understanding the global environment.[102] To this end, "activities in the Antarctic Treaty area shall be planned and conducted so as to limit adverse impacts on the Antarctic environment and dependent and associated ecosystems" and so as to avoid, *inter alia*:

> (i) detrimental changes in the distribution, abundance or productivity of species or populations of species of fauna and flora;
> (ii) further jeopardy to endangered or threatened species or populations of such species; or
> (iii) degradation of, or substantial risk to, areas of biological, scientific, historic, aesthetic or wilderness significance.

Prior assessment of the impact of activities upon the Antarctic environment and dependent and associated ecosystems is a general environmental principle, given further elaboration in Annex I of the Protocol. One of the factors to be taken into account in making such assessments is "whether there exists the capacity to monitor the key environmental parameters and ecosystem components so as to identify and provide early warning of any adverse effects of the activity and to provide for such modification of operating procedures as may be necessary in the light of the results of monitoring or increased knowledge of the Antarctic environment and dependent and associated ecosystems".[103] Regular and "effective" monitoring is to be carried out to detect, *inter alia*, unforeseen effects of activities carried on both within and outside the Antarctic Treaty area on the Antarctic environment and dependent and associated ecosystems.

One of the deficiencies of the ATS identified in the IUCN's *Strategy for Antarctic Conservation* is the absence of one body with Antarctica-wide interests and responsibilities in respect of Antarctic conservation.[104] This deficiency is adddressed in the Protocol which, once it enters into force, establishes a Committee for Environmental Protection (CEP) of which each party is entitled to be a Member.[105] Both CCAMLR's Scientific Committee and SCAR have observer status, and other relevant scientific, environmental and technical organisations may be invited to participate as observers.[106] The functions of the Committee are set forth in Article 12. It is in this respect that not all of the criticisms of the existing ATS are met, given the CEP's absence of legislative and enforcement functions.[107] It is an advisory body only with its functions confined to providing advice and formulating recommendations to the parties in respect of, *inter alia*: the functioning of the Protocol; its effectiveness; inspection procedures; the need for scientific research, including environmental monitoring; and the state of the Antarctic environment. The Committee must submit a report of each of its sessions to the ATCM which "shall reflect the views expressed".[108] This is made publicly available following considera-

---

[102] Article 3(1) EP.

[103] Article 3(2)(c)(v) EP.

[104] *Op. cit.*, n. 1, at p. 28, para. 77.

[105] See Articles 11 and 12 EP. Contracting parties to the Antarctic Treaty but not to the Protocol may have observer status in the Committee.

[106] With the approval of the ATCM: Article 11 EP. As Scully observes, the establishment of the Committee for Environmental Protection attempts to chart a middle course between duplicating the efforts of SCAR and other bodies and creating a costly and extensive new institutional machinery: R. Tucker Scully, "Protecting Antarctica: Progress in Chile" (1991) 26: 1 *Antarctica Journal of the United States* 4, at p. 8.

[107] See, for example, the Antarctica EPA advocated by Barnes in Joyner and Chopra, *op. cit.*, n. 2, at pp. 245–249.

[108] This reflects the outcome of an earlier battle fought by the Scientific Committee under CCAMLR, with the need for different points of view expressed in Committee to be reflected in the final report.

tion by the ATCPs. In anticipation of the establishment of the CEP, the XIXth ATCM in 1995 will consider those matters which would be dealt with by the CEP in a special Transitional Environmental Working Group (TEWG).[109] This may be seen as part of the general commitment of the Antarctic Treaty parties to the provisional application of the Protocol pending its entry into force.[110]

Enforcement is thus left to the contracting parties who are obliged to take "appropriate measures" within their competence – including administrative actions – to ensure compliance with the Protocol,[111] and to publish annually a report of such measures undertaken. A weak inspection system is provided for in Article 14, with reports and comments thereon ultimately made publicly available. This parallels the inspection system under the Antarctic Treaty and CCAMLR.

The Protocol is a framework agreement, supplemented by five annexes covering: environmental impact assessment (Annex I); Conservation of Antarctic Fauna and Flora (Annex II); Waste Disposal and Waste Management (Annex III); Prevention of Marine Pollution (Annex IV); and Area Protection and Management (Annex V). Of most relevance to the present discussion of the ecosystem approach to environmental protection in Antarctica are Annexes II and V.

Annex II

Annex II of the Protocol reproduces to a significant degree the provisions of the 1964 Agreed Measures for the Conservation of Antarctic Fauna and Flora. There is more emphasis than formerly upon the protection of the diversity of flora and invertebrates in the Antarctic Treaty Area and upon humane methods of taking. Annex II also bans the introduction of dogs onto land or ice shelves in the Antarctic Treaty Area from 1 April 1994. In general, however, the scope of the Agreed Measures is retained.

The Agreed Measures introduced into the ATS the concepts of specially protected species and specially protected areas, both found in Annexes to the 1991 Protocol, as well as prohibiting within the Treaty Area the "killing, wounding, capturing, or molesting of any native mammal or native bird" without a permit. Annex II contains a similar prohibition, with the addition of a prohibition in respect of native plants. Thus Article 1(g) of the Protocol defines "take" or "taking" to include removing or damaging such quantities of native plants that their local distribution or abundance would be significantly affected. The Agreed Measures attached certain general conditions to the issuing of permits, a pattern also carried through to Annex II. Thus, for example, the Annex sets forth a permit condition ensuring that "the diversity of species, as well as the habitats essential to their existence, and the balance of the ecological systems existing within the Antarctic Treaty area are maintained." The standard for activities with the Treaty Area is to minimize, not eliminate, harmful interference with the normal living conditions of any native mammal, bird or plant. Article 1(h) of Annex II of the Protocol defines "harmful interference" to include, *inter alia* "any activity that results in the significant adverse modification of habitats of any species or population of native mammal, bird, plant or invertebrate".

---

[109] Draft Final Report of the XVIIIth Antarctic Treaty Meeting, XVIII ATCM/WP 37, 22 April 1994, at p. 9, para. 42. Representatives of SCAR, CCAMLR and COMNAP will be invited to participate in the work of the TEWG.

[110] See the Final Act of the Eleventh Antarctic Treaty Special Consultative Meeting, where the Protocol was adopted, reproduced in 30 ILM 1455 (1991). Annex V was added at the XVIth ATCM in Bonn on 17 October 1991, Recommendation XVI-10.

[111] Article 13 EP.

Annex V

This Annex consolidates the measures taken in 1989 to subject SPAs to a Management Plan.[112] Previous SPAs are now redesignated as Antarctic Specially Protected Areas (ASPAs) with entry prohibited except by permit and the management thereof regulated by an adopted Management Plan. Proposals for designation of an ASPA may be made by any party, the Committee for Environmental Protection, SCAR, or the CCAMLR Commission (from whom prior approval is needed before designation of a marine ASPA), by submitting a proposed Management Plan to the ATCM.

In proposing an ASPA, parties are to identify "within a systematic environmental-geographical framework", and to include within the series of ASPAs, *inter alia*:

(a) areas kept free from human interference for future comparisons;
(b) representative examples of major terrestrial, including glacial and aquatic, ecosystems and marine ecosystems;
(c) areas with important or unusual assemblages of species...;
(d) the type of locality of only known habitat of any species;
(g) areas of outstanding aesthetic and wilderness value.

The area is to be of sufficient size to protect the values for which the special protection or management is required, an important shift in focus from the "minimum disruption to other Antarctic activities" requirement under the Agreed Measures. Designation is for an indefinite period, with Management Plans reviewed every five years. To assist in the designation and management of new protected and managed areas, SCAR is to produce a revised "ecosystem classification matrix" and new guidelines for the inspection of protected areas for the XIXth ATCM in 1995.[113]

Similar procedures apply to the second category of protected area in Annex V, Antarctic Specially Managed Areas (ASMA), where activities pose risks of mutual interference or cumulative environmental impacts or the area contains sites or monuments of recognised historic value. Entry into an ASMA does not require a permit, but it will be subject to a management plan. It was anticipated that the first ASMA would be designated at the XIXth ATCM in 1995 with the approval of a management plan drafted by Brazil and Poland in connection with Admiralty Bay.[114]

The Antarctic Treaty System and the Convention on Biological Diversity

The above discussion has focused on the measures taken within the Antarctic Treaty system to conserve biological diversity, with particular reference to the ecosystem approach. A key theme in Antarctic legal developments has been the preservation of the hegemony of the Antarctic Treaty system in regulating environmental protection there. This is clear from the preamble to the Protocol, which refers both to "the special legal and political status of Antarctica", with the special responsibility of the ATCPs to ensure activities there are consistent with the AT, and to "the interest of mankind as a whole" in the development of a comprehensive regime for the

---

[112] See further *supra*, text following n. 36.
[113] Draft Final Report of the XVIIIth ATCM, p. 23, para. 105.
[114] Clarification of the relationship between Articles 4 (ASMAs) and 5 (Management Plans) will be needed first. The United Kingdom delegation asked for clarification of the possible use of mandatory prohibitions in Management Plans for ASMAs, refered to in Article 5(3)(f). See the Draft Final Report of the XVIIth ATCM, p. 25, paras. 110–111.

protection of the Antarctic environment and dependent and associated ecosystems. Yet only States which are contracting parties to the Antarctic Treaty may become parties to the Protocol, underscoring its interlinked and subordinate status.[115]

But what of the relationship between the components of the Antarctic Treaty system and other international conventions, including the Biodiversity Convention? Turning first to the components of the Antarctic Treaty system, the Antarctic Treaty and the Seals Convention are silent on their relationship with other international conventions; CCAMLR explicitly names the Seals Convention and the International Convention for the Regulation of Whaling, the rights and obligations under which remain unaffected by CCAMLR; and, finally, the Protocol explicitly addresses its relationship with these other components of the Treaty system and does not derogate from the rights and obligations thereunder. This last point is not without difficulty, however, given the ambiguous phrasing of Article 4 of the Protocol.[116]

The issue of the relationship of the Antarctic Treaty system to a range of global environmental treaties, including the Biodiversity Convention, was raised in a Chilean working paper at the XVIIth ATCM.[117] The response of the ATCM was to agree that proper co-ordination was a good idea, but that "the requirements for co-ordination were specific to each of the agreements and that primary responsibility for ensuring such co-ordination lay with the parties to the Antarctic Treaty that were parties to the other agreements".[118] This means that for uniform rules of application in the Antarctica Treaty area, recourse will still be had to the components of the Antarctic Treaty system and the measures in effect thereunder. For States also party to global environmental agreements of application to Antarctica, their ATS obligations will be supplemented by these additional obligations. But what if these obligations conflict?

The first step is to examine the provisions of the conflicting treaties. If silent on the point, then the general rules of treaty law would ensure that as between States party to both treaties, the later treaty prevails to the extent that it is incompatible with the earlier treaty *where the treaties relate to the same subject matter*.[119] "Relating to the same subject matter" will be construed strictly.[120] However, the Biodiversity Convention is not silent. Article 22 addresses its relationship with other international conventions.

---

[115] This may be contrasted with CCAMLR and the Seals Convention, both of which are free standing Conventions. CCAMLR requires non-ATPs "to observe as and when appropriate the Agreed Measures .... and such other measures as have been recommended by the [ATCPs] in fulfilment of their responsibility for the protection of the Antarctic environment from all forms of harmful interference" (Article V(1)).

[116] For discussion, see n. 96.

[117] *Op. cit.*, n. 16.

[118] Draft Final Report, *op. cit.*, n. 15, at p. 12, para. 55.

[119] Article 30, Vienna Convention on the Law of Treaties, 1155 UNTS 331. Article 30(3) of the Convention provides: "When all the parties to the earlier treaty are parties also to the later treaty but the earlier treaty is not terminated or suspended ... the earlier treaty applies only to the extent that its provisions are compatible with those of the later treaty." The following paragraph considers the situation where the States parties to the later treaty do not include all the parties to the earlier one, in which case the above rule applies to States parties to both; whereas in the case of one State party to both and another party only to one of the treaties, it will be the treaty to which both States are party which governs their mutual rights and obligations (Article 30(4)).

[120] Sir Ian Sinclair, *The Vienna Convention on the Law of Treaties* (2nd edn., 1984), at p. 98. Where the later treaty is of a general character and the earlier treaty is specific in nature, then the issue is not one of successive treaties relating to the same subject matter, but of treaty interpretation involving consideration of the maxim *generalia specialibus non derogant: ibid.* Thus the compatibility of the framework Biodiversity Convention with the components of the ATS may be resolved through recourse to principles of treaty interpretation, rather than to what Sinclair calls "[a] particularly obscure aspect of the law of treaties", successive treaties: *ibid.*, p. 93.

Essentially paragraph 1 follows the familiar pattern of no derogation, with an important caveat where the exercise of the rights and obligations under another existing international agreement "would cause a serious damage or threat to biological diversity". Quite apart from the difficulty of interpreting when the threshold of "a serious damage or threat" has been reached, it is unlikely that the *exercise of rights and obligations* under the Antarctic Treaty system will meet this threshold. Indeed, it is striking how many provisions of the Protocol relate to the obligations of biological diversity conservation contained in the Biodiversity Convention. The same cannot necessarily be said for the other components of the ATS – CCAMLR for example. Inaction in the face of threatened species and populations could be a breach of Article 8(k), but any failure within CCAMLR to adopt conservation measures could be resolved through unilateral national legislation to avoid the possibility of breach. On the other hand, Article 8(k) could form the basis of an interesting challenge to the exercise of the opt-out procedures under CCAMLR where the species is a threatened one, and the State opting out is also a party to the Biodiversity Convention.

The CCAMLR example gives rise to further difficulties, however, since Article 22(2) of the Biodiversity Convention requires the contracting parties to implement the Convention with respect to the marine environment "consistently with the rights and obligations of States under the law of the sea". This subordinates the Biodiversity Convention to "the law of the sea" – presumably both customary and treaty law on the subject.[121] Does this include CCAMLR? If so, then the Biodiversity Convention is subordinated to it. If not, then a State party to both Conventions would resolve conflict by reference to Article 22(1).

A further and greater difficulty relates to the jurisdictional scope of the Biodiversity Convention. It applies to components of biological diversity in areas within the limits of national jurisdiction, and to processes and activities under the jurisdiction or control of States within or beyond the areas of national jurisdiction. Given the unresolved nature of sovereignty claims in Antarctica, the reluctance of the ATCPs to address *ensemble* the relationship of this and other treaties to the ATS is unsurprising.

Conclusion

Although negotiated subsequent to the Protocol on Environmental Protection, the Convention on Biological Diversity entered into force on 29 December 1993, barely eighteen months after it was opened for signature at the UNCED Conference in Rio. The formal negotiation of the Convention commenced in February 1991, at the time when the ATCPs were concluding their negotiation of the Protocol for adoption in Madrid on 4 October 1991. The Antarctic Treaty system is not immune from general developments in international environmental law, and it is possible to detect in the Protocol more than merely a consolidation of measures within the ATS. Moreover, given the emphasis upon environmental protection and an ecosystem approach, it is unlikely that serious conflicts will emerge between the Biodiversity Convention and the ATS at least in respect of obligations of conservation. Indeed, there is significant scope for the ATS to continue to make substantial contributions to the conservation of global biological diversity.

---

[121] See L. Glowka, F. Burhenne-Guilmin and H.Synge, in collaboration with J.A. McNeely and L. Gundling, *A Guide to the Convention on Biological Diversity* (1994), IUCN Environmental Policy and Law Paper No. 30, at p. 109.

7

The Role of *Ex Situ* Measures in the Conservation of Biodiversity

Lynda M. Warren

The Meaning of *Ex Situ* Conservation

The World Conservation Strategy[1] defines conservation as "the management of human use of the biosphere so that it may yield the greatest sustainable benefit to present generations while maintaining its potential to meet the needs and aspirations of future generations". This definition is reflected in the objectives of the Convention on Biological Diversity[2] which are the conservation of biological diversity, the sustainable use of its components, and the fair and equitable sharing of the benefits arising through the utilisation of genetic resources.[3]

Biological diversity is defined in Article 2 as meaning the variability among living organisms from all sources and includes diversity between species, within species and of ecosystems. According to Soulé[4] there are five levels in the biodiversity hierarchy: (a) whole systems such as landscapes or ecosystems; (b) assemblages such as associations or communities; (c) species; (d) populations and (e) genes. The Convention is concerned with all of these. The components of biodiversity for which *ex situ* conservation is appropriate must, given our present state of scientific knowledge and conservation expertise, be considered limited to variability between and within species, i.e. to levels (c)–(e).[5] *In situ* conservation, on the other hand, is appropriate for all five hierarchical levels.

Ex situ conservation is defined in the Convention as "the conservation of components of biological diversity outside their natural habitats".[6] The alternative, *in situ*

---

[1] IUCN/UNEP/WWF, *World Conservation Strategy: Living Resource Conservation for Sustainable Development* (1980), para. 1.4.
[2] Hereinafter referred to as the Convention.
[3] Article 1.
[4] M.E. Soulé, "Conservation: Tactics for a Constant Crisis" (1991) *Science* 253, pp. 744–750.
[5] Note, however, that *ex situ* conservation and reintroduction of so-called "keystone species" may prove critical in future for the conservation of some ecosystems. See conclusions to this chapter.
[6] Article 2.

International Law and the Conservation of Biological Diversity (C. Redgwell and M. Bowman, eds.: 90 411 0863 7: © Kluwer Law International: pub. Kluwer Law International, 1995: printed in Great Britain), pp. 129–144

conservation, is defined as "the conservation of ecosystems and natural habitats and the maintenance and recovery of viable populations of species in their natural surroundings and, in the case of domesticated or cultivated species, in the surroundings where they have developed their distinctive properties".[7]

The Convention is not original in separating types of conservation in this way. The World Conservation Strategy,[8] for example, makes a similar, but slightly more complicated, distinction with reference to the ways of preserving genetic diversity:

(a) on site – in which the stock is preserved by protecting the ecosystem in which it occurs naturally;

(b) off site, part of the organism – in which the seed, semen or other element from which the organism concerned can be reproduced is preserved;

(c) off site, whole organism – in which a stock of individuals of the organism concerned is kept outside its natural habitat in a plantation, botanical garden, zoo, aquarium, ranch or culture collection.

Ex situ conservation, therefore, involves the removal of one or more specimens or parts thereof from the wild and keeping them in a viable condition elsewhere. As a distinction from *in situ* conservation, however, this is not completely clear. Would it be *ex situ* conservation or *in situ* conservation to remove specimens from the wild and transfer them to another site in the wild that had all the qualities of the natural habitat of the species but from which the species was absent, perhaps through loss or simply because it had failed to colonise? One answer would seem to be that it would be depend on whether the specimens would survive in their new home unaided or whether they would need to be managed in some way such as by feeding them or fencing them in. If the new site happens to be a protected area, however, it may well be a managed habitat anyway, thus blurring the distinction further. The Convention defines *in situ* conditions as "conditions where genetic resources exist within ecosystems and natural habitats, and, in the case of domesticated or cultivated species, in the surroundings where they have developed their distinctive properties".[9] This suggests that any site that was not currently inhabited by the species in question would not be *in situ*.

The meanings of "natural" and "natural habitat" are also unclear in the absence of a precise definition. As Markham *et al.*[10] observe, some of the most valuable habitats, such as chalk grasslands, are dependent on some form of resource management. Even apparently wilderness areas are often the result of historical intervention by man.[11]

The categorisation of *ex situ* conservation into whole organism conservation and conservation of parts is useful because the techniques involved and the legal controls necessary are likely to be very different. It is also useful, in this respect, to distinguish between animals, plants and micro-organisms. Unfortunately the

---

[7] *Ibid.*

[8] *Op. cit.*, n. 1, para. 6.5.

[9] Article 2.

[10] A. Markham, N. Dudley and S. Stolton, *Some Like it Hot: Climate Change, Biodiversity and the Survival of Species* (1993), at p. 126.

[11] It is probable, for example, that no natural vegetation remains in Great Britain. Even apparently pristine places, such as the Southern Ocean, have been significantly altered by human activity.

Convention makes no distinction. Nor does it consider the different motives for *ex situ* conservation.

Motives for *Ex Situ* Conservation

The immediate purpose of *ex situ* conservation is, of course, to preserve an element of biological diversity. There may, however, be several reasons for wishing to do this as indicated in Table 7.1.

Conservation

Traditionally, the motive has been simply to ensure the survival of a species, population or variety. In most cases *ex situ* conservation has been the last resort without which the species would have become extinct. The immediate aim has been to increase numbers of individuals by captive breeding or artificial propagation. The possibility of reintroduction into the wild comes later if sufficient numbers can be produced and the circumstances giving rise to the original vulnerability in the natural habitat have been rectified. The first successful reintroduction of an animal species was the release of the Alpine ibex in the last century. Arzdorf[12] catalogues 129 reintroductions of which thirty are currently self-sustaining. The majority are of birds or mammals. Notable successes include Przewalski's horse, Père David's deer and the Hawaiian goose, all of which were extinct or virtually extinct in the wild. *Ex situ* conservation of plant species started later and has been applied most extensively to crop species and varieties. Successes with severely threatened plant species have been less spectacular but include the artificial propagation of cuttings from the one remaining wild specimen of café marron, a Rodrigues Island endemic.[13] Maunder *et al.*[14] refer to a number of species of plants that are extinct or critically endangered in the wild but which survive in commercial horticulture.

Table 7.1   Motives for *Ex Situ* Measures

Conservation
Extinct, or virtually extinct in wild e.g. café marron
Wild population non-viable e.g. Arabian oryx, Mauritius kestrel
Loss of genetic vigour e.g. red kite

Research and education
Conservation biology e.g. propagation techniques, behavioural studies, genetic finger-printing
Encyclopaedias i.e. as living museum collections
Public awareness

Commercial gain
Trade e.g. rare orchids
Synthesis e.g. medicinal plant products
Genetic engineering

---

[12] W. Arzdorf *Stand, Möglichkeiten und Grenzen von Zucht und Auswilderung bedrohter Tierarten als Beitrag zum Artenschutz* (1990).

[13] See V.H. Heywood, "Conservation of Germplasm of Wild Plant Species", in O.T. Sandlund, K. Hindar and H.D. Brown (eds.), *Conservation of Biodiversity for Sustainable Development* (1992), pp. 189–203. For an update on progress, see (1994) 28(1) *Oryx* 7.

[14] M. Maunder, M.F. Fay and T.M. Upson, "*Ex Situ* and *in Situ* Approaches to Plant Conservation: Two Distinct Options or Complementary Points on the Conservation Spectrum?", in M. Groves, M. Read and B.A. Thomas (eds.), *Species Endangered by Trade: A Role for Horticulture?* (1993), pp. 44–53.

One of the reasons why populations die out and species become extinct is that, as numbers decrease, the genetic diversity decreases also. Inbreeding occurs and the species or population loses its genetic vigour. Captive breeding programmes involving individuals from different populations can help to prevent this process. Even when numbers of individuals are extremely low, transfers of breeding individuals between zoos can maximise the chances of preventing inbreeding.

It is not just wild species and populations that can benefit from *ex situ* techniques. In a sense, of course, domestic varieties of plants and animals do not have a natural habitat[15] but this does not mean that they are not worth conserving. Because they rely on human activity for their very existence there is often a need for positive conservation to maintain varieties once they are no longer of immediate use.[16]

Emergency conservation measures as a last resort are not the ideal way to achieve the conservation of biodiversity, however, and policies are now geared more towards "just in case" preservation whereby stocks of commercially important and/or threatened species or varieties are maintained outside of the natural habitat. Because techniques for conserving parts of organisms work more successfully for plants than for animals, most of the effort has been on plants. One of the major issues for the future, as techniques become more refined, is likely to be the selection of the species.

Education and Research

Ex situ measures can be of indirect benefit to conservation in the wild by enabling research into conservation biology. Topics covered include research into the behavioural and nutritional requirements of species, the development of techniques for artificial fertilisation or plant propagation, and genetic finger-printing so that the viability of field populations can be evaluated. Maunder *et al.*[17] make the further point that zoos and botanic gardens are the equivalent of an encyclopaedia of information for research and educational purposes. Research is also an important element of commercial *ex situ* practices.

Commercial exploitation

By no means all of the species subject to *ex situ* conservation are in imminent danger in the wild. Many are kept because they are commercially important. Keeping them under *ex situ* conditions ensures a ready supply, facilitates research and cuts down on costs. This type of *ex situ* conservation can be divided into three types. First, the species itself may be of commercial value as a living organism. Rare plants, such as orchids and certain species of bulbs, fit into this category. Second, a species, usually a plant, may be grown artificially because of its pharmaceutical or other profitable qualities. Third, in some cases an experimental collection of living material, usually a micro-organism, will be needed as a basis for genetic engineering so as to maximise returns or to obviate the need for the wild species.

---

[15] Hence the reference in the Convention to "the surroundings where they have developed their distinctive properties" (Article 2).

[16] Within the UK, for example, the National Council for the Conservation of Plants and Gardens (NCCPG) maintains national collections of plant varieties and the Rare Breeds Survival Trust conserves domesticated animals.

[17] *Op. cit.*, n. 14.

Ex Situ Measures for the Conservation of Biodiversity as required under the Convention

The Convention provides for these different motives for *ex situ* conservation but does not treat them separately. The provisions of Article 9 are given in full in Table 7.2. Article 9(b) provides for the establishment and maintenance of facilities for *ex situ* conservation and research. It does not require that research to be for the purposes of conservation.[18] Indeed, in the case of commercially important species, it is just as likely that the *ex situ* conservation is for the purposes of research. In such a case, the *ex situ* measures are no more than a form of farming for wild species.[19]

Articles 9(c) and (d) are an acknowledgement that emergency measures may be necessary to prevent a species from becoming extinct. Article 9(c) requires contracting parties to take *ex situ* measures to collect specimens with a view to reintroductions into the wild at a later stage. The basic requirement of Article 9(d) is that collection from the wild must not threaten the natural population. In the case of very rare species, of course, collection will be a threat but this will be a calculated risk based on the expectation of being able to reintroduce individuals following a captive breeding programme. The Article refers to special "temporary" *ex situ* measures. Given the practical difficulties involved in building up an *ex situ* stock for reintroduction it is likely that any such *ex situ* programmes will be long term. Furthermore, given the risks, often unquantifiable, of reintroductions, it would be foolish not to keep a breeding stock in captivity "just in case". In this respect, therefore, the measures cannot be "temporary". If all goes according to plan, however, the threat to the *in situ* population will be temporary.

The fact that special measures are considered necessary to deal with emergency situations underlines the fact that these are to be exceptions. The *ex situ* conservation programme is aimed at the full range of biological diversity, not just those species that are under greatest threat.

Preference is to be given to measures taken in the country of origin of the material, which is defined as meaning the country which possesses the resources in

Table 7.2 Article 9 *Ex Situ* Conservation

Under Article 9 each Contracting Party shall, as far as possible and as appropriate, and predominantly for the purpose of complementing *in situ* measures:

(a) adopt measures for the *ex situ* conservation of components of biological diversity, preferably in the country of origin of such components;

(b) establish and maintain facilities for *ex situ* conservation of and research on plants, animals and micro-organisms, preferably in the country of origin of genetic resources;

(c) adopt measures for the recovery and rehabilitation of threatened species and for their reintroduction into their natural habitats under appropriate conditions;

(d) regulate and manage collection of biological resources from natural habitats for *ex situ* conservation purposes so as not to threaten ecosystems and *in situ* populations of species, except where special temporary *ex situ* measures are required under subparagraph (c) above; and

(e) co-operate in providing financial and other support for *ex situ* conservation outlined in subparagraphs (a) to (d) above and in the establishment and maintenance of *ex situ* conservation facilities in developing countries.

[18] It is true that *ex situ* measures taken under Article 9 are to be predominantly for the purposes of complementing *in situ* measures but this does not rule out other purposes.

[19] See also the discussion on indigenous propagation of bulbs below, text accompanying n. 55.

in situ conditions. Article 9(e) is an acknowledgment of the likely need for financial and practical support in order to give effect to this preference in developing countries.

The financial implications of *ex situ* conservation have influenced the development of the provisions in the Convention. The International Environmental Law Conference held at the Hague in 1991[20] concluded that it would be difficult to obtain financial compensation from foreign institutions for the commercial use of biodiversity when this could be avoided by growing species in their own *ex situ* facilities. It was recommended that preference be given to *in situ* conservation but that, where desirable, *ex situ* facilities should be provided in the state where the material originates. Where this is not possible it is recommended that the state of origin should have preferential access.[21]

These recommendations have been incorporated in the Convention and colour the approach to *ex situ* conservation. The result is a confusion between the scientific needs for *ex situ* conservation and the question of intellectual property rights over the use of the genetic material so conserved. The Convention does not take account of the fact that many species in need of *ex situ* conservation are not, in themselves, commercially important. Even in the case of commercial species, *ex situ* conservation may not provide a sustainable source of material for exploitation. Only in those cases where material can be mass produced through artificial propagation under laboratory conditions is there likely to be a danger of foreign institutions benefiting through *ex situ* conservation. The real problem is in the collection of wild material and the synthesis of natural products following laboratory analysis. Giving preference to *ex situ* conservation in the country of origin will not necessarily solve this.

There are, nevertheless, sound reasons for encouraging the establishment of *ex situ* collections within the same geographical area as the *in situ* habitat. Markham *et al.*[22] note that, of the 1,500 botanic gardens worldwide, less than 10% are located in the tropics, although this is where the majority of plant species occur. It makes sense to maintain collections of species in conditions that are as close to their natural state as possible.

Even so, *ex situ* conservation in the country of origin can have a number of drawbacks. While it may sometimes be better to maintain collections of native wildlife in the country of origin, there are many instances where it is more appropriate to house them elsewhere, for example to facilitate captive breeding programmes; to avoid unnecessary, very costly, duplication of effort; or simply to spread the risk by not putting all the eggs in one basket, as it were.

Part of the problem is that the Convention treats all types of *ex situ* measures alike. If the need is to maintain a stock of individuals in as near natural conditions as possible then a site in the country of origin is likely to be best. If, however, a species has become so rare that artificial propagation requiring technical expertise is necessary it may be more appropriate to export individuals. Similarly, if parts of organisms are to be conserved, it might make more scientific (and economic) sense to export the material to a state with existing expertise and facilities rather than to experiment in the country of origin.

---

[20] S. Bilderbeek (ed.), *Biodiversity and International Law* (1992), at p. 31.
[21] *Ibid.*, Recommendation 5a, at p. 31 and p. 198.
[22] *Op. cit.*, n. 10.

Techniques of *Ex Situ* Conservation

The measures available differ for animals, plants and micro-organisms. For animals, the need usually arises because natural population levels have become too low for maintenance of the population (or species), given the normal reproductive success and genetic vigour. The conservation response is to collect animals from the wild for captive breeding. This is a specialist exercise, however, and depends on a detailed knowledge of the species' nutritional and behavioural requirements. The problem of inbreeding is dealt with by cross-breeding animals collected from, or bred from animals collected from, different wild populations. The first co-operative and co-ordinated breeding programmes started in the early 1980s. There is now a comprehensive Animal Records Keeping System compiled from zoo records and amounting to an international stud book. The IUCN's Captive Breeding Specialist Group is concerned with the interactive management of *ex situ* and wild populations.

Biotechnological advances are of great value to captive breeding programmes. Examples include *in vitro* fertilisation techniques and DNA finger-printing which enables scientists to profile a natural population and determine the degree of interrelatedness. Most *ex situ* conservation of animal species takes place in zoos and is carried out with the intention of reintroduction into the wild, although not necessarily at the site of origin.

Ex situ measures for plant conservation are more diverse and include gene banks for seeds, tissue or pollen, and growing collections in plantations, field gene banks and botanic gardens. There have been several gene banks for commercially important plants developed, such as, for example, the Nordic Gene Bank for Agricultural and Horticultural Plants. In most cases the criterion for inclusion is the potential for exploitation rather than the conservation status or country of origin.[23] Collections of native wild plants under threat are less common.[24] Even botanic gardens do not necessarily carry representative collections of native flora. Wilcock,[25] for example, found that conservation of native Scottish plants was a low priority for Scottish botanic gardens.

Ex situ conservation of micro-organisms consists of the establishment and maintenance of cultures. These are often the only reliable source for research purposes and are of considerable commercial importance.[26]

What *Ex Situ* Conservation Involves

The main elements of *ex situ* conservation, whatever the methods employed and whatever the motives, are collection from the wild, transport, maintenance, propagation and, where appropriate, relocation. Although there is scant legal provision for *ex situ* conservation as such, its component activities are subject to legal control which may, in some cases, prove detrimental to the aims of *ex situ* conservation.

---

[23] For further discussion of this point see Heywood, *op. cit.*, n. 13.

[24] Developing countries are most likely to concentrate efforts on commercially important species and the developed countries with a history of *ex situ* conservation have generally built up a collection of exotic species. For example, even the Royal Botanic Gardens, Kew holds seeds of only about 40% of native British seed-producing plant species. For further details of the UK's *ex situ* programme see *Biodiversity: The UK Action Plan* (1994), Cm 2428, Chapter 5. This publication was produced as part of the UK's commitment to save biodiversity, made at UNCED.

[25] C.C. Wilcock, "Botanical Sanctuaries for Scotland's Flora" (1990) 45 *Trans. Bot. Soc. Edinb.* 509–517.

[26] *Op. cit,*. n. 24, para. 5.22.

Taking

There can be no *ex situ* conservation without the material to conserve and the ultimate source of such material, whether it be a group of individuals, a single individual, or a cutting or seed, must be the wild. It is ironic, therefore, that one of the legal measures employed in *in situ* conservation is a control over the taking of wild specimens. In the case of endangered species, which may be the very species for which *ex situ* conservation measures are needed, the control may amount to a complete prohibition. This apparent conflict between the desire to safeguard wild populations from all attempts to deplete numbers and the need to remove specimens from the wild for captive breeding or translocation purposes is usually dealt with by providing for exceptions to absolute prohibitions on conservation grounds.

One of the first international conservation agreements was the Convention on Nature Protection and Wildlife Preservation in the Western Hemisphere (the Western Hemisphere Convention)[27] which came into force in 1942. One of its stated aims is to "protect and preserve in their natural habitat representatives of all species and genera of ... native flora and fauna ... in sufficient numbers and over areas extensive enough to assure them from becoming extinct through any agency within man's control".[28] Article VIII refers to listed species needing urgent protection. Such species are to be protected as completely as possible and taking is only to be allowed with appropriate official permission. Permission is only to be granted under special circumstances "in order to further scientific purposes, or when essential for the administration of the area in which the animal or plant is found". There is no direct reference to *ex situ* conservation, nor is there any reference to propagation or restocking. This probably reflects the current thinking of the time, when *ex situ* measures were largely restricted to attempts at captive breeding in zoos. Such measures are presumably included within the term "scientific purposes".

Lyster[29] describes the Western Hemisphere Convention as "a visionary instrument, well ahead of its time in terms of the concepts it expresses" which paved the way for subsequent treaties on wildlife protection. Most of these subsequent agreements put special emphasis on habitat protection. Species protection measures are usually associated with the establishment of protected areas or relate to measures controlling exploitation. Some of the key agreements are considered here.

The African Convention on the Conservation of Nature and Natural Resources (the African Convention)[30] emphasises the need to establish "conservation areas" but also has, as one of its objectives, special protection of animal and plant species threatened with extinction. Such species are included in the Annex to the Convention. Protection for Class A species is strict. Collection is permitted only on the highest authority and only "if required in the national interest or for specific scientific purposes". As with the Western Hemisphere Convention, there is no specific reference to *ex situ* conservation.

The Convention on the Conservation of European Wildlife and Natural Habitats (the Berne Convention)[31] is primarily concerned with conservation measures for the habitats of listed species. It does, however, also require protection of listed plants

---

[27] 161 UNTS 193.
[28] Preamble.
[29] S. Lyster, *International Wildlife Law* (1985), at p. 97.
[30] 1001 UNTS 4.
[31] UKTS 56 (1982), Cmnd 8738; ETS 104.

and animals. Article 5 prohibits deliberate collecting of Appendix I plants and Article 6 prohibits all forms of deliberate capture of Appendix II animals. Exploitation of Appendix III species is to be regulated to keep populations out of danger but otherwise there is no prohibition on taking. Exceptions to prohibitions are permitted under Article 9(1) but are limited to socio-economic circumstances and do not refer to *ex situ* conservation or any associated measures.

The ASEAN Agreement on the Conservation of Nature and Natural Resources[32] does include direct reference to *ex situ* conservation. Article 5, which covers endangered and endemic species, provides that contracting parties shall, wherever possible, prohibit the taking of listed species except for exceptional circumstances by special allowance from the designated authorities. Article 3, which is concerned with genetic diversity, requires parties to endeavour to promote and establish gene banks and other documented collections of animal and plant genetic resources with a view to ensuring the survival of all species under their jurisdiction and control. Taken together, these provisions allow collection for *ex situ* conservation. Furthermore, Article 4, which deals with sustainable use of exploitable species, provides that special measures, such as restocking, shall be provided for whenever the conservation status of a harvested species so warrants.

The Convention on the Conservation of Migratory Species of Wild Animals (the Bonn Convention)[33] requires a contracting party that is the range state for an endangered migratory species listed in Appendix I to prohibit the taking of animals of such species.[34] Exceptions may be made only for limited purposes including taking for the purpose of enhancing the propagation or survival of the affected species. Exceptions must be clearly defined and limited in space and time and must not be to the disadvantage of the species. Parties must inform the Convention Secretariat of exceptions.

The European Council Directive on the Conservation of Natural Habitats and of Wild Fauna and Flora (the Habitats Directive),[35] continues the emphasis on habitats and, therefore, *in situ* conservation. However, it also provides for a system of strict protection for listed species of plants and animals within their natural range and requires Member States to prohibit all forms of deliberate capture or collection.[36] Further, Member States must prohibit the keeping, transport and sale and exchange of specimens of these species taken from the wild, except where these were taken legally before implementation of the directive.[37] Annex V of the directive lists plants and animals whose taking from the wild and exploitation may be subject to management measures. Under Article 14 these measures must be taken where Member States deem it necessary so as to maintain the favourable conservation status[38] of the species concerned. The measures that may be taken include temporary or local prohibition on collecting specimens from the wild and the establishment of a system of licences. The latter could, for example, be used to authorise the appropriate

---

[32] (1985) 15 EPL 64.

[33] (1980) 19 ILM 15.

[34] Article III(5). A range state is defined as any state exercising jurisdiction over any part of the range of the migratory species. Article I(1)(h).

[35] 92/43/EEC, OJ L206, 22.7.92, p. 7.

[36] Articles 12 and 13.

[37] *Ibid.*

[38] The conservation status of a species is considered favourable when population dynamics data indicate long-term viability; its natural range is not being reduced or likely to be reduced in the foreseeable future; and there is, and will probably continue to be, a sufficiently large habitat to maintain populations in the long-term (Article 1(i)).

scientific authority to take wild material where a species is in danger of over-exploitation in the wild and in need of *ex situ* conservation. Unfortunately there is no comparable measure available for the Annex IV species protected by Articles 12 and 13. Instead, Article 16(a) of the directive permits derogation from their provisions provided that there is no satisfactory alternative and the derogation is not detrimental to the maintenance of populations at favourable conservation status in their natural range. Member States must report biannually on derogations stating, *inter alia*, the species concerned, the reason for derogation and the risk involved. Derogations are permitted for the following purposes: to protect wild fauna and flora and to conserve habitats; to prevent serious damage to agricultural and other interests; in the interests of public health and safety or for imperative reasons of overriding public interest; for research and education; for repopulating and reintroducing species and for associated breeding programmes, including artificial propagation of plants; and to allow the taking or keeping of limited numbers of certain specimens of Annex IV species. It is unfortunate that the directive does not expressly provide for a licensing system for specimens of endangered species to be taken from the wild for conservation purposes necessitating *ex situ* measures. Lumping *ex situ* conservation measures with socio-economic purposes as a reason for derogation detracts from the positive aspects of *ex situ* measures as a tool for achieving favourable conservation status, rather than as a reason for going against it.

There are no formal international agreements designed to provide for collections made for the purposes of establishing gene banks for commercial purposes. Clearly, some of the provisions in the above instruments relating to taking for the purposes of establishing gene banks or more generally for the purposes of scientific research could include such activities but the context is not really concerned with cultivated varieties. Rose[39] refers to the code of conduct for plant germplasm collecting and transfer endorsed by the Commission on Plant Genetic Resources established by FAO.

Transporting

Despite the emphasis in the Convention on *ex situ* measures being undertaken in the country of origin, most activity to date has involved transporting material from one state to another. The Convention on International Trade in Endangered Species of Wild Fauna and Flora (CITES),[40] although primarily concerned with regulating the import and export of wildlife for commercial purposes, is also of relevance to transboundary movements of wildlife for *ex situ* conservation.[41]

The provisions of CITES apply to specimens of listed species.[42] Trade is regulated through a system of import and export permits. For species included in Appendix I, which is limited to species threatened with extinction, transboundary movement for commercial purposes is, in effect, prohibited. No export is allowed without an export permit which can only be given if, *inter alia*, the specimen has been legally obtained and its export will not, in the opinion of the designated scientific authority

---

[39] See the chapter by Greg Rose in this volume.

[40] 993 UNTS 243; UKTS 101 (1976), Cmnd 6647.

[41] Note that "trade" is defined in CITES Article I(c) as meaning export, reexport, import and introduction from the sea. There is no reference to financial reward.

[42] Article I(b) defines "specimen" as any animal or plant, alive or dead, and any readily recognisable parts of derivatives thereof. There is no specific reference to micro-organisms and none is listed in any of the appendices.

of the export state, be detrimental to the survival of the species. Legal import requires presentation of the export permit and an import permit issued by the import state. The issue of the latter is conditional on the advice of the scientific authority of the import state that the import will be for purposes that are not detrimental to the survival of the species and on the management authority being satisfied that the specimen is not to be used for primarily commercial purposes. For Appendix II species, for which limited exploitation is deemed acceptable, specimens cannot be imported by CITES states unless they are accompanied by export certificates from the country of origin.

There are, however, some exceptions to the rigours of CITES control. Article VII lists exceptions, included among which are animals bred in captivity and artificially propagated plants. Article VII.4 provides that specimens of Appendix I species that have been captive bred or artificially propagated for commercial purposes shall be deemed to be specimens of Appendix II species so that no import permit is needed. Article VII.5 provides that, where the appropriate authority of the export state is satisfied that the specimens have been captive bred or artificially propagated, certificates to that effect are acceptable in lieu of permits. This applies to specimens of all listed species. Furthermore, Article VII.6 provides that no permits are required for the non-commercial loan, donation or exchange between registered scientists of living, appropriately labelled, plant material.

"Bred in captivity" has been defined[43] as applying to "offspring, including eggs, born or otherwise produced in a controlled environment, either of parents that mated or otherwise transferred gametes in a controlled environment if reproduction is sexual, or of parents that were in a controlled environment when development of the offspring began, if reproduction is asexual". Parental breeding stock must be established in a manner that is not detrimental to survival of the species in the wild and must be maintained without augmentation from the wild except for occasional additions from wild populations to prevent deleterious inbreeding. The breeding stock must be managed in a sustainable way, that is it must be maintained indefinitely so that at least a second generation can be produced in a controlled environment.[44]

Conference Resolution 8.17, which replaces Conference Resolution 2.12(c) recommends that "artificially propagated" is interpreted to refer only to plants grown by man from seeds, cuttings, divisions, callous tissues or other plant tissues, spores or other propagules under controlled conditions. The artificially propagated stock must be established and maintained in a manner not detrimental to the survival of the species in the wild and managed in a manner designed to maintain the artificially propagated stock indefinitely.

The transport of living specimens from state to state can also incur concerns over the spread of disease, so that quarantine regulations may need to be complied with in addition to any CITES regulations. The International Plant Protection

---

[43] Proceedings of the Second Meeting of the Conference of the Parties, Conf. Res. 2.12.

[44] Conf. Res. 2.12 defines a controlled environment for animals as "an environment that is intensively manipulated by man for the purpose of producing the species in question, and that has boundaries designed to prevent animals, eggs or gametes of the selected species from entering or leaving the controlled environment". The CITES Plants Committee, meeting for the first time in 1989, suggested that controlled conditions for plants means being under an environment that is intensively manipulated by man for the purpose of producing selected species. General characteristics of controlled conditions may include, but are not limited to, tillage, fertilisation, weed control, irrigation, or nursery operations such as potting, bedding, or protection from weather.

Convention 1951 provides for phytosanitary certificates to be issued for the export of plants. Conference Resolution 4.16 recommended that phytosanitary certificates be used instead of certificates of artificial propagation for the export of Appendix II species in order to facilitate international transactions.

While CITES is the main convention to deal with transboundary movements of wildlife, it is not the only one. The African Convention[45] requires special provisions for the export of protected species. In addition to permits required to collect the specimens there must be a further authorisation indicating their destination. The purpose of export is not covered but, in the case of the most protected species, authority for collecting in the first place will be limited to requirements of national interest or for scientific purposes.[46] The Western Hemisphere Convention[47] contains provisions similar to those in CITES. The import, export and transit of protected flora and fauna is to be regulated by issuing permits to authorise exports and by prohibiting imports from other contracting parties unless accompanied by such a certificate. With the success of CITES,[48] however, both of these rather imprecise and narrowly applicable sets of provisions have been largely forgotten.

Storage and Manipulation

The fate of living material collected from the wild will depend on the type of material and on the motive behind the *ex situ* conservation. There are no binding international legal agreements controlling the use of such material, although improper use may be in breach, for example, of import conditions under CITES. Instead, use will be regulated by any national laws concerning animal experimentation, standards of animal welfare, plant breeding rights, access to environmental information and quarantine.

For flowering plants, seed collection and storage provides a relatively straightforward way of safeguarding against future threats to species. Because the techniques involved are relatively simple and, given the size of a seed, the storage requirements minimal, seed banks are increasing in popularity. There is a strong element of risk involved, however, in that future viability cannot be assured. Alternative or, in some cases, additional methods are employed, such as growing specimens in botanic gardens. Both types of *ex situ* conservation amount to controlled captivity. The main issue of relevance to the Convention is the access to material. Most botanic gardens operate an open access policy and the international network of gene banks provides open access in principle. As Rose[49] points out, however, where cultivated plants are concerned commercial sensitivities may limit the extent of this access in practice.

In the case of highly endangered plant species, artificial propagation techniques may be used to increase numbers under controlled conditions for relocation into the wild or to produce cultivated specimens of highly desirable species so as to take away the need for wild collections. The most dramatic instance of such activity is

---

[45] Article IX.

[46] Article VIII.1(1).

[47] Article IX.

[48] Lyster describes CITES as "perhaps the most successful of all international treaties concerned with the conservation of wildlife", *op. cit.*, n. 29, at p. 240. Burgess, however, is less complimentary. She cites evidence that suggests that some signatories fail to abide by CITES requirements: J. Burgess, "The Environmental Effects of Trade in Endangered Species", in OECD, *The Environmental Effects of Trade* (1994), at pp. 123–152.

[49] In this volume.

shown by the artificial propagation of rare orchid species from seeds using a special medium, which obviates the need for the fungal mycorrhiza normally essential for seed development. These are grown and can be transported in flasks. One of the problems with artificially propagated material is that it ultimately derives from wild material. This can give rise to problems under CITES. The European Community, for example, has experienced considerable difficulties in agreeing a new regulation to implement the plant provisions of CITES because of the impracticalities of registering all propagules.[50]

For the sort of commercial *ex situ* measures that are of most concern to the developing world, the collection of wild material is just the first step in the development of a commercial product, be it a new drug synthesised following analysis of collected material or a new variety of micro-organism derived by artificial selection or, even, genetic engineering. The issues of relevance to the Convention in such instances are the financial reward to the country of origin for supplying material and the transfer of technology developed as part of the commercial enterprise. One recent solution has been the co-operative venture between the drug company MERCK and the government of Costa Rica whereby a Costa Rican company carries out the experimental work and provides MERCK with the end results. Whether such a policy of keeping producer and market at arm's length will prove practical in the long term is uncertain given the huge sums of money involved.

For animals, the main reason for *ex situ* conservation is to undertaken captive breeding, further details of which are given in the section on techniques, above.

Return to the Wild

The ultimate aim of any *ex situ* measures should be to promote the survival of the species or population. The return of specimens to the wild is often the ultimate goal. The various types of release have been categorised as follows.[51] "Restocking" or "reinforcement" is the release of individuals to reinforce an existing population so as to increase the viability of the population. "Reintroduction" or "re-establishment" is the deliberate release of individuals of a species into an area from which it has been lost with the aim of establishing a self-sustaining, viable population. "Introduction", on the other hand, is the intentional or accidental dispersal by human agency of living organisms outside their historically known native range. Maunder and Ramsay[52] are critical of the implications of the distinction between reintroduction and introduction because they believe that it could be detrimental to conservation. Introductions are usually treated with great caution because of the potential for introducing disease and/or of upsetting the ecological balance of a community and many states have national legislation to control them. The Botanical Society of the British Isles categorises a release as a re-establishment if it is within 1 km of a recorded natural site. Maunder and Ramsay believe that this is too restrictive and is dependent on detailed information that will not usually be available. It is their view that the reintroduction of an extirpated population on a site more than 1 km from a recorded natural population but in a piece of contiguous habitat should

---

[50] Note, however, that all orchid flasked seedlings are excluded from CITES controls.
[51] The definitions are based on those given in a review by M. Maunder and M. Ramsay, "The Reintroduction of Plants into the Wild: An Integrated Approach to the Conservation of Native Plants", in A.R. Perry and R.G. Ellis (eds.), *The Common Ground of Wild and Cultivated Plants* (1994), at pp. 81–88.
[52] *Ibid.*

not be termed an introduction. That term should be reserved for the establishment of exotic or non-native species outside their natural range. The distinction could be important in interpreting the Convention. Article 9(c) requires parties to adopt measures for the reintroduction of threatened species into their natural habitats but does not define the terms "reintroduction" and "natural habitats".[53] It does, however, distinguish it from the introduction of alien species which must be prevented if there is a threat to ecosystems, habitats or species.[54]

A good example of the blurred distinction between reintroductions and introductions is provided by the indigenous plant propagation project instigated in Turkey by the Fauna and Flora Preservation Society.[55] Many bulbs on the commercial market have been collected from wild populations in Turkey and then exported to Holland for growing on prior to sale.[56] The project is designed to protect the wild populations by translocating a few specimens to field sites in Turkey where they can be propagated under field conditions. The selected sites must, obviously, possess the qualities of the natural habitat but do not necessarily support any wild specimens. In terms of the Convention, therefore, this might or might not be a reintroduction. It is also unclear whether or not it can be classified as *in situ* or *ex situ* conservation. If the release amounts to a restocking, albeit under controlled conditions, it could be interpreted as an *in situ* measure under Article 8(f) or an *ex situ* measure under Article 9(c). If it is an introduction or reintroduction it will not be an *in situ* measure. It would not fit the conservationist's normal understanding of *ex situ* conservation because the bulbs are being cultivated for commercial purposes rather than for their conservation. But, as is shown above, the Convention does not rule out *ex situ* measures of this kind and the end result of the enterprise, should it prove successful, will be conservation of an *in situ* population.

McNeely presents criteria to be satisfied before reintroductions should be sanctioned.[57] These include the permission and involvement of appropriate government agencies in the recipient and donor states; the identification and elimination or control of the original causes of extinction; the establishment of a recognised protocol involving a feasibility study, the reintroduction itself and subsequent monitoring; and the restriction of reintroductions of critically endangered species to agencies with proven ability to plan, undertake and follow up projects.[58]

Conclusions

For years *ex situ* conservation has been regarded as a cul-de-sac and most conservation effort has concentrated on habitat protection as the main device for main-

---

[53] "Habitat", however, is defined in Article 2 as "the place or type of site where an organism or population naturally occurs".

[54] Article 8(h).

[55] For further details of the project see M. Read, "The Indigenous Propagation Project: A Search for Co-operation and Long-term Solutions", in M. Groves, M. Read and B.A. Thomas, *op. cit.*, n. 14, at pp. 54–62 and M. Read and B.A. Thomas, "Propagation not collection", in A.R. Perry and R.G. Ellis, *op. cit.*, n. 51, at pp. 101–110.

[56] Note, in connection with transport of collected bulbs, that Turkey is not a member of CITES.

[57] J.A. McNeely (ed.), *Parks for Life*, Report of the IVth World Congress on National Parks and Protected Areas (1993), Recommendations 11.4 and 13. These are the draft conclusions of workshops and are based on policy guidelines drafted by the IUCN's Reintroductions Specialist Group of the Species Survival Commission.

[58] This last criterion provides a good example of the opportunities for cooperation in supporting conservation in developing countries as it is generally the developed countries that have acquired the necessary expertise.

taining species. The view promoted by the Convention, that *in situ* and *ex situ* measures are complementary, is one that is now widely accepted although there is still some disagreement as to the relative importance of each. The relegation of *ex situ* measures to subsidiary importance in the Convention probably has more to do with financial expectations than conservation. There is evidence to suggest that the developed countries were prepared to see greater prominence attached to *ex situ* measures.[59]

The relative importance of *ex situ* and *in situ* conservation on a global scale is irrelevant. What matters is that the optimum blend of measures is adopted to deal with each individual problem.[60] Nevertheless, Markham *et al.*[61] see three particular problems in relying too much on *ex situ* measures. First, they believe that *ex situ* conservation can act as an excuse for further habitat destruction; second, only a tiny proportion of a species' genetic material is preserved and thirdly many previous programmes of captive breeding and seed banks[62] have failed to maintain the species concerned. To this may be added the problems of selecting those species to conserve.

The Workshop on Ecological Restoration at the IVth Congress on National Parks and Protected Areas concluded[63] that priority for reintroductions into protected areas should be given to keystone species upon which the integrity of the protected area depends. Heywood[64] defines a keystone, or target, species as one whose disappearance would significantly decrease the biodiversity of an area or on whose presence the stability or very existence of the ecosystem depends. In practice, however, it may not always be easy to identify such species.[65]

Even if keystone species can be identified, the reintroduction of captive bred or artificially propagated specimens into the wild is not without scientific difficulties, only some of which are covered by any sort of legal control. First there is the question of success rates. As is acknowledged in McNeely,[66] the decline of a wild population cannot simply be redressed by adding to its numbers. The reason for the decline must also be considered. If it is through habitat loss then adding to the population size may simply serve to hasten the demise. If it is because of persecution, there must be laws to protect the species and these must be enforced.[67] There is also the danger that the individuals will be unable to cope in the wild either

---

[59] According to W. Burhenne, "Necessary Elements of a Biodiversity Convention", in Bilderbeek, *op. cit.*, n. 20, at p. 62, many developing countries wished the Convention to give equal weight in importance to *ex situ* conservation as to *in situ* conservation. See also the Working Group reports referred to in P.W. Birnie and A.E. Boyle, *International Law and the Environment* (1992), at pp. 485–486 and n. 135.

[60] See, for example, Maunder *et al.*, *op. cit.*, n. 14, and *Biodiversity: The UK Action Plan*, *op. cit.*, n. 24, at para. 5.3. The latter describes *in situ* and *ex situ* conservation as opposite ends of a spectrum with no absolute distinction. It goes on to say that a "seamless blend of both in an integrated programme is therefore appropriate to conserve biodiversity".

[61] *Op. cit.*, n. 10.

[62] Seed banks are not suitable for all species. Furthermore, although costs of storage are not high, seeds cannot be kept indefinitely and fresh collections must be made or seeds grown up to give new seeds.

[63] McNeely, *op. cit.*, n. 57, Recommendation 13.

[64] *Op. cit.*, n. 13.

[65] The Workshop on Reintroductions of Extirpated Species recommended that further research be carried out into the identification of key species. See McNeely *op. cit.*, n. 57, Recommendation 11.4.

[66] *Op. cit.*, n. 57.

[67] It is no coincidence that many of the species that are subject to captive breeding programmes are top predators that interfere with human activity or are commercially desirable to the extent that illegal capture is still worth the risk.

because they are incapable of fending for themselves or because they have become too familiar with people.

If sites offering *in situ* conditions or known to have been recent sites for the species are not available it may be necessary to introduce individuals to a new comparable site. Specimens bred in captivity might be supplemented by individuals translocated from an *in situ* site. There are considerable dangers in manipulating nature in this way. There are numerous examples of ecological disasters resulting from accidental or deliberate introductions into the wild. For this reason most countries have strict national laws controlling the introduction of alien species into the wild but most do not regulate the release of individuals of native species. This is a dangerous loop-hole in any system designed to protect biodiversity rather than species. This is because introducing "foreign" individuals may result in a weakening of the native genetic stock especially if the introduced organisms exhibit greater reproductive vigour. The problem becomes even more acute with genetically modified organisms. Here there are usually strict controls on the release. The conditions attached, however, may presume a much greater understanding of the natural ecosystem involved than is currently available.

According to Markham *et al.*[68] the "main aim of ex situ conservation for the future must be to provide a store of species and genetic material so that in future they may be returned to nature". Such reintroductions must be seen as part of the wider concept of restoration ecology, that is, the establishment of a group of species in numbers and proportions comparable to those occurring under natural conditions.[69] This is much easier said than done, however. Olwell[70] states the general view of experts that translocations should be regarded as experiments. He concludes that we know little about how to restore most endangered plants to natural habitats. Yet 25% of the US Fish and Wildlife service's plant recovery plans recommend the use of reintroductions.

As the techniques of artificial breeding and gene manipulation become more elaborate and more successful there are signs of complacency creeping into official thinking. Such thinking fails to appreciate that restoration ecology is very much in its infancy. Olwell[71] cites the opinion of experts that we are not yet ready, scientifically, to trade viable habitat of an endangered plant species for a reintroduced population. If the idea is that sensitive habitats do not have to be protected absolutely because they can be recreated elsewhere using translocation techniques backed up by releases from *ex situ* propagation or breeding, biodiversity will not be conserved. There are, unfortunately, strong signs that this is exactly what is intended.[72]

---

[68] *Op. cit.*, n. 10, at p. 129.

[69] Modified from the definition of "complete restoration" cited in Maunder and Ramsay, *op. cit.*, n. 51.

[70] P. Olwell, "Restoring Diversity: Strategies for Rare Plant Reintroductions" (1993) 7(3) *Plant Conservation* 1–2.

[71] *Ibid.*

[72] Article 6(4) of the Habitats Directive (*op. cit.*, n. 35), for example, will permit specially protected areas to be degraded for reasons of socio-economic importance provided that there is no satisfactory alternative and that compensatory measures are provided.

8

International Regimes for the Conservation and Control of Plant Genetic Resources

Gregory Rose

Introduction

Plant Genetic Resources

Animal life relies on plant life. Plants are the basis on which other living organisms build. The term "plant genetic resources" refers to the seeds, spores and other reproductive material by which plants propagate. This chapter examines legal aspects of the international conservation and control of these plant genetic resources (PGRs). It commences with an explanation of how human beings prioritise the conservation of particular PGRs in accordance with the need for human security – in terms of availability of food, essential materials and biosphere life support functions, as well as the rarity and substitutability of the particular PGRs.

The total amount of genetic resources increases gradually as species diversify. This is the natural process of evolution, a diversification amongst genes in living organisms, increasing the total number of unique genes. Maintaining biodiversity, widely construed, means maintaining the size of the pool of unique genes or even the natural rate of growth of the pool of unique genes.[1] These genes occur at all the taxonomic levels into which biologists classify living organisms; i.e. family, genus, species, subspecies (races) or individuals. However, for human society to value each individual living organism equally and strive for its preservation and reproduction is not economically possible.

Functional Priorities

All genetic resources may be considered as intrinsically valuable for their own sake as unique forms of life. However, other than within certain religious and ethical systems, plant genetic resources have usually been viewed from the economic

---

[1] J.H. Vogel, *Genes for Sale: Privatisation as a Conservation Policy* (1994).

International Law and the Conservation of Biological Diversity (C. Redgwell and M. Bowman, eds.: 90 411 0863 7: © Kluwer Law International: pub. Kluwer Law International, 1995: printed in Great Britain), pp. 145–169

perspective of their value to humans. This anthropocentric, rather than biocentric, biodiversity conservation ethic allocates animal organisms also dependent upon the maintenance of various PGRs a secondary importance.

Identification of the value of particular PGRs to humankind would be useful in order to allow a selective approach to conserving genetic resources to be adopted, as practical economic considerations might require that conservation efforts be prioritised. A corollary of this thinking is that the most useful PGRs are likely to be identified as those most necessary in order to ensure that security for humanity is maintained. An obvious priority is world food security. Other priority use values are to provide agricultural sources for industrial products, such as fibres, and construction materials, such as timber, as well as sources for medicines. From a more informed perspective, paramount importance might go to the less understood and obvious services provided by PGRs in maintaining biosphere functions, such as biogeochemical cycles, soil balances and the gaseous mix of the atmosphere. However, the dominant perspectives on the utility of PGRs, adopted by the applied science, commercial and political communities, emphasise food and agricultural security.

Substitutability

Those PGRs which are the most useful to humankind and which are the rarest or the least substitutable by other PGRs are the most valuable and deserving of conservation. In many cases it is difficult to know which plant genetic resources are the rarest or least substitutable around the world. There is little firm data available at a global level to enable prediction of the comparative usefulness of particular PGRs. Similarly, poor inventories concerning the rarity or substitutability of PGRs impede their relative evaluation. However, it can be said that those which are rare specimens higher up the taxonomic chart, e.g. are the sole representatives of a family rather than a genus, or of a species rather than a subspecies, are likely to be less substitutable as they have the fewest genetic characteristics in common with other organisms. Individual specimens in a healthy race are likely to be most easily substitutable and therefore have the least value as living organisms, from a utilitarian perspective.[2]

Control

Maintenance of as much diversity as possible will generally provide a better buffer against potential changes in environmental conditions which could detrimentally affect human security. For example, the Irish potato famine was the result of a blight on one plant genetic resource upon which a human population was singularly dependent. Despite hortatory statements concerning the need for conservation of PGRs, most international instruments and institutions dealing with PGRs are primarily concerned with the conditions of their control, distribution and utilisation, rather than with conservation.

For example, United Nations agencies, such as the UN Food and Agriculture Organisation, originally looked to conserve PGRs and distribute them amongst the peoples of the world for their common benefit. Commercial biotechnology companies have been concerned to ensure that plants which have been technologically improved upon, as compared to those in their natural state, are respected as private

[2] *Ibid.*

property. The divergence in approaches to control, distribution and utilisation remains a source of political tension requiring resolution. Unfortunately, it deflects attention from the essential common concern, which is the conservation of PGRs needed by all.

General Approaches

General, global approaches to conservation of PGRs are primarily those which deal with the conservation of biodiversity as a whole. Prominent organisations in this work are the UN Environment Programme (UNEP), the International Union for the Conservation of Nature (IUCN) and the UN Conference on Environment and Development (UNCED).

Agenda 21

In terms of legal product, the most remarkable achievement is that of UNCED, which concluded Agenda 21, an action plan for sustainable development in the twenty-first century. This forty-chapter instrument includes several chapters relevant to the conservation of PGRs. Its mention here is merely cursory; however, it must be noted that Chapter 14, Programme G, deals with "Conservation and Sustainable Utilisation of Plant Genetic Resources for Food and Sustainable Agriculture". Chapter 14 sets out many of the activities necessary for the preservation of PGRs which are discussed in this paper and estimates an annual cost for these activities of about US$600 million, including US$300 million on grant or concessional terms. Several inter-State programmes have emerged relatively recently, during the same period that Agenda 21 was being developed. For example, the USA–Japan Environmental Partnership provides US$20 million annually from 1994 to 1997 to establish Natural Resource Conservation and Management Centres in Asia, some of which may undertake biodiversity prospecting.[3] Similarly, in 1992 the USA National Institutes of Health, National Institute of Mental Health, National Science Foundation and Agency for International Development established a programme to fund "International Cooperative Biodiversity Groups" designed to "promote conservation of biological diversity, through the discovery of bio-active agents from natural products and to ensure that equitable economic benefits from these discoveries accrue to the country of origin".[4]

Also relevant is Chapter 15 of Agenda 21, which relates to biodiversity conservation and calls on governments to develop measures and arrangements to help biodiversity source countries and their peoples share the benefits of the commercialisation of genetic resources.

Agenda 21 is a "soft law" instrument, not legally binding on signatories and its effect on State behaviour is yet to be fully assessed. It might be described as a reflection of, as much as a catalyst for, general efforts to conserve PGRs at the international level. It indicates that increasingly important activities in the conservation of PGRs are being undertaken at the inter-State level.

---

[3] W.V. Reid, S.A. Laird, C.A. Meyer, R. Gomez, A. Sittenfeld, D.H. Janzen, M.A. Gollin and C. Juma, *Biodiversity Prospecting: Using Genetic Resources for Sustainable Development* (1993), at p. 26.

[4] The National Institutes of Health, National Institute of Mental Health, National Science Foundation and Agency for International Development, "International Cooperative Biodiversity Groups: Request for Applications", 12 June 1992; cited *ibid.*, at p. 51.

UN Convention on Biological Diversity

Negotiations for a Convention on Biological Diversity[5] began in November 1990 under the auspices of the UNEP, with the assistance of the IUCN, and resulted in a final version agreed in May 1992. It was signed by 157 nations at the United Nations Conference on Environment and Development (UNCED) in Rio in June 1992 and came into force ninety days after the thirtieth ratification, on 29 December 1993. At the end of September 1994, there were eighty-four parties and the first meeting of the Conference of the Parties is due to be held from 28 November to 9 December 1994. The Convention is to be supplemented by protocols as and when necessary. A protocol on biosafety, dealing with the release of genetically modified organisms, and a protocol based on the International Undertaking on PGRs (discussed below) are under discussion.

Conservation

The Convention's overall objectives are stated in Article 1 as being the conservation of biological diversity, the sustainable use of its components and the fair and equitable sharing of the benefits arising out of the utilisation of genetic resources. Parties are obliged, in broad terms, to develop national strategies for conservation and sustainable use of biodiversity, promote *in situ* and *ex situ* conservation, introduce environmental impact assessments for projects likely to have significant adverse effects on biodiversity and periodically submit reports on the measures taken to implement these obligations.[6] Overall, the Convention promotes free trade in genetic resources as part of an arrangement that also provides a mechanism for financing biodiversity conservation and obliges developed countries to finance those activities and to promote the transfer of technology. It is important to note that the Convention is directed towards the conservation of all genetic resources, not only plants, and that its source is in a conservation ethic, rather than in the search for agricultural security.

Access

The Convention rejects the notion of genetic resources being a common good and Article 3 recognises a nation's sovereign right to exploit them as national resources, subject to the proviso that these activities do not affect the environment of other countries.[7] Article 15 restates this principle of national sovereignty but commits nations to facilitate access to their domestic resources for environmentally sound uses. It stipulates that access should be on "mutually agreed terms" and will generally require a nation's "prior informed consent". Articles 15.1, 15.2, 15.4 and 15.5 provide that:

1. Recognising the sovereign rights of States over their natural resources, the authority to determine access to genetic resources rests with the national governments and is subject to national legislation.

---

[5] (1992) 31 ILM 822.

[6] Articles 6, 8, 9, 14 and 26 of the Convention on Biological Diversity.

[7] Article 3 provides:

> States have, in accordance with the Charter of United Nations and the principles of international law, the sovereign right to exploit their own resources pursuant to their own environmental policies, and the responsibility to ensure that activities within their jurisdiction or control do not cause damage to the environment of other States or areas beyond the limits of national jurisdiction.

2. Each Contracting Party shall endeavour to create conditions to facilitate access to genetic resources for environmentally sound uses by other Contracting Parties and not to impose restrictions that run counter to the objectives of the Convention.
3. Access, where granted, shall be on mutually agreed terms and subject to the provisions of this Article.
4. Access to genetic resources shall be subject to prior informed consent of the Contracting Party providing such resources, unless otherwise determined by that Party.

A potential source of confusion is that Article 15 refers only to "genetic resources". This term, as defined in Article 2, appears not to extend to derived chemical extracts. Such extracts could therefore be construed as being excluded from the access provisions despite this interpretation being contrary to the Convention's objectives.

Another concern is that the Convention covers only the future removal of genetic resources, either directly from the country of origin or via other countries that have acquired the genetic material in accordance with the Convention. It excludes the actions of countries that have previously acquired genetic resources (e.g. seeds held in established gene banks) or that go on to acquire such resources from countries that are not parties to the Convention. Article 15.3 states:

> For the purpose of this Convention, the genetic resources being provided by a Contracting Party ... are only those that are provided by Contracting Parties that are countries of origin of such resources or by the Parties that have acquired the genetic resources in accordance with the Convention.

To address this hiatus, Resolution 3 of the Conference for the Adoption of the Agreed Text of the Convention on Biological Diversity (the "Nairobi Final Act"),[8] concerning the interrelationship between the Convention and the promotion of sustainable agriculture, requested that the matter be dealt with within the context of the Global System for Plant Genetic Resources (discussed below) and that complementarity and co-operation be sought between the Convention and that Global System.[9]

Owing mainly to pressure from the USA during the final negotiations over the text of the Convention, there were important qualifications made to the technology transfer provisions of Article 16. Phrases such as "as appropriate", "where mutually agreed" and "consistent with the adequate and effective protection" of existing rights seem severely to qualify the obligations on commercial seed development companies in developed countries to give up patented technologies and trade secrets as part of a reciprocal deal for access to PGRs in developing countries. In addition, a number of countries have used interpretative statements to clarify and qualify their positions on the meaning of ambiguous articles in the Convention. These statements tend to give restrictive meanings to technology transfer and funding provisions. For example, the USA has expressed its intention to ratify the Convention by the end of 1994 and to file an accompanying interpretative statement concerning technology transfer, intellectual property rights, research, funding, sovereign immunity and sovereign control over PGRs.

---

[8] (1992) 31 ILM 822.
[9] *Ibid.*

Despite its shortcomings, the Convention on Biological Diversity is having a far reaching effect on the management of PGRs. By establishing in binding legal terms at global level the principle of State sovereign control over access to its PGRs, the Convention has focused attention on the universal value of those PGRs for food, agriculture and medicine. This encourages countries which control such resources to seek ways to conserve them, so as to realise their value in monetary terms, and encourages other countries which might find their access to such resources limited by restrictive regulations or resource erosion to co-operate in management regimes, so as to ensure the future availability of and access to such resources. In particular, the Convention has stimulated work in international organisations dealing with management of PGRs, such as the Food and Agriculture Organisation of the United Nations.

UN Food and Agriculture Organisation

The Food and Agriculture Organisation (FAO) is now the primary organisation carrying responsibility for global conservation of plant genetic resources. Historically, it had been bypassed for reasons of political convenience in favour of non-UN organisations. However, as developing countries express greater interest in asserting control over their PGRs, and as they express this principally through the UN system and in relation to PGRs important to food and agriculture, PGR conservation activity in the FAO has been reinvigorated.

The FAO first addressed genetic resources as early as the opening meeting of its Committee on Agriculture in 1946. However, little action was taken in those early days.[10] In 1957 the first specialised international newsletter on crop genetic resources was published. In 1961 a conference on plant genetics was held in conjunction with the International Biological Programme (IBP) and an International Technical Conference on Plant Genetic Resources was held in 1967. In 1968 a crop ecology unit was created and, a little later, the Plant Production and Protection Division became active in the area of plant genetic resource conservation.[11] The 1970s saw a great increase in global activity, but a great deal of it sidestepped the FAO, taking place in parallel developments outside the UN system (these developments are discussed below under the headings Gene Banks and Botanic Gardens). Nevertheless, the FAO held Second and Third International Technical Conferences on PGRs in 1973 and 1981. It was only in the 1980s that significant international legal measures to stem the erosion of PGRs came into the FAO fold and the FAO Global System on Plant Genetic Resources was established.

Since its biannual peak Conference in 1983, the FAO has been developing a Global System on Plant Genetic Resources aimed at promoting the conservation and sustainable utilisation of PGRs by the international sharing of the benefits and burdens of conservation. The system consists essentially of: (a) a flexible legal framework (the International Undertaking); and (b) an inter-governmental forum (the Commission on Plant Genetic Resources). By November 1993, 140 countries were part of the Global System either as members of the Commission[12] or through adhering to the International Undertaking.[13]

---

[10] C. Fowler and P.R. Mooney, *The Threatened Gene: Food, Politics and the Loss of Genetic Diversity* (1990), at p. 149.

[11] *Ibid.*

[12] *Report of the 27th Session of the FAO Conference*, FAO Doc.C93/REP/5, at p. 6.

[13] In March 1994 107 countries were signatories to the International Undertaking.

FAO Commission on Plant Genetic Resources

The intergovernmental forum under the FAO Global System is the Commission on Plant Genetic Resources (CPGR). It was established by the FAO Conference in 1983.[14] It reviews international developments, promotes the progress of the Global System, and develops international agreements and instruments necessary to facilitate PGR conservation and use. The CPGR is open to all FAO members and associate members and makes its decisions by consensus, although a "one country, one vote" approach to decision making can be taken where necessary.[15] The CPGR meets biannually and operates intersessionally through a Working Group of twenty-three members who are elected on a geographically representative basis. The terms of reference of the Working Group are presently being reviewed but are, broadly, to consider the progress made in implementing the Commission's programme of work and any other matters referred to it by the Commission.[16]

The CPGR is the expert body undertaking preparatory information collection, research, formulation and adoption of documents preliminary to FAO deliberation. The distribution of work on PGRs between the CPGR and the FAO itself is not always very clear. Much of the CPGR work is orientated around information collection and redistribution. However, it also has a major role in catalysing and co-ordinating PGR conservation activities. The FAO Conference itself is then the forum within which controversial political matters are played out. The CPGR has some determinative authority of its own but is a subsidiary body. It is responsible for the periodic publication of a report on the state of erosion of the world's PGRs, early warning of major genetic erosion problems, development of a global plan of action for adoption by the FAO concerning the co-ordination of conservation of PGRs, and preparation for a Fourth FAO International Technical Conference on PGRs.

Periodic Publication on the State of the World's PGRs

Signatories to the International Undertaking on Plant Genetic Resources (see below) are obliged to report annually to the FAO on measures they have taken or propose to explore, preserve, evaluate and make available PGRs.[17] The CPGR Secretariat has also prepared questionnaires concerning the national situation of PGR conservation and sent them to all FAO members for responses.[18] Based largely on these responses and annual reports, an FAO report on *The State of the World's Plant Genetic Resources* will be compiled.[19] Covering all aspects of conservation and the utilisation of PGRs, at the request of the CPGR, such periodic reports are to identify programmes being carried out by regional, international and non-governmental organisations.[20] The reports are intended to expose gaps, constraints and emergency situations so as to guide the CPGR's future discussions and provide an authoritative base for the preparation of an FAO Global Plan of Action.[21] The first

---

[14] Resolution C 9/83, *Report of the 22nd Session of the FAO Conference*, FAO Doc.C/83/REP (1983).

[15] J. Esquinas-Alcazar "The Global System on Plant Genetic Resources" (1993) 2 RECIEL 153.

[16] CPGR/94/WG9/7, *Draft Terms of Reference of the Working Group*.

[17] See Article 11, International Undertaking on PGRs, below.

[18] CPGR 91/5.

[19] CPGR 91/REP, *Report of the 4th Session of the Commission on Plant Genetic Resources* (1991).

[20] Esquinas-Alcazar, *op. cit.*, n. 15, at p. 157, n. 14.

[21] CL 103/16, *Report of the 5th Session of the Commission on Plant Genetic Resources* (1993), at pp. 3–4.

State of the World's PGRs report will be prepared during the preparatory process for the Fourth International Technical Conference in 1996, under the guidance of a regionally balanced group of experts and in co-operation with other relevant organisations, such as the International Plant Genetic Resources Institute.[22]

Early Warning System

Flowing from the information synthesis described above, an "Early Warning System" is intended to be put in place. As indicated in Agenda 21, Chapter 14 Programme G, on conservation of PGRs, in some instances the loss of plant genetic diversity is as great in gene banks as it is in the field, owing to inadequate security of the PGR holdings.[23] The Early Warning System is to draw rapid attention to specific hazards threatening the operation of gene banks and to threats of extinction of plant species.[24]

Global Plan of Action on PGRs

The CPGR Global Plan of Action on PGRs is intended for adoption by the FAO in 1996. It will ensure co-ordination within a clear global framework of the conservation activities for PGRs of major national and international agencies as well as non-governmental organisations. Thus, responsibilities will be divided so as to avoid duplication of effort and to promote efficient use of funds amongst the organisations involved in the implementation and financing of the Plan. Financing will take place in accordance with provisions outlined in the International Undertaking on PGRs (see below) and Agenda 21. Although the Plan of Action will be prepared and adopted by the FAO, the CPGR requested the FAO to prepare it and the CPGR will periodically review and update it, as well as supervise its implementation.[25] The Plan itself is intended to be a political instrument and will not be legally binding on any member of the FAO or CPGR.

Fourth International Technical Conference

Following on from the FAO 1967, 1973 and 1981 technical conferences, a Fourth International Technical Conference on PGRs is planned for 1996. The CPGR has stated that the aims of the Conference are to make the Global System fully operative and to develop the relevant parts of Agenda 21 and the Convention on Biological Diversity into concrete, costed programmes, projects and activities that would be integrated into the Global Plan of Action.[26]

The first report on the *State of the World's PGRs* and the Plan of Action assembled during the preparatory process will be presented at this Conference. Solutions to the questions of how to treat *ex situ* collections not acquired in accordance with the Convention on Biological Diversity are to be sought. Of particular note is the "bottom-up" approach to be adopted. The Conference is to be intended to be "country driven" and based on broad participation from governments, scientific institutes, non-governmental organisations, and local and farmers' communities.[27]

---

[22] Esquinas-Alcazar, *op. cit.*, n. 15 at p. 157, n. 16.
[23] See also C. Fowler and P.R. Mooney, *op. cit.*, n. 10, at pp. 163–165.
[24] Esquinas Alcazar, at p. 155.
[25] *Ibid.*
[26] *Op. cit.*, n. 21, at p. 13.
[27] Esquinas-Alcazar, at p. 55; CPGR/94/WG9/5, *Progress report on the International Conference and Programme on the Conservation and Utilization of Plant Genetic Resources.*

In Situ Conservation

The FAO, in co-operation with other organisations, is developing a network of *in situ* conservation areas. This work focuses primarily on conserving wild relatives of cultivated plants, the promotion of "on farm conservation" and the utilisation of land races. A major priority is seen as the training of national experts.[28]

Legal Codes

In April 1993 the CPGR endorsed a draft International Code of Conduct for Plant Germplasm Collecting and Transfer.[29] This was presented to the 27th Session of the FAO Conference in November 1993 for final approval and adopted there.[30] In April 1993, the CPGR also discussed a preliminary draft Code of Conduct for Biotechnology as it affects the conservation and utilisation PGRs.[31] (The Codes are considered below, in relation to the legal framework.)

FAO Legal Framework for PGRs

The International Undertaking on Plant Genetic Resources was adopted, like the decision creating the CPGR, by the FAO Conference in 1983.[32] Unlike the Convention on Biological Diversity, which is directed towards the preservation of all biological resources, plant and animal, irrespective of their socio-economic function, the International Undertaking is concerned with the preservation of plants for agriculture. It is aimed at ensuring that PGRs, especially those species of present or future economic and social importance, are conserved, utilised and made available for plant breeding and other scientific purposes. The International Undertaking contains provisions dealing with the exploration and collection of genetic resources (Article 3), preservation, evaluation and documentation *in situ* and *ex situ* (Article 4), international co-operation in conservation, exchange and plant breeding (Article 6), international co-ordination of gene bank collections and information systems (Article 7), conservation and management activities funding (Article 8), activities monitoring by the FAO (Article 9) and the maintenance of phytosanitary measures for plant protection (Article 10). Another fundamental difference between the Convention on Biological Diversity and the International Undertaking on PGRs is that the Convention is a legally binding instrument, whereas the Undertaking is not.

The Undertaking has been dogged by controversy since its adoption as it deals with the politically explosive issues of PGR control. Its effectiveness has been hampered by the competing interests of the (mostly developing) countries which have a natural abundance of PGRs and wish to maintain control over them, and, on the other hand, the (mostly developed) countries which have made capital investments in refining PGRs and wish to exercise control themselves. Each also desires unhindered free access to the others' holdings of PGRs.

---

[28] *Op. cit.*, n. 21, at p. 12
[29] CL 103/16-Sup. 1, "Agreed Text on Draft International Code of Conduct for Plant Germplasm Collecting and Transfer".
[30] Resolution 8/93, *Report of the 27th Session of FAO Conference*, FAO Doc.C/93/REP/5, (1993).
[31] *Op. cit.*, n. 21, at p. 10
[32] Resolution 8/83, *op. cit.*, n. 14.

Access

The Undertaking was originally based on the principle that PGRs are part of the shared "heritage of mankind". Article 1 expresses the Objective of the Undertaking as follows: "The Undertaking is based on the universally accepted principle that plant genetic resources are a heritage of mankind and consequently should be available without restriction." Whilst "heritage of mankind" is not defined, the Undertaking makes it clear that this means that the world's PGRs should be freely available to all. Article 5 (Availability of Plant Genetic Resources) states:

> It will be the policy of adhering Governments and institutions having PGRs under their control to allow access to samples of such resources, and to permit their export, where the resources have been requested for the purposes of scientific research, plant breeding or genetic resource conservation. The samples will be made available free of charge, on the basis of mutual exchange or on mutually agreed terms.

In Article 2, plant genetic resources are defined as the reproductive or vegetative propagating material of the following categories of plants:

i.   cultivated varieties (cultivars) in current use and newly developed varieties;
ii.   obsolete cultivars;
iii.   primitive cultivars (land races);
iv.   wild and weed species, near relatives of cultivated varieties;
v.   special genetic stocks (including elite and current breeders' lines and mutants).

In this original form, the Undertaking seems to imply that all genetic resources (including the refined elite lines of plant breeders in developed countries) are the property of all and should therefore be freely accessible. Whilst most developing countries signed the Undertaking, many developed countries refused to sign since they felt that their commercial plant breeders would be harmed by the loss of property rights in the elite lines.[33]

To deal, *inter alia*, with these concerns, three Resolutions were later adopted unanimously at the FAO Conferences of 1989 and 1991 and added to the Undertaking as Annexes.[34] These qualify the principle of PGRs being common heritage in several ways, i.e. by: (a) asserting sovereign rights of countries over their PGRs; (b) clarifying that free access does not necessarily mean free of charge; (c) limiting the benefit of free access to those adhering to the Undertaking; and (d) limiting the scope of free access to exclude breeders' lines and farmers' breeding material.[35]

A balance was sought to be achieved between exclusive property rights and general access to, on the one hand, modern commercial biotechnological products, such as breeders' lines, and traditional farmers' and wild varieties on the other. As described below, legal systems for the protection of commercial biotechnological property are already well established, but not for the protection of traditional farmers' varieties and wild varieties in developing countries.[36]

---

[33] Developed countries which did not adopt the International Undertaking in 1983 were Canada, France, Germany, Japan, New Zealand, Switzerland and USA.

[34] Resolutions 4/89 (1989), 5/89 (1989), and 3/91 (1991). These resolutions were negotiated by both members and non-members of the CPGR, including also countries which had and had not signed the International Undertaking. (CPGR/94/WG/2, *Revision of the International Undertaking: Background and the Step by Step Process*, at p. 2.)

[35] *Ibid.*

[36] *Ibid.*

On the matter of access, it should also be noted that the International Undertaking and the Convention on Biological Diversity deal with access to PGRs in similar but divergent ways. Both provide for access on mutually agreed terms and leave open the question of whether the terms should be agreed on a bilateral or multilateral basis. However, the emphasis in the International Undertaking is on multilateral solutions and international institutional mechanisms, whilst the Convention is geared more towards bilateral agreements.[37] Further, the International Undertaking deals with access to all PGRs for food and agriculture, whereas the Convention on Biological Diversity does not address access to existing *ex situ* collections except in the country of origin, i.e. it does not address existing collections in foreign gene banks. As noted above, Resolution 3 of the Conference for the Adoption of the Agreed Text of the Convention on Biological Diversity calls for a solution to this gap in the Convention's subject treatment within the FAO Global System.

Farmers' Rights

To meet the demands of developing countries and traditional farmers to property rights as the originators of most of the earth's current stock of PGRs, the International Undertaking Annexes recognise and loosely define the notion of Farmers' Rights. Resolution 4/89 recognises the "enormous contribution that farmers of all regions have made to the conservation and development of plant genetic resources" whilst Resolution 5/89 defines their rights as "arising from the past present and future contributions of farmers in conserving, improving and making available plant genetic resources, particularly those in the centres of origin/diversity". In essence, Farmers' Rights are a collectively held benefit, allocated to traditional farmers mostly in developing countries, in reward for their past work in the conservation and development of PGRs.

Farmers' Rights are quite different from modern, private commercial property rights in that they are not allocated directly to individuals. They are for the common benefit of farmers and farming communities in all areas of the world, especially in centres of origin/diversity of PGRs, and are to benefit them by ensuring the protection and conservation of PGRs and of the biosphere, and by ensuring that they participate fully in the improved use of PGRs through plant breeding and other scientific methods.[38] Responsibility for administration of this common benefit is "vested in the International Community, as trustee for present and future generations of farmers"[39] and is to be implemented by it, in particular, through an International Fund for PGRs monitored by the CPGR.[40] Resolution 3/91 provides that the International Fund resources should be substantial, sustainable and based on principles of equity and transparency.[41]

---

[37] *Ibid.*, at p. 3. Suggested complementary options for provision of access to PGRs are (a) international framework agreements to facilitate bilateral exchanges; and (b) multilateral agreements on the availability and conditions of access to *in situ* and/or *ex situ* collections. (CPGR/94/WG9/4, *Issues for Consideration of "Stage II": Access to Genetic Resources and Farmers' Rights*, at p. 5.)

[38] Resolution 5/89, *Report of the 24th Session of the FAO Conference*, FAO Doc.C89/REP (1989).

[39] Resolution 5/89, *ibid.*

[40] Resolution 4/89, *ibid.*

[41] The Fifth Session of the CPGR (April 1993) agreed that questions which remain to be answered in relation to the International Fund include: the nature of funding (voluntary or mandatory); linkage between financial responsibility and benefit from the use of PGRs; whether countries, uses or consumers should bear financial responsibilities; how the relative needs of beneficiaries (especially developing countries) are to be estimated; and how farmers and local communities would benefit from the funding: *op. cit.*, n. 21.

The emphasis of the work of the International Fund is to be on development of capacities in developing countries' PGR conservation and management. Agenda 21 Chapter 14 gives an idea of the amount required for the International Fund for this purpose, i.e. US$600 million per annum, half of which would be on grant or concessional terms. Thus, when operative, the Global Plan of Action could be financed through the International Fund, which would redistribute the benefits of PGR utilisation back to developing countries but in a manner ensuring that the benefits go to conservation and management, rather than highways and power stations.[42]

Unfortunately, the political accommodations restricting access to commercial biotechnology and creating complementary Farmers' Rights have resulted in a document which is highly ambiguous. Annex III acknowledges that many consequential questions remain unanswered and notes that: "conditions of access to plant genetic resources need further clarification".

Article 11 of the International Undertaking requires States to provide information to the FAO at yearly intervals on the measures that they have taken or propose to take to achieve the objectives of the Undertaking. However, as it is a vague and non-binding agreement with no other enforcement mechanism, many of the provisions in the Undertaking have been simply ignored or breached. For example, there has been no progress in setting up the International Fund and very little progress in ascribing definite Farmers' Rights.

Nevertheless, the UN Conference on Environment and Development and the adoption of the Convention on Biological Diversity have accelerated the pace of developments in the FAO. In April 1993 a resolution agreed by the Commission on Plant Genetic Resources called for the revision of the Undertaking, beginning with integration of the three Annexes into the main text. The CPGR agreed that the Undertaking should be made consistent with the UN Convention on Biological Diversity 1992 and noted that the FAO might, if requested, revise the Undertaking as a legally binding document which might take the form of a protocol to the Convention.[43] This resolution has since been endorsed by the FAO biannual conference in November 1993 and work is currently under way through the Working Group of the CPGR.[44]

Code of Conduct for Plant Germplasm Collecting and Transfer

The Code of Conduct for Plant Germplasm Collecting and Transfer aims to regulate the collection and transfer of PGRs so as to facilitate access and sustainable utilisation, and prevent genetic erosion. It also includes reporting provisions to allow the CPGR to monitor its implementation. It was formally adopted in November 1993 at the FAO biannual conference.[45]

---

[42] The Fifth Session of the CPGR noted that no agreement had been reached on the nature of contributions to the International Fund (*ibid.*). However, progress has been made in another forum. The Keystone International Dialogue, in which governments, industry, international organisations and non-governmental organisations participated, supported the concept of an International Fund to implement Farmers' Rights. It agreed to a mandatory fund of a minimum of US$500 million pa, or US$1.5 billion from 1993–2000. (*Report of the Third and Final Session of the International Dialogue*, Oslo 1991.)

[43] *Op. cit.*, n. 21.

[44] *Report of the 27th Session of the FAO Conference*, FAO Doc.C/93/REP/5 (1993).

[45] *Ibid.*

The Code of Conduct recognises the rights of local farming and indigenous communities to the PGRs they maintain. It reiterates the view expressed in the modified International Undertaking that nations have sovereign rights over their PGRs, but that PGRs should be made readily available. Article 3.2 provides:

> The code recognises that nations have sovereign rights over their PGRs in their territories and it is based on the principle according to which the conservation and continued availability of plant genetic resources is a common concern of mankind. In executing these rights, access to plant genetic resources should not be unduly restricted.

At the same time, the Code of Conduct on Plant Germplasm Collecting and Transfer re-emphasises the need to share the benefits derived from PGRs between the source States and the collectors or users of germplasm by suggesting ways in which the collectors or users may pass on a share of the benefits to the donors, recognising the rights and needs of local communities and farmers.

Codes of Conduct are non-binding instruments that merely set standards which States may adhere to as they wish. They do not normally require any formal adherence process. However, it is common practice for States to inform the FAO of their intention to apply one of its Codes of Conduct. In some cases (such as the FAO Code of Conduct on the Distribution and Use of Pesticides), implementation mechanisms have involved the FAO asking governments to monitor and periodically report back to it concerning their observance of the Code. The Code of Conduct for Collecting and Transfer of PGRs provides that in the case of non-observance by a collector of the principles of the Code or the rules and regulations of the host country, the host country can report the matter to the CPGR. Any such offence against the Code would jeopardise an offending State's entitlement to a certificate from the FAO saying that there are no unresolved complaints outstanding against it.[46]

It should be noted that related regional developments outside the FAO may promote the implementation of the Code of Conduct. The Manila Declaration Concerning the Ethical Utilisation of Asian Biological Resources[47] recommends the development of adequate national legislation to exercise control over the collection and export of biological material. Appendix 1 to the Manila Declaration is a Code of Ethics for Foreign Collectors of Biological Samples which recommends a set of actions by foreign collectors to ensure that the developing country signatories which provide biological samples are not disadvantaged by the transfer of PGRs.[48]

Code of Conduct for Biotechnology

An FAO Code of Conduct for Biotechnology is currently being drafted. It promotes the development of appropriate biotechnologies for the conservation and sustainable utilisation of PGRs, the promotion of bio-safety standards and the equitable sharing of the benefits between the developers of the technology and the donors of the germplasm. The legal status of this Code of Conduct as an instrument of soft law is the same as that for the Code of Conduct on Collecting and Transfer, discussed above.

---

[46] E. Canal-Forgues, "Code of Conduct for Plant Germplasm Collecting and Transfer" (1993) 2 RECIEL 171.

[47] Declared at the Seventh Asian Symposium on Medicinal Plants Spices and other Products (ASOMPS VI), Manila 2–7 February 1992, attended by 283 scientists from thirty-one countries.

[48] Appendix 2 is a set of Contract Guidelines designed to promote minimum standards to achieve equity in partnerships concerning transfer of genetic resources between developed countries and the country of origin.

FAO Agreements with Gene Banks

Draft basic agreements between the FAO and its Member States are being prepared to establish a network of gene banks (see below). These will bring gene banks previously under national control within the auspices of the FAO. The effect of the agreements will be to improve international co-ordination in conservation efforts and also to ensure that collections of PGRs are maintained on the basis of unrestricted international access.

Gene Banks

The principal method for *ex situ* conservation of PGRs is through gene banks, which store plants in their dormant form, as seeds. Gene banks range in sophistication from well catalogued collections in climate-controlled storage to poorly sorted bags held in humid barns. The most important tend to be public institutions such as dedicated international centres, national centres and public research institutes. Private collections can also play a role.

International Gene Banks

Prominent international gene banks are those in the network of International Agricultural Research Centres (IARCs) supported by the Consultative Group on International Agricultural Research (CGIAR). The Washington-based CGIAR was established in 1971 by sponsoring agencies such as the Ford and Rockefeller funds[49] to co-ordinate the work of four already established research centres and to extend the scope, reach and effectiveness of agricultural research, promoting the "green revolution".

The structure of the CGIAR itself is an informal association of forty public and private sector donors, research centres and non-donor representatives from developing countries (where much of the research is carried out). The donors consist of a consortium of countries, private foundations and regional development banks, the World Bank, UN Development Programme and FAO.[50] The CGIAR is chaired by the World Bank.

Today there are, in all, eighteen IARCs with a total annual budget of over $300 million in the CGIAR network. Their work is broadly divided into six categories: (a) productivity research, (b) management of natural resources, (c) improvement of the policy environment in developing countries, (d) research institution building in developing countries, (e) germplasm conservation, and (f) building of linkages between agricultural research institutions around the world. In 1990 and 1991, CGIAR decided to reorient towards more work on sustainable development. In order to maintain the productivity of natural resource bases it therefore decided to focus on protecting the diversity of PGRs, dealing with soil degradation, addressing climate change, promoting growth in less productive areas, and management of pests and nutrients in ways that would reduce dependence on agricultural chemicals. This new approach will result in expansion of the system to include new international centres.[51] For example, a Centre for International Forestry Research was established in Bogor, Indonesia in early 1994.

---

[49] J.R. Kloppenberg, *First the Seed* (1988), at p. 160.
[50] *Ibid.*, at p. 164.
[51] CGIAR, *Geneflow: A Publication About the Earth's Plant Genetic Resources* (1992).

Located within the FAO in Rome, but still constituted outside of the UN system, is the International Plant Genetics Resource Institute (IPGRI, formerly known as the International Board for Plant Genetic Resources). It was established in 1974 and is perhaps the most important of the IARCs, being the most concerned with active conservation of PGRs. IPGRI co-ordinates the activities of the other IARCs and the establishment of regional centres for the conservation of PGRs, as well as providing financial assistance for other non-CGIAR conservation facilities.[52]

Most IARCs have specific local or regional responsibilities for germplasm conservation, with a few also collecting specific crops on a worldwide basis.[53] The CGIAR has also provided expertise and financial assistance to institutions which are not CGIAR research centres. Eighteen IARCs and 227 non-IARC seed banks in ninety-nine countries now hold 90% or more of the known landraces for important commercial crops such as corn, oats and potatoes. The IARCs have an estimated 61,000 accessions in storage[54] (which amounts to 16% of unduplicated world holdings)[55] and the world's largest collection of genetic resources.[56]

The work of the CGIAR has, however, been subject to criticism from some developing countries and conservationists as tending to wrest control of PGRs from developing countries and traditional farmers and delivering it to developed countries and commercial agricultural interests.[57] As the CGIAR stands outside of the UN system, it can be accused of being answerable to donor countries and their agri-industrial interests as much as to international food and agricultural security concerns.

In theory, the CGIAR network recognises the principle of open access, and requires that genetic material held in its network should be available on a free and unrestricted basis for all bona fide users. The CGIAR describes itself as trustee of global germplasm holdings with the whole world community as beneficiaries.[58] Accordingly, CGIAR members state that they will not take out intellectual property protection over any germplasm in their possession. CGIAR also declares that plant genetic material will be released to private users only on the basis of agreements that require that the users negotiate with the CGIAR if derived varieties or genes isolated from the material were to be protected and used commercially. These transfer agreements are designed to ensure that any useful plant genetic material could not be withheld from the source country and that the CGIAR could not be prevented from using that material for the benefit of developing countries.[59]

However, in practice, there are a number of problems with this policy. The meaning of CGIAR trusteeship is not clear and a number of commentators have questioned whether it has the legal capacity to hold anything on trust.[60] Its trusteeship is further thrown into doubt by the nature of the relationship of every IARC

---

[52] S. Johnston, "Conservation Role of Botanic Gardens and Gene Banks" (1993) 2 RECIEL 175.

[53] Ibid., at p. 176.

[54] D. Van Sloten, "IBPGR and the Challenges of the 1990s: a Personal Point of View" 6 Diversity 38; cited in B. Groombridge, Global Biodiversity: Status of the Earth's Living Resources (1992), at p. 557.

[55] Op. cit., n. 52.

[56] Rural Advancement Fund International (RAFI), A Report on Germplasm Embargoes, (RAFI Communiques, September 1988), at p. 1.

[57] Fowler and Mooney, op. cit., n. 10, at p. 150.

[58] CGIAR, On Intellectual Property, Biosafety and Plant Genetic Resources, CGIAR Discussion Paper (1992), at p. 2; cited in Johnston, op. cit., n. 52, at n. 30.

[59] Ibid.

[60] P. Mooney, "The Law of the Seed: Another Development and Plant Genetic Resources", in Development Dialogue 1–2, (1983), at p. 72; cited in Johnston, op. cit., n. 52, at n. 30.

with its host government. Further, the IARC legal constitutive documents tend not to make explicit provisions governing the ownership of the PGRs held. Typically, if an IARC's existence is terminated, its assets (including plant genetic material) become the property of the host national government.[61] Problems also arise from CGIAR gene banks not possessing sovereign immunity and so being subject to the laws of the country of location. Thus, if a country of location decides to restrict access to gene banks, the IARC is obliged to comply. The US Government, for instance, has stated that any material so received would become national property and it has admitted that political considerations have dictated a US policy of exclusion from access for a few countries.[62]

Article 7 of the International Undertaking requires the development of an international network of base collections in gene banks under the auspices and/or jurisdiction of the FAO to hold PGRs for the benefit of the international community on the basis of unrestricted access. At its second session (in 1985) the CPGR considered legal arrangements to establish the international network required under Article 7 of the Undertaking.[63] In 1988 the FAO invited the IARCs of CGIAR to place their collections under its auspices.[64] In April 1993, at the fifth session of the CPGR, the IARCs agreed to do so. A Basic Agreement for International Agricultural Research Centres is now being formulated and seeks to clarify matters concerning the concepts of trustee and beneficiary as they relate to the concept of ownership, obligations with respect to conservation of germplasm and its availability, the policy role of the CPGR and the duration of these Basic Agreements.[65]

National Gene Banks

In addition to the international co-ordination of conservation through the IARCs of CGIAR, there are important national systems for the conservation of PGRs.

The US National Plant Germplasm System is a sophisticated and diffuse network managed by the US Department of Agriculture together with State and private institutions, private industry and individuals. The system is overseen by the National Plant Genetic Resources Board which co-ordinates the acquisition, preservation, evaluation and distribution of national germplasm resources. However, the Department of Agriculture's policy on germplasm acquisition tends to limit the collection of PGRs to those of interest to US scientists and so the collection is predominantly composed of species of recognised economic value.[66]

An FAO Legal Office Study has shown that ownership of PGRs in governmental or public institutional gene banks is vested in the State in which the gene bank is located. Whilst such national gene banks fall under explicitly national controls and many of these restrict access, the scope and method of the controls varies greatly. Some examples of the controls placed upon access to their collections by national gene banks are given below. By the time of the fifth Session of the CPGR in 1993,

---

[61] J.R. Kloppenburg (ed.), *Seeds and Sovereignty: the Use and Control of Plant Genetic Resources*, at p. 251.

[62] Communication between T.W. Edministor, Administrator, Agriculture Research Service, US Department of Agriculture and R.H. Demuth, Chairman, IBPGR (19 January 1977), cited in Mooney, *op. cit.*, n. 60, at pp. 30–31.

[63] CPGR/85/6.

[64] CPGR Circular State Letter 1988.

[65] CPGR/94/WG/2, *Revision of the International Undertaking: Background and the Step by Step Process*, at p. 2.

[66] Johnston, *op. cit.*, n. 52, at pp. 178–179.

thirty-one countries had indicated their willingness to make their gene banks part of the International Network.[67]

(a) Ecuador[68] – cocoa germplasm export from Ecuador is restricted.

(b) Germany (FAL, Braunschweig)[69] – practises a policy of reciprocal access to its gene bank collection.

(c) Finland[70] – whilst professing a policy of open access, restricts access to 78% of its germplasm collection (according to IPGRI databases).

(d) Other European countries[71] – IPGRI databases suggest that six European countries are on record as restricting availability of more than 10% of their national holdings, with another twelve restricting it to a lesser extent.

(e) Turkey[72] – restricts access to certain parts of its collection (including fig, grape, hazelnut and pistachio) because of their economic importance to national agricultural development.

(f) Thailand[73] – exportation of mango, durian, grape and pomelo fruits is banned by the Thai Ministry of Agriculture.

(g) Uganda[74] – sugar cane germplasm export from Uganda is restricted.

(h) USA[75] – whilst policy towards public collections is based on the goal of free and unrestricted access for bona fide scientists, a succession of political embargoes have been imposed by the USA, the most recent of which are against Cuba and Serbia.

Botanical Gardens

International Gardens

Botanical gardens have served, particularly in the eighteenth and nineteenth centuries, as centres for the collection, propagation and distribution of plant genetic resources of great commercial value to the sea trading powers of the day, such as Holland and Britain. Today, botanic gardens are the single most important type of institution involved in *ex situ* conservation of wild plants. There are over 1,550 botanic gardens in the world, about 800 of which are currently active in plant conservation.[76] Together they manage over 3 million accessions and also have at least 528 seed banks, of which 220 are reported as having wild seed available for distribution.[77]

---

[67] CPGR/94/WG9/4, Annex 1, *Review of the International Undertaking: Issues for Consideration of Stage II: Access to Genetic Resources and Farmers' Rights.*

[68] According to FAO description reported in Rural Advancement Fund International (RAFI), *op. cit.*, n. 56, at p. 5

[69] FAO, *Seeds Review 1984–85* (1987), at p. 482.

[70] R. Vellve, *Saving the Seed: Genetic Diversity and European Agriculture* (1982), at p. 88.

[71] *Ibid.*

[72] J.J. Hardon, "Crop Genetic Resources Conservation in Western Europe", paper presented to the Workshop on opportunities for European Cooperation in the Conservation of PGRs, (1989), at p. 10.

[73] RAFI, *op. cit.*, n. 56, at p. 5.

[74] According to FAO description reported in RAFI, *ibid.*

[75] US Office of Technology Assessment, *Technologies to Maintain Biological Diversity* (1987), p. 233, cited in Johnston, *op. cit.*, n. 52, at n. 23.

[76] Groombridge, *op. cit.*, n. 54, at p. 549.

[77] C. Heywood, V. Heywood and P. Jackson, *International Directory of Botanical Gardens V* (1990), extracted in B. Groombridge, *op. cit.*, n. 54, at p. 559.

The activities of botanic gardens at an international level are co-ordinated by the Botanic Gardens Conservation International (BGCI) (formerly the Botanic Gardens Conservation Secretariat). It was established in 1979 and currently has a membership of 317 botanic gardens. It aims to disseminate information and co-ordinate *ex situ* conservation of endangered wild plants and produces annual reports and inventories of each member garden.[78]

Member gardens are encouraged to distribute and propagate species, especially endangered ones. In 1987, the BGCI published *The International Transfer Format for Botanic Gardens Records* with the intention of developing a standard for the storage and transfer of electronic information for botanic gardens and other institutions maintaining living plant collections. At present the BGCI has received over 20,000 International Transfer Format plant records from botanic gardens for incorporation into the database.[79]

Many botanic gardens have avowed policies of unrestricted access. However these policies are adopted through decisions of the gardens themselves rather through international agreement and will be discussed in the following section.

National Gardens

In addition to co-ordination at an international level by the Botanical Gardens Conservation International, there are also a number of important botanic gardens co-ordinated at a national level. These include, for example, the Brazilian Botanic Garden Network (Rede Brasileira de Jardins Botanicos – RBJB), established in January 1991, and the Australian Network for Plant Conservation, established in March 1991.[80]

Whilst, in theory, many botanic gardens recognise the principle of open access, in practice there are a number of restrictions. For example, Argentina requires an undertaking from botanic gardens wishing to export wild genetic stock that they will inform them to whom they will pass on the material and for what purpose the stock will be used.[81] Similarly, the famous Kew Royal Botanic Gardens in Britain require assurances from recipients that if the accessed seed stock is commercialised then Kew will receive a share of the profits. Kew would then retain a portion of these royalties and pass the remainder on to the donor country.[82]

Property Rights in Genetic Resources

Naturally occurring or landrace PGRs may be variously described as the common property of all humankind, or as owned by the sovereign State within whose territory they occur, or as the property of the persons who control the areas of land or water on or in which the PGRs occur. In the international context, rendering PGRs the common property of all mankind gives rise to the "tragedy of the commons" at a global level, as there is no international authority to impose conservation imperatives on the international community.

Resources characterised as common property, i.e. belonging to no one and notionally the responsibility of all, tend in fact not to be protected by anyone. The

---

[78] Groombridge, *ibid.*, at p. 549.
[79] Johnston, *op. cit.*, n. 52, at p. 174.
[80] *Ibid.*, at p. 179.
[81] *Ibid.*, at p. 174.
[82] *Ibid.*, at p. 179.

net individual benefit of taking responsibility for the protection of the commons is usually calculated by individuals to be less than that of maximising their exploitation.[83] Thus, the argument is made that, in order for naturally occurring and landrace PGRs to be conserved, there must be economic value in conservation of such PGRs for those who can exercise control over and conserve them.[84]

At first reckoning, the solution might seem to be to render these PGRs as State property. However, the same "tragedy" problem arises here as the most biodiverse States are developing countries with few infrastructural resources to effect possession of their PGRs, i.e. to inventory, control and protect their wild or landrace PGRs from erosion. (The same problem may also arise in the less biodiverse developed countries, as the infrastructural resources required are indeed huge.) In addition, the allocation of property rights in naturally occurring PGRs to individual States raises the difficulty of distributing the economic benefits equitably among those States which have the same varieties of naturally occurring PGRs.

An alternative approach is to allocate property rights in naturally occurring or landrace PGRs at the sub-State level, rendering them the private property of small communities or persons who have custody over them. However, this approach again raises practical infrastructure and equity problems concerning inventory, protection and fair distribution of the benefit.

Despite the problems inherent in making wild and landrace PGRs the property of States, communities or persons, a trend of "proprietising" has emerged. In developing countries, States are in the process of claiming national property rights over their wild and landrace PGRs. In developed countries, the same phenomenon is taking place and, in addition, private property rights over technologically improved PGRs are being consolidated. Thus, two kinds of property rights are being claimed. One is over the tangible property contained in PGRs, i.e. the actual organic material. The other kind of right is over the intellectual property incorporated into technologically improved PGRs.

Tangible Property

State Rights

In response to the commercialisation of PGRs by seed companies, measures are being taken to assert State ownership of PGRs naturally occurring or traditionally bred within State jurisdiction and control. These measures are designed to ensure that there is an economic return to the State from which the commercialised genetic material originates. The measures consist of laws declaring the physical property of such organic material to be the property of the State. Like hydrocarbons, these PGRs occurring within sovereign territory and the maritime exclusive economic zone cannot be appropriated without the permission of the State, which may specify conditions for appropriation, such as royalties and technology transfer.

The Costa Rican government has established a scheme based on the assertion of State rights over the tangible property of PGRs under its jurisdiction by establishing an organisation, INBio (see below), to which it transferred all rights to Costa Rican naturally occurring and traditionally bred PGRs.[85] Indonesia is considering, with

---

[83] G. Hardin, "The Tragedy of the Commons" (1969) *Science* 1243.

[84] Vogel, *op. cit.*, n. 1.

[85] R. Gomez, A. Piva, A. Sittenfeld, E. Leon, J. Jiminez and G. Mirabelli, "Costa Rica's Conservation Programme and National Biodiversity Institute – INBio", in Reid *et al.*, *op. cit.*, n. 3.

the assistance of the Asian Development Bank, the establishment of a Biodiversity Marketing and Commercialisation Board. It is envisaged that this Board would play a similar role to that of INBio in Costa Rica.[86]

Similarly, the recognition of a State's sovereign rights over its PGRs in the UN Convention on Biological Diversity (see above), prompted the government of the Australian state of Queensland to propose an amendment to its Nature Conservation Act so as to give the state outright ownership of its flora and fauna and "guarantee that it shares in any profits made from exploiting them".[87] Countries such as China, Ethiopia and Turkey strictly regulate the export of national germplasm, particularly of economic crops. In further recent developments, the Manila Declaration recommends that "national governments ... develop adequate legislation to exercise control over the collection and export of biological material".[88]

Although States which have commercially attractive PGRs within their jurisdiction are willing to claim proprietary rights over them, their ability to exercise effective control in order to conserve PGRs and to regulate foreign access to them is not at all apparent. In respect of regulation of foreign access, a great deal depends upon the attitude adopted by those foreign persons seeking access. These persons may be private entrepreneurs or foreign governmental institutions and are often not even the end-user of the PGR. Many pharmaceutical, agricultural or biotechnology research institutions and industries in developed countries receive PGRs through intermediaries – enterprises which prospect for PGRs.

For example, Biotics Ltd, a private firm based in Britain, works as a broker providing pharmaceutical companies with PGRs. Biotics buys samples from source country institutions and through a contract agrees to share any royalties with the source country institution. Similar contracts are drawn up between Biotics and the pharmaceutical company that ultimately holds the patent. Glaxo Group Research Ltd and SmithKline Beecham are examples of companies with contracts with Biotics.[89] Public sector intermediaries also gather and supply PGRs. For example, in the USA, Monsanto Inc. and the National Cancer Institute have agreements with the Missouri Botanical Garden for the supply of PGRs, whilst the Pfizer and Boehringer Ingelheim corporations have agreements with the New York Botanical Garden.[90] Obviously, State regulation over access to its PGRs will depend to a great extent on a co-operative attitude to such regulation being adopted by those persons seeking access.

In the cases of prospecting by the enterprises of foreign governments, it seems that a responsible attitude may not yet be prevalent. At the beginning of 1993, the US National Cancer Institute (NCI) was prospecting for drugs in twenty-five nations. There were formal agreements with only four of these countries, although others are being drawn up. Cameroon recently negotiated a deal with the NCI providing that if the NCI develops a commercially successful anti-cancer drug from a chemical present in a rare vine recently discovered in Cameroon, NCI will try to ensure that Cameroon receives a fair and equitable share of the profits. However, NCI has merely promised to "make its best effort" to ensure that Cameroon receives royalties on drug sales, as NCI would patent any effective drug and then

---

[86] Reid *et al. ibid.*, at p. 26
[87] "Queensland Sets Out Rights Over Native Species" *New Scientist*, 1 May 1993, p. 7.
[88] See n. 47 above.
[89] Reid *et al., op. cit.*, n. 3, at p. 25
[90] *Ibid.*

license companies to market it. The NCI argues that it cannot promise a specific royalty to Cameroon as it cannot preordain the future rights of a licensee. This arrangement is vague and makes little mention of technology transfer or the use to which any royalties will be put.[91] Another major economic and technology power, Japan, has recently launched a biodiversity research programme in Micronesia. Twenty-four Japanese companies (including Suntory and Nippon Steel) and the Ministry of International Trade and Industry have established the Marine Biotechnology Institute in Micronesia. Again, however, no arrangements have been made to share royalties with Micronesia or to pay an exploration fee.[92] These circumstances reflect the efforts still required in order to meet the commitments made by signatories to the International Undertaking (above) to implement the notion of Farmers' Rights.

Plant Breeders' Rights

Plant breeders' rights are property rights for the benefit of innovators of plant varieties. Such innovators may be State institutions or private individuals. The property in which the right is held consists of the organic material which forms the new plant breed.

As a means of protecting plant breeders' rights at the international level, the International Union for the Protection of New Varieties of Plants (UPOV), an intergovernmental organisation, established the International Convention for the Protection of New Varieties of Plants (known as the UPOV Convention) in 1961. The 1961 text, as amended in 1978, forms the basis of the current laws on plant variety protection in the member States of UPOV.[93] In 1991, UPOV comprised nineteen countries. These were predominantly developed countries, although some others do have their own plant breeders' rights.[94]

Under the UPOV Convention, a breeder may obtain exclusive rights to a novel plant variety if it is distinctive, uniform and stable.[95] The Convention is geared towards cultivars that breed true-to-type for the desired trait, which is a difficult standard for a wild variety to meet (certainly before several generations of breeding). Thus, whilst it may be possible to describe some wild plants as property under the UPOV Convention, it would not be possible to protect many others. Nor would it be possible to protect micro-organisms.[96]

Revisions made to the UPOV Convention in 1991 (which are not yet in force) will lead to a strengthening of the legal protection system by extending the period of protection for the originating breeder and by narrowing the rights for other breeders and farmers.[97] The 1991 amendments make clear that the term "breeders" includes not only those who have bred a plant but also those who have "discovered and developed it". It also makes clear that widespread knowledge of the existence of a new variety is not a bar to a claim.[98] These changes could increase the opportunity for village farmers and indigenous peoples to use the UPOV Convention to claim

---

[91] "High hopes hanging on a useless vine" *New Scientist*, 16 January 1993.

[92] Reid *et al.*, *op. cit.*, n. 3, at pp. 2, 26.

[93] N.J. Byrne "Plant Breeding and UPOV" (1993) 2 RECIEL 136.

[94] Reid *et al.*, *op. cit.*, n. 3, at p. 171.

[95] International Convention for the Protection of New Varieties of Plants (UPOV) 815 UNTS 89, TIAS 10199.

[96] Reid *et al.*, *op. cit.*, n. 3, at p. 172.

[97] N.J. Byrne, *op. cit.*, n. 93, at p. 139.

[98] UPOV Convention, Article 6.

plant breeders' rights in relation to landraces. In an ideal world, this opportunity might be harnessed to conservation efforts, providing an incentive to local peoples to preserve wild varieties for their potential economic value. However, the effect of the 1991 amendments is yet to be seen.

Intellectual Property

State rights controlling access to PGRs under their jurisdiction and plant variety rights each concern the tangible organic material of plants. Intellectual property rights, on the other hand, concern the ideas involved in inventions. Inventions relevant to this discussion of PGRs may be particular technical processes, such as processes for the manipulation of genetic material so as to achieve a desired trait in a plant. At a much more legally controversial level, the invention may be the dis-covery of information encoded in genetic sequences which is sought to be patented.

A utility patent gives an inventor the right, for limited time, to exclude others from making, using or selling the patented invention in a particular country. Control is exercised at the national level, and national governments co-ordinate their efforts within the framework of international agreements. A survey of the role of intellectual property rights in biodiversity conservation is dealt with else-where in this work. However, for the sake of completeness, some important inte national agreements and decisions relating to patents and PGRs are noted here.

Paris Convention

The Paris Convention of 1883 (as revised) provides that the utility patents of any party to the Convention are to be recognised and given effect by other parties, who are to provide the same protection for such foreign patents as they would for patents under their own jurisdiction. The Paris Convention imposes requirements for minimum rights under relevant national laws – although these are far from demand-ing. In 1991, there were ninety-nine parties, including sixty from less technologi-cally active countries.[99] In the USA, the scope of patentable subject matter is wider than elsewhere and includes not only purified compounds but also genetically altered animals, plants and micro-organisms.[100] The USA has already allowed plant varieties to be protected under patent.[101] However, most other countries preclude or restrict patents or inventions over PGRs.[102]

European Patent Convention

The European Patent Convention of 1973 applies to fourteen nations, including all European Union Countries. It effects the harmonisation of national patent laws by the adoption of similar statutes in each participating State. The European Convention has an important exclusion clause, Clause 53B, which excludes "plant and animal varieties" and "essentially biological processes" from patentability, (although not microbiological processes and products thereof). The meaning of

---

[99]   W. Lesser, *Equitable Patent Protection in the Developing World: Issues and Approaches* (1991), at p. 8.
[100] I. Walden, "Intellectual Property in Genetic Sequences" (1993) 2 RECIEL 126, 128.
[101] CGIAR, *Intellectual Property Issues for the International Agricultural Research Centres*, Issues in Agriculture 4, April 1992.
[102] Reid *et al.*, *op. cit.*, n. 3, at p. 168.

these terms has been subject to differing interpretations concerning whether they prohibit patents on organisms *per se*, or refer only to plant and animal varieties produced by traditional breeding processes.[103] However, the granting of a European patent for the Harvard Oncomouse (a genetically engineered mouse containing a human oncogene) suggests that genetically engineered plants and animals can be patented. The judgment in a European Court of Justice case on the matter seems to suggest that a balance should be struck in each instance between the public benefits and disadvantages when deciding whether to grant a patent. But the more recent *Plant Genetic Systems* case suggested that environmental considerations should not be taken into consideration amongst these public interest concerns.[104]

Despite an apparent trend towards legal recognition of patents over PGRs, they remain controversial. Questions are raised as to their legitimate extent, especially in terms of their benefit to the public interest. A possible conservation benefit in allowing some patenting of PGRs is demonstrated in the case of the agreement between the Merck Pharmaceutical Corporation and Costa Rica's National Institute of Biodiversity in Costa Rica (INBio). Merck is the largest pharmaceutical company in the world. INBio is a non-profit, scientific organisation, created in 1989 on the recommendation of the Costa Rican government with the aim of identifying and classifying the diverse biological species found in the rainforests and protected areas of Costa Rica. The agreement between them has been widely viewed as ground breaking, since it marks commercial recognition of a State's rights over PGRs found under that State's jurisdiction and its rights to benefit from others' use of those PGRs.[105]

In September 1991, Merck agreed to pay INBio US$1,135,000 for the opportunity to screen thousands of plants, insects and micro-organisms. The payment to INBio over two years is to cover the costs of national conservation activities, with Costa Rica's Ministry of Natural Resources receiving US$100,000. In addition, INBio and the Costa Rican government will each receive a quarter of the royalties on the sales of any commercial products that are developed as part of the agreement (with Merck receiving the other half). As part of the deal, Merck are also providing equipment to INBio for the establishment of a biodiversity sample processing laboratory within the University of Costa Rica and INBio scientists are being trained in Merck laboratories.[106]

The Merck–INBio arrangement gives some force to the suggestion that there is a role for intellectual property rights in the conservation of PGRs, i.e. the commercial value of intellectual property rights in products derived from naturally occurring or landrace PGRs can lead to investment in the conservation of those PGRs. Commercial interests might work in partnership with States (or others controlling areas where PGRs are located) so as to secure such benefits.

---

[103] D. King "Legal, Socio-Economic and Ethical Issues in the Patenting of Plants and Animals", January 1992.

[104] D. Alexander, "Some Themes in Intellectual Property and Environment" (1993) 2 RECIEL 113, 114.

[105] Reid *et al.*, *op. cit.*, n. 3, at pp. 2, 53.

[106] *Ibid.*, at p. 53.

Conclusion

There is a discrete body of international institutional effort and of soft law concerning the conservation and control of plant genetic resources. This institutional effort and soft law predates and intermeshes loosely with more general biological diversity conservation institutions and laws. It also evolved contemporaneously and connects closely with agricultural property protection mechanisms, such as plant variety and intellectual property rights.

In the late 1950s, the problems of PGR erosion began to be addressed at the public international level by the UN Food and Agriculture Organisation. However, its efforts were limited by the unwillingness of its members to compromise the commercial interests of local agricultural corporations' access to foreign PGRs, to relinquish private property in technologically improved PGRs, or to cede national control over PGRs naturally occurring under their jurisdiction. Thus, progress in establishing binding legal regimes for plant genetic resources conservation was poor. Such progress as was made was in the area of *ex situ* conservation in international gene banks, rather than *in situ* conservation. The institutions put in place avoided legal questions of responsibility for and control over those *ex situ* collections and, instead, focused on technical work.

The Convention on Biological Diversity reinvigorated progress towards a legally binding regime specifically for the conservation of PGRs. It gave widespread and legally binding recognition to existing State sovereign rights over genetic resources naturally occurring within State jurisdiction, including State rights to govern the terms of foreign access to those resources, and to benefit from foreign use of such resources. Thus, the Convention on Biological Diversity broke an impasse between developed and developing countries over international rights to PGRs. As developing countries now have clearly recognised rights, there are improved opportunities for them to press ahead with international conservation arrangements which they perceive as equitable and co-operative rather than disadvantageous and coercive. Conversely, developed countries, recognising that terms for their access to foreign PGRs need to be negotiated, perceive mutual advantage in terms which incorporate conservation obligations. Therefore, progress towards a binding international regime for conservation of PGRs is now being made.

The FAO serves as the vehicle for this progress. It is the main international institution responsible for inter-State arrangements for conservation of those PGRs which the international community values most highly, i.e. those useful for food and agriculture. Since 1983 it has been working towards a Global System for PGRs, the two main elements of which are institutional – the Commission on PGRs – and legal – the International Undertaking on PGRs and related legal instruments. Other elements in the Global System include the FAO periodic publication *State of the World's PGRs*, as well as its Early Warning System, Plan of Action, *In Situ* and *Ex Situ* Networks and Technical Conferences. However, whilst consensus and institutional foundations for the Global System have gradually been built, the System is not yet truly operational. Critical to its success is establishment of the International Fund for PGRs and the implementation of Farmers' Rights. These are essential to provide the resources needed for conservation and to bring those resources to the traditional farmers who daily control many PGRs. It is encouraging to note that the activities and perspectives generated through the Convention on Biological Diversity are stimulating progress on this front. For example, the International Undertaking, which introduces the International Fund and Farmers' rights, is being revised and clarified, and it may also be given binding force as a Protocol to the Convention. Conservation efforts mandated by the Biological Diversity Convention and financial conditions for access to PGRs may also be directed towards the implementation of Farmers' Rights.

In contrast to the intensified dialogue and co-operation concerning PGRs at public international level, there is a growing tendency at national and private commercial levels to emphasise exclusive ownership of PGRs. Naturally occurring and landrace species, as well as technologically developed PGRs, are increasingly recognised as subject to legal rights of exclusive control by States and agricultural companies. Remarkably, a self-correcting counterbalance to the tendency toward exclusive control, which might make many PGRs unavailable to potential users, seems to be emerging from new international arrangements being put in place to ensure that useful PGRs remain available. For example, the legal status of *ex situ* collections acquired prior to the Convention on Biological Diversity coming into force is being revised so that national gene banks and gene banks under the umbrella of Consultative Group on International Agricultural Research will formally place their germplasm collections under the auspices of the FAO. Access agreements between the agricultural or pharmaceutical industry sectors and PGR source countries tend to provide that ownership of the ultimate product of the prospecting effort must be shared. These agreements will be closely examined to see if they direct some of the benefits to PGR conservation, complementing the public international process by providing new finance and technology for PGR conservation.

The Convention on Biological Diversity has stimulated a great deal of interest in new ways to prevent erosion of PGRs and the work of the FAO on conservation of plant genetic resources seems to be coming to fruition. Yet, the tensions between the approaches of the Convention on Biological Diversity and the International Undertaking on PGRs are critical. The Convention emerges from a conservation ethic and is concerned with all biological resources, it places emphasis on controls on future foreign access to each State's biological resources. However, the finance and technology benefits gained by PGR source countries under the Convention on Biological Diversity could be used for any purpose, such as logging, land clearance and mono-cropping. In contrast, the International Undertaking is based upon a regime of uncontrolled access, although it does introduce a notion of historic reward for such access, known as Farmers' Rights. Assurances that extension of the rights of exclusive control over access to PGRs will promote their conservation and utilisation are still needed. Whilst the International Undertaking is concerned solely with plant genetic resources and has an agricultural application orientation, it does provide that the rewards for access, i.e. the Fund established to meet Farmers' Rights, will be directed towards global conservation of PGRs. The FAO Code of Conduct for Plant Germplasm Collecting and Transfer also makes some provision for these concerns but the evidence is not yet available to indicate whether either mechanism will be effective.

Whilst the link between control of access to plant genetic resources and conservation of them is evident, it is not yet apparent whether the many parties involved will successfully realise this link. The FAO and persons concerned that current efforts should promote conservation may soon find their hopes either realised or dashed, as the pace of institutional and legal developments has certainly become energetic. This dynamism is reason enough to hope for positive results.

9

Intellectual Property Rights and Biodiversity

Ian Walden

Introduction

The accelerating depletion of our natural resources can be expressed along a continuum from species to individual segments of functional genetic code. It is the potential value of genetic material, however, that is at the core of our attempt to preserve biodiversity: the preservation of genetic material which has, as yet, undiscovered beneficial properties. Increasingly it has been recognised within the environmental community that one approach to encouraging nations to preserve their natural genetic resources, particularly among the developing nations, is through the provision of economic incentives. This viewpoint has been most forcefully expressed by Wilson:

> The only way to make a conservation ethic work is to ground it in ultimately selfish reasoning ... an essential component of this formula is the principle that people will conserve land and species fiercely if they foresee a material gain for themselves, their kin, and their tribe.[1]

One manifestation of this approach is the concept of "debt swaps", where creditor nations agree to reduce the debt burden on developing countries in return for the implementation of environmental policies. Another possibility has been the area of "property" rights, particularly intellectual property rights.[2] One argument, put forward by certain commentators within the environmentalist lobby, is that if developing countries were able to establish intellectual property-style rights over the valuable genetic material found within their biota, they would then have an economic incentive to preserve the habitat which sourced such material.[3]

---

[1] E.O. Wilson, *Biophilia* (1984).
[2] As a general legal definition, an item of "property" is simply something in which an individual or legal entity can assert rights against others.
[3] See, for example, C. Juma, *The Gene Hunters* (1989) and T.M. Swanson, "Economics of a Biodiversity Convention" (1992) 21:3 *AMBIO* 250–257.

International Law and the Conservation of Biological Diversity (C. Redgwell and M. Bowman, eds.: 90 411 0863 7: © Kluwer Law International: pub. Kluwer Law International, 1995: printed in Great Britain), pp. 171–189

It was the Biodiversity Convention,[4] however, which brought intellectual property issues to the centre of the political debate on biodiversity, when the United States government decided not to sign the Convention. One of the main grounds for objection was the perceived attitude of the Convention towards intellectual property rights (IPRs). The USA believed that the Convention's provisions failed to recognise the positive role that intellectual property rights could play in the conservation of biodiversity, while the countries of the developing world were portrayed as wanting to restrict the application of IPRs in order to "share in profits and technology".[5]

This chapter examines the nature of intellectual property rights and reviews the extent to which they are used to protect biological material. Intellectual property rights already play a central role in the rapidly growing biotechnology industry. The second part of the chapter considers the wider issue of preserving biodiversity through the assertion of tangible property rights, primarily through controlling access to the land upon which such material is found. The possibility of establishing some form of *sui generis* intellectual property right for wild, unmodified genetic material is examined in the context of extending the range of property rights available to developing countries rich in genetic material. Such a proposal gives rise to substantial theoretical and practical issues which extend beyond the objective of providing countries with an economic incentive to conserve biodiversity. Finally, it is concluded that although intellectual property rights need to be a component within a wider policy framework for biodiversity conservation, such policy should be based primarily upon legal mechanisms, such as contractual agreements, controlling the use of tangible property.

Intellectual Property Laws

Intellectual property rights (IPRs) are a particular aspect of property covering "all things which emanate from the exercise of the human brain".[6] The major intellectual property rights are patents, plant breeding rights, trade secrets, trade marks and copyright. The general principle behind IPR protection is that the "right holder" is given some form of monopoly control over the economic exploitation of the material concerned. The overriding economic justification behind such protection is as a reward and incentive for the efforts of those involved in the creation of the "property", as well as to prevent unfair competition from others.

Intellectual property laws historically distinguish between the treatment given to human creations as opposed to creations of nature. This section reviews the use of patents, plant breeding rights and trade secret laws within the biotechnology industry as a means of protecting the commercial exploitation of genetic material.

Patent Law

The patent system has grown up over a long period of history. Patents protect ideas and their expression within new products and processes. Patents confer upon the inventor of a new process and/or product *exclusive monopoly rights* with regard to its

---

[4] 31 ILM 818 (1992).

[5] M. Chandler, "The Biodiversiversity Convention: Selected Issues of Interest to the International Lawyer" (1993) 4 Col JTL at p. 162.

[6] J. Phillips and A. Firth, *Introduction to Intellectual Property Law* (1990), at p. 3.

economic exploitation for limited periods of time (up to twenty years) in exchange for the *public disclosure* of the details of the invention.[7] The patent system developed to promote inventive activities which were seen to benefit economic growth.[8]

Patent legislation generally requires that the claimed process and/or product meets four requirements:[9]

(a) Is it an invention?
(b) Is it new?
(c) Does it involve an inventive step?
(d) Does it have industrial applicability (i.e. is it useful)?

These requirements have been significantly harmonised through a number of international treaties.[10]

"Process" patents have traditionally been viewed as being of less economic value than "product" patents, partly because they are difficult to police since new technical methodologies can also often be found to produce a particular product. A third option is an all encompassing "product-by-process" patent claim, which gives rise to the most profitable monopoly rights. Such a claim can be used to extend legal protection to substances (products) that would not by themselves satisfy the necessary requirements.[11] However, "product-by-process" patents are more difficult to obtain, unless it can be shown that the product could not be described adequately by any other means.

Article 53(b) of the European Patent Convention[12] states that *no* protection is available for:

> plant or animal *varieties or essentially biological processes* for the production of plants or animals; this provision does not apply to microbiological processes or the products thereof.

Historically, plant varieties were excluded because of the existence of an international legal regime for plant breeding rights (see below); while animal varieties were generally considered to be ineligible. However, subsequent case law has adopted liberal interpretations of the words in italics: *varieties* has been held not to cover all

---

[7] Such information should enable the invention to be "performed by a person skilled in the art". This aspect of patent law can raise significant problems for genetic material since it can be particularly difficult to describe such material in a technically and legally unambiguous fashion. To mitigate against such difficulties, the Budapest Treaty on the International Recognition of the Deposit of Micro-organisms for the Purposes of Patent Protection of 28 April 1977, and other national and international laws, permit deposition of the material with a recognised depository institution.

[8] Indeed, the US courts have been seen as supporting the biotechnology industry in particular; see R.A. Armitage, "The Emerging US Patent Law for the Protection of Biotechnology Research Results" [1989] 2 EIPR 47, at p. 57.

[9] *Guidelines for Examination in the European Patent Office*, para. 1.1 of Chapter IV, and para. 2.2.

[10] For example, the Paris Convention for the Protection of Industrial Property, first concluded in 1883 and periodically revised; and, most recently, the TRIPS Agreement within the GATT, discussed further *infra*.

[11] See, for example, *Cochrane v. Badische Anilin und Soda Fabrik* 111 US 293 (1884) at 311. The EC's draft "Proposal for a Council Directive on the Legal Protection of Biotechnological Inventions", OJ C10, 13 January 1989, at p. 3, provides in Article 12(2) that animal and plant varieties could be patented by subsuming them within "product-by-process" patents.

[12] The Convention on the Grant of European Patents, concluded October 1979, entered into force on October 1977. The provisions of the Convention are enacted in substantially similar form in the patent laws of the Member States, of which there are currently fourteen.

plants and animals per se;[13] while *essentially biological* has been held not to cover those processes where interference by man constitutes a "substantial part".[14] In addition, the exclusion from Article 53(b) of "microbiological processes or the products thereof" would seem to extend patentability to a significant range of genetic material, such as bacteria and other micro-organisms.[15]

The *European Patents Handbook* states that, apart from those exceptions noted above, all biological inventions are in principle patentable. It goes on to note that "[d]ifficulties normally arise only in relation to novelty, inventive step and (if the invention is based on preliminary experiments) possibly industrial applicability" (1991, at 18/3). In addition, the Guidelines for Examination issued by the European Patent Office state that:

> To find a substance freely occurring in nature is also mere discovery and therefore unpatentable. However, if a substance found in nature has first to be isolated from its surrounding and a process for obtaining it is developed, that process is patentable. Moreover, if the substance can be properly characterised either by its structure, by the process by which it is obtained or by other parameters ... and it is "new" in the absolute sense of having no previously recognised existence, *then the substance per se may be patentable*. An example of such a case is that of a new substance which is discovered as being produced by a micro-organism.[16]

From this statement it would seem that under European patent law a pre-existing distinction is not necessarily made between the patentability of naturally as opposed to artificial genetic material. Naturally occurring genetic material is potentially patentable provided that it can be isolated, characterised and is "new".

In the USA, the first significant case in the area of patenting genetic material was *Diamond v. Chakrabarty* in 1980.[17] This was the first occasion on which a patent had been granted for a live organism and therefore overturned the "product of nature" principle which had previously dominated this area. The successful application concerned bacterium which were capable of breaking down crude oil (potentially useful for cleaning purposes). In the decision, the Court stated that:

> the patentee has produced a new bacterium with markedly different characteristics than any found in nature ... *His discovery is not nature's handiwork, but his own; accordingly, it is patentable subject matter under patent law.*

As the quote makes clear, the key element in the granting of the patent was the need for human action upon the genetic material. Since that initial application, the most publicised biotechnology patent has been that awarded to the "Onco mouse" in 1988, concerning the successful implantation of cancer causing genes.[18]

---

[13] No general exclusion on inventions in the sphere of animate nature can be inferred from the European Patent Convention": *Ciba-Geigy* [1984] OJ EPO 112, at p. 114.

[14] *Lubrizol Genetics* [1990] OJ EPO 71.

[15] R. Teschemacher, "Patentability of Microorganism per se" (1982) 13: 1 *International Review of Industrial Property and Copyright Law* 27.

[16] *Guidelines for Examination in the European Patent Office*, Part C, Chap IV, at 2.3.

[17] [1980] 477 US 303, 206 USPQ 193.

[18] See *Ex parte Allen* 2 USPQ 2d 1425 (1987). Subsequently, the European Patent Office followed suit and gave approval to the "Oncomouse" patent application (EPO T19/90, 3 October, 1990). In the UK, biotechnology products have faced more substantial problems in obtaining patent protection than in the US: see *Genentech Inc's Patent* [1989] RPC 147; however, see also the recent decision of the UK Patents Court in *Biogen Inc. v. Medeva plc* of 4 November 1993.

Patents and Tangible Property Rights

In *John Moore* v. *Regents of the University of California*,[19] a dispute arose between the University Hospital and a patient over the use of some of his "cells" that had been removed during an operation and subsequently used in the production of a patented medical product. The patient took legal action to recover some of the monies that had been received from sales of the product. The use of human tissue, as the basis upon which a biotechnology patent was awarded, gave rise to conflicting "property" rights between the patent owner and the individual tissue donor. The Court eventually concluded that the common law tort of unlawfully converting personal property had taken place, and therefore the patient was entitled to compensation.

In terms of patent law, this decision violated some of its underlying principles. First, it granted economic rights to the source material (ie. the patient), when under current patent regimes only inventors who contribute to the intellectual conception of an invention are entitled to a patent. Second, it breached the existing principle that "naturally occurring, unmodified cells are products of nature and free to all".[20]

In terms of biodiversity, the result of this case should be seen as significant. One of the major problems of using patent law as a means of protecting natural genetic material is the fact that historically "products of nature" have been viewed as "free to all". This case illustrates, however, that this principle can be breached in circumstances where an alternative, more fundamental, right is recognised to exist.

The Human Genome Project

Over recent years, one of the most significant developments in the area of genetic sequences is the "human genome" project. This international project is designed to identify and map all the sequences within human DNA genetic material. The size of the task means that it is expected the project will take many years to complete, involving participants around the world.

As part of the "human genome" project, the National Institute of Health in America (NIH) recently applied to the US Patent and Trade Mark Office for patent protection for 2412 identified sequences, "before the function of products encoded by an associated gene are known".[21] In a preliminary decision, the Office has decided that the sequences lack novelty, inventiveness and usefulness, and therefore the applications were rejected. The lack of information with regard to the functions of the gene sequences creates a clear problem in terms of proving "usefulness". In addition, the methodology for discovering the gene sequences involved a primarily computerised process based around seven automatic DNA sequencing robots. It is likely that such a process lacked the necessary inventive step.

This is an important case in terms of providing intellectual property protection for natural genetic sequences, since mere discovery of a gene sequence will increasingly become a computerised process which, if given legal protection, would primarily benefit developed countries with the ability and resources to utilise significant computing power to the "churning through" of such information. The "human genome" project also raises even more fundamental questions concerning the right to "own" the natural components of the human body.

---

[19] 202 Cal.App.3d 1230, 249 Cal. Rptr. 494 (1988) and 7 *Biotechnology Law Report* 355.
[20] W. Noonan, "Ownership of biological tissues" (1990) 72: 2 JPTSO 110. See also *Funk Bros. Seed Co.* v. *Kalo Inoculant Co.*, 333 US 127, 130 (1948).
[21] S.B. Maebius, "Novel DNA Sequences and the Utility Requirement" (1992) 74: 9 JPTSO 651.

In summary, patent law would not appear to distinguish clearly between natural and artificial gene sequences. Any patent claim involving a genetic sequence is required to meet the four basic requirements. It is the first requirement, whether "discovered-in-nature" substances are excluded patentable subject matter, that gives rise to major difficulties when making a claim involving a natural genetic sequence.[22] A successful patent application would be possible depending on the extent to which subsequent development work is carried out (e.g. to isolate and purify the material), the nature of the claim, and the position of the sequence within that claim (e.g. product-by-process). The other three requirements, novelty, obviousness and application, need to be met in *any* claim based on a genetic sequence.

Plant Breeders' Rights

Plant breeders' rights (PBRs) are a *sui generis* intellectual property right for plant varieties that arose during the nineteenth century to satisfy the demands made by the plant breeding industry. Such rights were treated in a legislatively distinct manner from the patent system because of the conception that animate inventions should be treated separately from inanimate inventions.[23]

Within the United Kingdom, a new plant variety can gain protection if it falls within a category of one of the "schemes" laid down by the government. The purpose behind the schemes is to vary the extent of protection to suit the commercial requirements of the variety. The criteria for obtaining protection under a scheme are that a plant variety be new (in the sense that no prior commercialisation has occurred), distinct, uniform and stable.[24] The period of protection varies between twenty and thirty years according to the scheme in which the variety is included. Within the United States, plant varieties are also protected under *sui generis* legislation. In a 1985 decision, however, the US Patent Office decided to admit plant varieties within the scope of the patent system, such that they can now obtain dual protection.[25]

At the international level, plant breeders' rights have been the subject of international agreement, namely, the International Convention for the Protection of New Varieties of Plants 1961. The Convention has subsequently been amended in 1972, 1978 and, most recently, in 1991.[26] This latest revision was designed to ensure the Convention's continued relevance as a form of legal protection in the face of the trend towards patenting plant varieties.[27] It therefore removes the previous obligation upon signatory nations not to grant both patent and plant breeders' rights to the same species. The 1991 revision also extends the scope of the Convention to all plant genera and species and will give exclusive rights covering "harvested" material to the "breeder".[28] The Convention allows countries to give protection on the basis of "discovery", since a "breeder" is defined as "the person who bred, or *discovered*

---

[22] The World Intellectual Property Organisation (WIPO) has examined the protection of biotechnological inventions and recommended that biological matter occurring in nature should be patentable: WIPO Meetings, Paris Union, *Industrial Property* (1986), at pp. 251–274.

[23] Juma, *op. cit.*, n. 3, at p. 149.

[24] Phillips and Firth, *op. cit.*, n. 6, at para. 25.6.

[25] *In re Hibberd*, 227 US Patents Quarterly *(BNA)* 443 (PTO Bd. App. & Int. 1985).

[26] UPOV Document DC/91/138, 19 March 1991.

[27] N. Byrne, *Commentary on the Substantive Law of the 1991 UPOV Convention for the Protection of Plant Varieties*, Centre for Commercial Law Studies, 1991.

[28] However see also the issue of "farmers' rights" to use farm-saved seed: N. Bryne, "Plant Breeding and the UPOV" (1992) 2: 2 RECIEL 139.

and developed, a variety" (Article 1(iv)). The level of subsequent development that must be associated with the discovery is generally recognised as being minimal.[29]

Despite the trend towards the patenting of plant varieties, explicitly accepted in the 1991 revision of the Convention, plant breeding rights continue to be of significant relevance in the area of protecting plant genetic material.

Trade Secrets

The biotechnology industry relies upon trade secret protection in a number of circumstances:

(a) to protect information prior to an application for a patent;
(b) to protect peripheral, undisclosed know-how related to the patent, and
(c) to protect information that is unpatentable, or for which patent law provides ineffective commercial security.[30]

Indeed, it has been noted that trade secret protection is increasingly being used by the biotechnology industry as an effective method of protection.[31] Part of the reason behind this trend would seem to be the extensive amount of litigation which has grown up in the field of biotechnology patents due to difficulties and uncertainties regarding the scope and effectiveness of patent protection. One key advantage of trade secret protection is that the length of protection is not limited to a certain number of years; it simply depends on the information remaining a "secret". A disadvantage of trade secret protection for genetic material is the fact that the commercial exploitation of the material will usually involve its introduction into the public domain, thereby losing its nature as a secret.

The first step in any action a breach of trade secrets is for the courts to decide if the information involved can be categorised as a trade secret or confidential information. In Australia, the courts have shown themselves willing to view plant genetic material as having the appropriate qualities of trade secrets. In *Franklin* v. *Giddin*[32] the defendant stole cuttings from the plaintiff's genetically unique nectarine trees. An action was taken for the improper acquisition of confidential information embodied within the trees' genetic code. The judge accepted this position, declaring that:

> The parent tree may be likened to a safe within which there are locked up a number of copies of a formula for making a nectarine tree with special characteristics ... when a twig of budwood is taken from the tree, it is as though a copy of the formula is taken out of the safe.

Such secrecy does not have to be absolute, rather it is a question of objective fact. Although the courts will consider all forms of misappropriation of trade secrets, an action will usually only succeed if the plaintiff can show that adequate precautions were taken to protect the secret nature of the information.

Another disadvantage when using trade secrets law are its limitations in providing international protection. Unlike patents and copyright, trade secrets protection has

---

[29] *Ibid.*
[30] I.P. Cooper, *Biotechnology Law* (1992), at 11–1.
[31] R.W.J. Payne, "The emergence of trade secret protection in biotechnology", (1988) 6 *Bio/Technology*, at pp. 130–131.
[32] [1978] 1 QdR 72.

not traditionally been the subject of international agreement.[33] The use of confidential information abroad can therefore circumvent and render the "right" ineffective domestically. Protection can usually only be enforced internationally where the confidentiality is contractually based.

The Biodiversity Convention and the Uruguay Round

The Biodiversity Convention has given rise to significant controversy in the area of intellectual property rights. Article 16, "Access to and Transfer of Technology", was cited by the United States government as one of the reasons behind their refusal to sign the Convention: "The Convention focuses on [intellectual property rights] as a constraint to the transfer of technology."[34] The language of the Convention regarding technology transfer has been seen as potentially permitting countries to restrict the intellectual property rights of companies that develop products based on resources obtained from the country.[35] Article 16, paragraph 5 provides:

> The Contracting Parties, recognizing that patents and other intellectual property rights may have an influence on the implementation of this Convention, shall cooperate in this regard subject to national legislation and international law *in order to ensure that such rights are supportive of and do not run counter to its objectives* [emphasis added].

In addition, the Convention has also been seen as enabling countries to enact compulsory licensing regimes. Article 16, paragraph 4 provides:

> Each Contracting Party shall take legislative, administrative or policy measures ... with the aim that the private sector facilitates access to, joint development and transfer of technology ... for the benefit of both governmental institutions and the private sector of developing countries ...

The Convention also includes a statement, at Article 22(1), that *preceding* international intellectual property agreements shall not be affected, "except where the exercise of those rights and obligations would cause a serious damage or threat to biological diversity".[36]

In early 1994, the Uruguay Round of the General Agreement on Tariffs and Trade (GATT) was completed. For the first time the multilateral trade negotiations and the resultant Agreement extended to trade in services, as well as goods. Within the GATT process, agreement was reached on the Trade-Related Aspects of Intellectual Property Rights (TRIPS).[37] The TRIPS Agreement represents a major move towards international harmonisation in the area of intellectual property rights. In terms of developing countries, one of the potentially most significant aspects of

---

[33] However, the TRIPS agreement, within the recently agreed GATT Treaty (see the discussion of the Biodiversity Convention and the Uruguay Round, below) does include the "protection of undisclosed information", at Article 39.

[34] State Department Press Memorandum, 29 May 1992 and *Report of the Intergovernmental Negotiating Committee for a Convention on Biological Diversity*, UNEP, 7th Negotiating Session, 5th Session of the International Negotiating Committee, at p. 35, UN Doc. UNEP/Bio.Div/N7-INC.5/4 (1992). See also the "Declaration of the United States of America", attached to the Nairobi Final Act, 31 ILM 848 (1992).

[35] "PTO, Biotech Group explain objections to Earth Summit's biodiversity treaty", (1992) 44 *Patent, Trademark & Copyright Journal (BNA)*, at pp. 120–121.

[36] See Chandler, *op. cit.,* n. 5, at p. 150.

[37] Agreement on trade-related aspects of intellectual property rights, including trade in counterfeit goods", MTN/FA II-AIC. There are 117 signatories to the agreement. See generally, J. Worthy, "Intellectual Property Protection after GATT" (1994) 5 EIPR 195–198.

the Agreement was the establishment of a framework of enforcement procedures for intellectual property rights, a key concern of developed countries.[38]

With respect to the nature of intellectual property rights, the Agreement states that:

> The protection and enforcement of intellectual property rights should contribute to the *promotion of technological innovation and to the transfer and dissemination of technology*, to the mutual advantage of producers and users of technological knowledge and in a manner conducive to social and economic welfare, and to a balance of rights and obligations.[39]

This statement reasserts the technological basis of IPRs and does not give any recognition to a potential role for IPRs within an environmental context, except to the extent that such a role is seen as an issue of "economic welfare". Article 8(2) permits countries to adopt measures which prevent abuses of intellectual property rights which "adversely affect the international transfer of technology". This could be conceived widely by developing countries with respect to biotechnology products, and seems to echo the sentiment contained within Article 16 of the Biodiversity Convention.

With regard to the patenting of biotechnology products, the TRIPS Agreement attempts to find a balance between the broad protection found under the US system, and the public interest concerns of developing countries, which found expression in the Biodiversity Convention. Members are permitted to exclude from patentability:

(a) diagnostic, therapeutic and surgical methods for the treatment of humans or animals;
(b) plants and animals other than microorganisms, and essentially biological processes for the protection of plants and animals other than non-biological and microbiological processes...[40]

With respect to compulsory licensing, the TRIPS Agreement places procedural limits upon the ability of governments to provide for such licensing. Authorisation can only be given on a case-by-case basis, and only where the proposed user has failed to reach reasonable commercial terms with the right holder.[41]

Despite the initial concern expressed by the USA over the Biodiversity Convention, the subsequent TRIPS Agreement seems to refute the suggestion that the Convention represents a significant threat to the rights of intellectual property right holders. Indeed, a reinterpretation of the Convention's provisions, in the light of the TRIPS Agreement, may mean that the Convention will actually strengthen the intellectual property regimes in the developing countries.[42]

---

[38] *Ibid.*, Part III.
[39] *Ibid.*, Article 7 "Objectives" (emphasis added).
[40] Article 27(3). Subparagraph (b) also provides for a review of this provision after a period of four years. This seems to reflect concerns that such distinctions may be increasingly unworkable.
[41] Article 31 "Other Use Without Authorization of the Right Holder".
[42] See M.A. Gollin, "An intellectual property rights framework for biodiversity prospecting", in World Resources Institute, *Biodiversity Prospecting* (1993), at p. 191.

Overview

Intellectual property rights were established to encourage human creativity by protecting the processes and products of such activity. Such rights are described as "intangible property". What are the key economic characteristics of intangible property rights?

(a) Within the legal jurisdiction, the right-holder has an *exclusive* monopoly right to restrict the use (e.g. copying, adaptation and distribution) of the *information* embodied within the subject matter;[43] and
(b) The right to use this *information* can then be sold outright, or particular uses can be "licensed".

Where the intangible property right is sold outright, the nature of the investment return is the same as if it were an item of tangible property. The key economic advantage that accrues to intellectual property, in terms of value, is the ability to licence for "royalties" multiple and continued use of the *same* piece of information.

When considering the protection of natural genetic material, it is their potential as a source of useful "information" which is the key to their economic value, not their physical manifestation. In terms of a biodiversity conservation policy, this characteristic of intellectual property rights may have important consequences. Developing countries could, for example, perceive that although they should preserve collections of species for research purposes, wider conservation policies would not yield economic value.

Property Rights

In nearly all legal systems, some form of property ownership forms the underlying basis upon which the economic system operates.[44] Individuals and legal persons (e.g. companies) are usually able to acquire rights in property and gain economic returns for their investment in such property through, for example, sale, lease or licence.

Land ownership, historically one of the main categories of property right, will usually convey an array of rights upon the owner, such as the right to extract any valuable minerals contained within the earth.[45] Such rights will also usually extend to the material existing on the surface of the land. A natural genetic material discovered on an area of land, which for example has an identified medicinal use, could therefore be exclusively harvested and marketed by the land owner.[46]

This fundamental right over property is recognised in the Biodiversity Convention in Article 3:

> States have, in accordance with the Charter of the United Nations and the principles of international law, the *sovereign right* to exploit their own resources...[47]

---

[43] Copyright, another form of time-limited intellectual property right, is usually said to protect *expression* rather than underlying ideas, although the distinction is not always clear. In the biotechnology field, copyright is not a significant means of protection, except perhaps for gene sequence diagrams.
[44] A primary distinction is made between real (immovable) property and personal (movable) property.
[45] Unless such rights are vested in the State, e.g. UK Coal Act 1938, s. 3.
[46] E.g. under the UK Theft Act 1968, s. 4(3), a distinction is made based on the final use of the material: "A person who picks mushrooms growing wild on any land, or who picks flowers, fruit or foliage from a plant growing wild on any land, does not (although not in possession of the land) steal what he picks, *unless he does it for reward or for sale or other commercial purpose.*"
[47] This Article is derived from, *inter alia*, Principle 21 of the Stockholm Declaration (emphasis added).

This is reasserted at Article 15(1) with respect to genetic resources:

> Recognising *the sovereign rights of States over their natural resources*, the authority to determine access to genetic resources rests with the national governments and is subject to national legislation [emphasis added].

The alternative viewpoint, that biodiversity should be classified as the "common heritage" of humankind, was rejected by the drafting committee at an early stage in the proceedings.[48]

The declaration in the Convention has already prompted governments to amend their legislation appropriately. The Australian State of Queensland, for example, has proposed an amendment to its Nature Conservation Act, "to give the state outright ownership of its flora and fauna and guarantee that it shares in any profits made from exploiting them". Such an action is seen as being essential to "halt a 'systematic search of our biota' by foreign laboratories and pharmaceuticals companies".[49]

During the 1970s and 1980s, the developed nations saw a significant rise in the value of land as an investment asset. This in turn led to the adoption of increasingly complex legal arrangements concerning the sale, leasing and subsequent use of land, such as joint ventures agreements between the land owner and the developer. It may be that developing countries could benefit greatly from the adoption of some of these legal mechanisms to ensure economic returns for investments in biodiversity which subsequently result in the "discovery" of a useful genetic sequence on a particular piece of land.

INBio[50]

A scheme based on the assertion of tangible property rights has been established by the Costa Rican government. The government established a quango, INBio, which is a research organisation composed of scientists working on developmental projects. INBio has recently signed an exclusive contractual agreement with the US pharmaceutical company, Merck, under which Merck is awarded all rights to develop and manufacture any "useful" genetic resources discovered by INBio.[51] In return, Merck has paid an up front fee of around US$1 million for the exclusivity arrangement, and has agreed to pay royalties upon any resultant commercial product.

The type of scheme outlined above, although not designed primarily as a mechanism for controlling the economic exploitation of national genetic material, does provide a useful case study of the type of policy that LDCs could pursue. In legal terms, the scheme is based on an assertion of legal ownership in the natural habitat as "tangible" property. In addition, it permits the "property owner" (in this case the Costa Rican government) to establish a range of subsidiary legal arrangements, such as providing for *contractual* rights to carry out prospecting activities in a particular territory. Under the INBio scheme, access to genetic material is therefore controlled by legal agreement and by the practical separation of the party who "prospects" for material from the party who develops any subsequent product.

---

[48] See IUCN, *A Guide to the Convention on Biological Diversity* (1993), Environmental Law and Policy Paper No. 3, at p. 3.
[49] Quoted in *New Scientist*, 1 May 1993, at p. 7.
[50] For a general review of the INBio scheme and its potential role in future environmental policies, see *op. cit.* n. 42.
[51] INBIO has been given a non-exclusive concession for prospecting. Independent prospecting is still permitted under licence from the Wildlife Department.

Seed Banks

In addition to countries appropriately asserting their tangible property rights, it is also critically important that the policies of such organisations as the International Plant Genetic Resources Institute (IPGRI) are reviewed.[52] The IPGRI co-ordinates the activities of seventeen International Agricultural Research Centres (IARCs) which operate as *ex situ* gene banks to maintain copies of plant germplasm.

In terms of legal protection, one of the key issues raised by these *ex situ* seed banks is the access that is provided to the stored germplasm. For example, the IARCs provide free and open access to the material being held, "for the benefit of the world community".[53] Where germplasm is provided to private organisations, however, the organisation is required to sign a "material transfer agreement".[54] This agreement states that in the event that the germplasm is to be commercially exploited under a protective legal regime such as patent law, then the organisation is required to enter into negotiations with the relevant IARC, "to ensure that any useful genes discovered in the material could not be withheld from the country from which the material originated". Such provisions also envisage the payment of monies to the country of origin.[55]

Although the centres have been established to conserve plant germplasm, the manner under which they operate their "open access" policy is seen as having "essentially disenfranchized the host country from their genetic assets".[56] In order to protect countries against such disenfranchisement, genetic databanks will need to operate on commercial grounds, otherwise the opportunity for the host/land owner to gain economic returns for exploitation rights would be removed. One possibility would be a royalty-style charge levied on users. This type of arrangement is currently being operated by some *ex situ* seed banks. For example, the Royal Botanic Gardens (UK) requires royalty returns on any commercialised product, half a which are returned to the donor country.[57] An alternative, or supplementary, solution would be to adopt some form of "expert's solution" whereby samples are only provided to persons who give some form of legal (i.e. contractual) undertaking not to use the sample except for experimentation nor to make it available to third parties.[58]

Compensation Right

The creation of a "compensation-style" *sui generis* right is a potential alternative mechanism to the promotion of intellectual property-style protection. Such a right could be classified as a "negative" property right, existing somewhere between positive tangible land rights and intangible intellectual property rights, as illustrated in Figure 9.1.

---

[52] Formerly the International Board for Plant Genetic Research (IBPGR).
[53] International Plant Genetic Resources Institute, 1993.
[54] The Commission on Plant Genetic Resources' document "Model Agreements for the International Agricultural Research Centres", CPGR/93/11.
[55] See S. Johnston, "Conservation Role of Botanic Gardens and Gene Banks" (1992) 2: 2 RECIEL 171–181.
[56] T.M. Swanson, "The Role of Wildlife Utilization and other Policies in Biodiversity Conservation", T.M. Swanson and E. Barbier (eds.), *Economics for the Wilds: Wildlife and Wildlands, Diversity and Development*, at p. 87.
[57] *Op. cit.*, n. 55, at p. 179.
[58] See, for example, Rule 28 of the European Patent Convention.

Figure 9.1   Property Rights Continuum

In its simplest form, a "compensation-style" right would operate through international recognition that the country which sourced a particular type of genetic material would have a "right" to a royalty-style payment from the organisation which had developed and marketed the end product. A comparable principle can be seen to operate under copyright law. Under the Berne Convention on international copyright, authors have a set of "moral rights" which are retained even after the economic rights have been transferred,[59] and can indeed continue to have certain economic rights.[60] A "compensation-style right is categorised as "negative", because it would be designed to provide economic returns only if a product resulted; in contrast, returns on "positive" property rights are usually asserted by the rightholder from "first-use".

In 1987, a UN Food and Agricultural Organisation (FAO) meeting of the Commission on Plant Genetic Resources adopted a call for the right to compensation of source states for donated germplasm.[61] To date, this concept (International Fund for the Conservation and Utilization of Plant Genetic Resources) has yet to be fully established as a practical proposition, since a method of funding the scheme has not been agreed.[62] However, if widely adopted, it would address the issue of providing countries with an economic incentive to preserve *in situ* genetic resources.

Bearing in mind the political issues surrounding the creation of an international compensation arrangement, it should be noted that the compensation-style solution could be enforced at an individual level, through contractual agreement. Such a scheme would again be based on controlling access to the genetic material and assertion of "land rights" with respect to its collection and use. The contractual agreements under which organisations would be permitted to prospect within a habitat could also include provisions stating that, in the event that a marketable "economic product" resulted from the source material, then the "land owner" is entitled to a percentage payment from the sale of that product. The amount could be the subject

---

[59] The Berne Convention for the Protection of Literary and Artistic Works 1886 (the most recent revision was Paris, 1971), states at Article 6(1): "Independently of the author's economic rights, and even after the transfer of the said rights, the author shall have the right to claim authorship of the work and to object to any distortion, mutilation or other modification of, or other derogatory action in relation to, the said work, which would be prejudicial to honour or reputation."

[60] The concept of "resale royalty rights" or *droit de suite*, at Article 14(1): "The author ... shall, with respect to original works of art and original manuscripts of writers and composers, enjoy the inalienable right to an interest in any sale of the work subsequent to the first transfer by the author of the work."

[61] The Commission (CPGR) was established at the 22nd Session of the FAO Conference in November 1983, which also adopted the "International Undertaking on Plant Genetic Resources", Resolution 8/83. This Undertaking was based on "the universally accepted principle that plant genetic resources are a heritage of mankind and consequently should be available without restriction" (Article 1). See further D. Cooper, "The International Undertaking on Plant Genetic Resources" (1992) 2: 2 RECIEL 158-166.

[62] See C.M. Correa, "Biological resources and intellectual property rights" (1992) 5 EIPR, at p. 154. See also Commission on Plant Genetic Resources, Report on the Fifth Session, April 1993, at paras. 16–18.

of a pre-agreed mechanism, such as use of an independent expert. Similar such "compensation" provisions are regularly used in property law when land is sold for development purposes.

Overview

It can be seen that the assertion of property rights, in conjunction with contractual arrangements, can act as an effective control over the exploitation of genetic material. Such control enables the owner of the material to gain economic rewards from the successful commercialisation of such material. In the past, developing countries have not made adequate use of such rights, thereby potentially disenfranchising themselves from those resources which have already passed to *ex situ* seed banks. The Biodiversity Convention contains specific provisions obliging countries to implement procedures for the control of access to biological material. Article 9(d), for example, states that Contracting Parties should:

> Regulate and manage collection of biological resources from natural habitats for ex-situ conservation purposes so as not to threaten ecosystems and in-situ populations of species...

In this regard, Article 15(5) of the Convention provides additional protection for source countries by requiring that those seeking access to genetic resources obtain the "prior informed consent" of the source country.[63]

The debate that has grown up in this area would seem, however, to be altering such perceptions; for example, the US National Cancer Institute, which prospects for genetic material in some twenty-five countries, has currently only entered into contractual agreements with four of them, but expects that number to substantially increase in the future.[64] It can be expected therefore that countries will increasingly make use of legal mechanisms, such as the INBio project, to preserve their control.[65]

Contrary concerns can be raised with regard to a property rights approach to conserving biodiversity. When countries assert their property rights, it tends to operate at a national level, such as INBio. However, much of the genetic material currently being prospected by the biotechnology industry currently comes from traditional materials used by indigenous tribes situated within, or spanning, national boundaries. The pursuance of a property right approach therefore needs to consider the diverse legal entities that should be encouraged to assert such ownership. During the preparation of the Biodiversity Convention, the Brazilian government successfully opposed an earlier draft which referred to the "common concern of all peoples", on the grounds that this could be used as a means of conferring rights on indigenous peoples. It can seen, therefore, that the assertion of property rights by one entity can mean the disenfranchisement of alternative groups.

A second potentially conflicting concern arises between the legal principle of property rights and alternative legal approaches that recognise the right to free access or the common ownership of natural resources. The policy of the IPGRI can be praised as well as criticised since, as has already been noted, the policy of free access

---

[63] The concept of "prior informed consent" finds echoes in the intellectual property field in the case of *John Moore* v. *Regents of the University of California, op. cit.,* n. 19.

[64] See S.K. Miller, "High Hopes Hanging on a 'Useless' Vine", *New Scientist,* 16 January 1993, at pp. 12–13.

[65] With respect to the implementation of Articles 16 and 19 of the Biodiversity Convention, the government of Switzerland has stated that it will make use of contractual agreements "between Swiss companies and private companies and government bodies in other Contracting Parties": Chandler, *op. cit.,* n. 5, at p. 166, n. 79.

is based on the "universally accepted principle that plant genetic resources are a heritage of mankind and consequently should be available without restriction" (FAO International Undertaking on Plant Genetic Resources). In addition, other international treaties covering natural resources, such as those covering the sea-bed, outer space and Antarctica prohibit the assertion of national territorial claims. However, any such "common ownership" approach is objected to both by developing and developed countries due to the current possibility of asserting intellectual property rights over genetic material.

A *Sui Generis* Right?

Interest in the potential role of intellectual property rights in biodiversity conservation can be demonstrated in two separate but related issues. The first issue is that by establishing legal protection for unmodified genetic resources, those responsible for investing resources in biodiversity conservation will have an economic incentive to preserve the natural habitat in which the genetic material is embodied. The second issue arises from the perception that unmodified genetic material is currently being expropriated by biotechnology companies from the developed world and converted into a product, which is then sold back to the developing nations at a significantly higher value. The establishment of some new form of intellectual property-style right is seen as one mechanism by which such issues can be tackled and "the fair and equitable sharing of the benefits arising out of the utilization of genetic resources" can be achieved.[66]

The Biodiversity Convention provides a possible legal basis upon which such a new right could be promoted. Article 11 of the Convention states:

> Each Contracting Party shall, as far as possible and as appropriate, adopt economically and socially sound measures that act as incentives for the conservation and sustainable use of components of biological diversity.

One possible "measure" is the creation of a *sui generis* "intellectual property-style" right in *the discovery of wild genetic material*.[67] Such *sui generis* legal regimes have been adopted in other fields to protect items that were not seen to have been adequately protected under the traditional law.[68] The creation of a *sui generis* "intellectual-property style" right gives rise, however, to a range of legal, policy and practical questions, some of which are discussed below.

Ownership

Who is responsible for discovering such material? Under current practices, useful genetic material is either found among the traditional materials used by indigenous peoples, or through deliberate prospecting for useful genetic resources.[69] The latter

---

[66] Biodiversity Convention, Article 1.
[67] See Correa, *op. cit.*, n. 62, at p. 154.
[68] E.g. semiconductor topographies and electronic databases.
[69] It should be noted that investment by biotechnology and pharmaceutical companies in such prospecting was minimal during the 1960s and 1970s, when synthetic chemistry was the primary means of developing new products. Although interest in natural products research has risen over recent years (partly due to improvements in screening techniques), it still represents a minor part of research efforts. Such interest could decline again as genetic engineering techniques develop and in response to rising prospecting costs. See, generally, Reid *et al.*, "A New Lease on Life", in *Biodiversity Prospecting, op. cit.*, n. 42, at pp. 1–52.

is primarily being carried out on behalf of organisations based in the developed countries, such as the US National Cancer Institute, a recognised exception being the INBio project. The creation of an intellectual property right for the individual discovering the genetic material would currently tend to favour the biotechnology companies from the developed world that fund such "prospecting", thus continuing and legally enhancing the existing situation. Such an imbalance is unlikely to alter until developing countries are able to devote sufficient resources to carrying out such basic research work themselves. The recognition of an "economic right" in the discovered material could, however, give developing countries the necessary incentive to invest the appropriate resources.

Under the Biodiversity Convention reference is made to the "Country of origin of genetic resources".[70] Under a *sui generis* system, would the "property" right accrue exclusively to this country, where the useful genetic material was originally "discovered"? This could create a complex system of conflicting claims between nations which would require some form of international arbitration procedure. In terms of preserving biodiversity, the "single right" issue could mean that countries that failed to register an original claim for a particular genetic resource would have little incentive to preserve their stock of the resource.[71] To avoid this result, and borrowing from the principles underlying copyright law, the existence of multiple different *in situ* "discovery" rights could be permitted. The existence of multiple rights could, however, undermine the economic benefits intended by the system. For example, would a biotechnology company be required to make payments to all those countries which "own" (possess) the germplasm, or only to that country from which it was specifically obtained? The former arrangement could make the costs prohibitive; while the latter would require some means by which origin could be ascertained. In addition, where payment is only made to one of the "right-holder" countries, a competitive market would be likely to arise between countries eager to sell the use of their genetic resources. Such a market could potentially drive the price (royalties) down to a level which may remove the opportunity cost benefits of maintaining biodiversity, rather than engaging in other forms of economic activity. In general with property, the owner has the right to fix the level of reward that he receives (e.g. licence fee or outright purchase price). However, any new international right system may need to specify the minimum level of royalties payable (e.g. on a percentage basis).

Scope

Would the property right attach to a particular "identified usefulness" of the unmodified genetic material,[72] or to the material as a whole? One key legal issue concerns the definition of "use". Under traditional patent law principles, product claims are held to extend to all its various uses, both recognised and potential. Over recent years, decisions from the European Patent Office have begun a trend to amend this principle by allowing claims, particularly in the area of medicines, to be made for discovered alternative "uses" of a particular compound.[73] Part of the

---

[70] Article 2 "Use of Terms".

[71] A range of "sustainable development" issues also arise when considering property rights. An alternative version of the problem would be where a landowner destroys natural habitat to farm a crop which was found to contain a useful compound; see Miller, *op. cit.*, n. 64, at p. 13.

[72] Swanson, *op. cit.*, n. 56, at p. 256.

[73] Second medical indication/EISAI, Decision Gr 05/83, [1985] OJ EPO 64, [1979–85] EPOR 241.

justification for this development was the recognition that the "first use" principle had a potentially detrimental effect in terms of scientific research, to the extent that investment in further research into the patented compound was seen as merely enhancing the value of the compound to the original claim holder.

In economic terms, however, the recognition of multiple uses each of which can give rise to separate "rights" could significantly reduce the expected returns for each right-holder; or alternatively, "first" claims could be significantly delayed until the claimant has satisfied himself that all potential marketable uses have been examined. The issue of "use" therefore raises further complexities. In many cases, any claim based on "identified use" would need to distinguish between the "genetic material" and the specific compound which creates the "useful" effect. For example, a vine recently discovered in the Cameroon contains an alkaloid, michellamine B, which appears to inhibit HIV's action upon human cells. In this case, the isolation and purification of the "useful" compound will be the subject of a patent application.[74] To cover such situations, any *sui generis* right may need to distinguish between the genetic material (the vine) and the potentially patentable compound (michellamine B); or, alternatively, the compound would be subject to two separate legal regimes.

Another critical issue concerns the *duration* of any right in genetic material. As has been mentioned previously, except for trade secrets, intellectual property rights are time-limited economic monopolies. Should a *sui generis* right be similarly time limited, or would a perpetual right be more suitable in terms of meeting the object-ive of incentives to maintain biodiversity conservation?

Formalities

What formalities regarding ownership would be appropriate? The information-based nature of intellectual property creates obvious difficulties with regard to protection. Historically, it has therefore proved necessary for formalities to be imposed upon right-holders, either at the moment at which the "property" is recognised, or as part of the enforcement process. The complexities and cost of such formalities are important considerations in terms of a *sui generis* IP right for unmodified genetic sequences.

Under patent law, the written statement is the primary basis upon which rights are asserted. Patents are unique and therefore the application which records the claim is the critical evidential document. When a patent application is made, part of the approval process involves extensive checking against existing patents to establish "uniqueness". In addition, it should also be noted that the manner in which a patent application is drafted, in terms of the scope and nature of the claim, is critical in determining the range of material that is granted a patent. Attempts to draw a clear distinction between the degrees of human intervention required to bring natural genetic sequences within the scope of the law therefore partly depends on the skill of the person drafting the claim.

Under copyright law, however, protection generally arises with no need for formal acceptance such as registration. Notices of authorship are usually attached to the "work" to notify the user that copyright exists, as well as assisting enforcement in some jurisdictions.

If a *sui generis* right were to be established for "identified use" in an unmodified genetic material, the question of what formalities are necessary would turn primar-

---

[74] Miller, *op. cit.*, n. 64.

ily on whether "ownership" was to be unique or multiple. The former, patent-based system, is extremely expensive to maintain and operate, but does provide for certainty. Significant litigation would be expected to result from entities disputing each other's claims, particularly where individual "uses" are the subject of separate claims. On the other hand, the latter, copyright-based system, is more flexible but gives rise to significant litigation at the point of proving infringement. In either case, the costs of enforcement and dispute resolution need to be considered, especially since this aspect would tend to disadvantage the developing countries. The establishment of meaningful property rights requires that suitable mechanisms exist to enable the right holder (e.g. a national government) to demonstrate "ownership", as well as proving that use has been made of a particular germplasm "owned" by the right-holder.

Conclusion

The general perspective adopted within this chapter is the possibility of granting countries some form of property right over their natural genetic material in order to provide countries with an economic incentive to institute environmental policies preserving biodiversity. From this standpoint, however, a distinction has been drawn between the existence or creation of a "positive" property right, which provides the owner with control over the economic exploitation of his germplasm, and the establishment of some form of international mechanism for compensating countries for the use of their natural genetic resources. The latter is outside the scope of intellectual property law, except to the extent that "information" is the underlying subject matter of the "right" and similar payment mechanisms could be adopted.

When considering the former option, property rights in natural gene material, a second critical distinction has been drawn between legal restrictions over access to, and removal of, the *physical* material which contains the expressed genetic information, the area of tangible property rights; and legal protection over the use of the *information* represented within the genetic material, the area of intangible intellectual property rights. In terms of preserving control over a habitat's natural biota and preventing the significant net loss of genetic material to the biotechnology industry of the developed world, the former area of "land rights" is the most important area for assertive environmental law.

National governments and individual landowners have the legal right to restrict access to organisations to prospect for genetic material on their territory. In the past, countries have failed adequately to enforce such rights, partly due to a lack of awareness of the potential value contained within their genetic resources. In addition, the policies of organisations such as the IPGRI will have to be altered to reflect the demand for controlled access.

In terms of overall *economic value*, legal protection of the information represented within genetic material must also be recognised as a concern; this is the area of intellectual property rights. Over recent years, patent law has become the major legal mechanism for protecting gene material. This chapter has considered intellectual property rights in relation to genetic material and biodiversity from two perspectives:

(a) Do intellectual property laws distinguish between genetic material which have been created or altered through human intervention from unmodified genetic material?

(b) What features would a *sui generis* "intellectual-property" style right for unmodified genetic material possess?

It has been shown that generally intellectual property laws do not necessarily make a clear or definite distinction between natural and artificial genetic material. To be information capable of being patented, for example, genetic "inventions" have to satisfy the requirements of novelty, obviousness and usefulness. Neither ideas nor discoveries are patentable in isolation, although both can be patentable when applied. Intellectual property rights are intimately linked with human activity and creativity; therefore it is the nature of the claim made in respect of genetic material which is the key determinant, not the source of the information.

If the first question has a qualified answer, then the need for a new *sui generis* right in natural genetic sequences would appear more limited. The extension of legal protection to mere discoveries would seem to be an unacceptable restriction of the principle of freedom of information and subsequently all forms of research and development. Some additional characteristics therefore need to be present. Borrowing from patent law, the concept of "usefulness" seems the most relevant requirement; although in a less stringent form than required under patent law. However, before promoting the establishment of such a *sui generis* right, it is necessary to consider and resolve a complex range of issues, from definitional problems connected to "use", to practical issues such as formalities and the cost of enforcement.

As with existing intellectual property rights, when considering the nature and extent of any *sui generis* right for unmodified genetic material it is necessary to have a clear focus on the policy issues that gave rise to the right. Patent laws are seen primarily as encouraging innovation of benefit to economic development, copyright laws are designed to reward human creativity, while trade secret laws primarily protect the holder from unfair competition. A *sui generis* right in unmodified genetic material would be designed to encourage investment in the preservation of biodiversity. When evaluating the nature of such a right, it is always necessary to measure its potential operational impact against the achievement of this policy aim.

Property rights are increasingly being asserted by owners of genetic material, whether at a governmental, organisational or individual level. Both tangible and intangible property protection are available to developing countries as a means of protecting their investment in biodiversity. The latter method is simply more complex and expensive, and to date has primarily been exploited by organisations from the developed world.

Property law protection for genetic material should not focus on the nature of the material itself, but on concerns of ownership and access. Restrictions on access, either through property or intellectual property laws, do enable economic returns to be achieved. Conversely, it should also be noted that inhibiting access to information has opportunity costs which need to be carefully weighed by the international community.*

---

* The author gratefully acknowledges the comments of Alison Firth of the Centre for Commercial Law Studies, Queen Mary and Westfield College, University of London, during various stages in the writing of this chapter.

10

Biodiversity Conservation in the United States

Kristina Gjerde

Introduction

The Global Convention on Biological Diversity (Biodiversity Convention) requires Contracting Parties to develop comprehensive national strategies, plans or programmes for the conservation and sustainable use of biological diversity.[1] To date, approaches at the national level to protection of biodiversity generally have been haphazard, developed in response to a particular threat, and not part of any comprehensive scheme. Thus, Contracting Parties now have the opportunity, as well as the obligation, to systematise their approaches to biodiversity conservation.

National strategies must confront the chief threats to biodiversity: habitat loss, over-exploitation, pollution and introduction of alien species.[2] It is widely acknowledged that the most effective way to preserve biodiversity is to maintain ecosystems capable of responding to natural environmental change.[3] This was recognised in the Preamble to the Biodiversity Convention, which states:

---

[1] Convention on Biological Diversity, June 1992, 31 ILM 818, Article 6(a). Biological diversity is defined in Article 2 as "the variability among living organisms from all sources including, *inter alia*, terrestrial, marine and other aquatic ecosystems and the ecological complexes of which they are part; this includes diversity within species, between species and of ecosystems".

[2] See e.g. J.A. McNeely, K.R. Miller, W.V. Reid, R.A. Mittermeier and T.B. Werner, *Conserving the World's Biodiversity* (1990), at p. 38; E.A. Norse (ed.), *Global Marine Biological Diversity: A Strategy for Building Conservation into Decision Making* (1993) at pp. 88–149. All such disturbances disrupt ecosystems. Over-exploitation, pollution and introduction of alien species simplify ecosystems, by killing only certain species, or allowing other species to proliferate and dominate. Habitat loss can split ecosystems into isolated patches which become too small to sustain species, or to support sufficient numbers of individuals for a diverse gene pool. Because the components of ecosystems are so intertwined, alteration in one component of ecosystem often results in disruption of the others. J. Williams and R.M. Novak, "Vanishing Species in our own Backyard: Extinct Fish and Wildlife of the United States and Canada", in L. Kaufman and K. Mallory (eds.), *The Last Extinction* (1986), at p. 114.

[3] See e.g. R.B. Costanza, B. Norton and B. Haskell (eds.), *Ecosystem Health: New Goals for Environmental Management* (1992). Healthy ecosystems are also better equipped to tolerate human-induced environmental changes such as global warming, sea level rise and ozone depletion: *ibid*.

International Law and the Conservation of Biological Diversity (C. Redgwell and M. Bowman, eds.: 90 411 0863 7: © Kluwer Law International: pub. Kluwer Law International, 1995: printed in Great Britain), pp. 191–210

A fundamental requirement for the conservation of biological diversity is the in-situ conservation of ecosystems and natural habitats, and the maintenance and recovery of viable populations of species in their natural surroundings.

Maintaining ecosystems requires a diverse suite of strategies – from limiting exploitation, managing habitat, acquiring land, to controlling activities and sources of pollution.[4] Such a broad-ranging approach necessarily relies on a well-developed legal system.

As host to the world's first national park, Yellowstone National Park, the United States is often considered to be in the forefront of nature conservation and environmental legislation. Since 1872, the United States has developed a comprehensive system of legislation designed both to protect the natural environment and to preserve nationally significant species. Indeed, the United States has taken the position that it requires no new legislation to implement the conservation obligations of the Biodiversity Convention.[5] Nevertheless, no system is perfect, and there may still be room for improvement. Thus, the United States may serve as an useful case study on legislative approaches to implementing the Biodiversity Convention.

This chapter will therefore describe and analyse the major US laws and programmes in four areas that most impact the conservation of biodiversity:

(a) species protection;
(b) habitat preservation;
(c) environmental impact assessments; and
(d) pollution prevention and control.

Land-use controls and zoning regulations are not specifically addressed in this chapter, as in the United States they are an area of law traditionally left to the states.[6]

---

[4] D. Jensen, "Protection and The Nature Conservancy", *Nature Conservancy*, January/February 1994, at p. 21. The Preamble to the Biodiversity Convention also highlights the importance of a proactive, precautionary approach to biodiversity conservation: "it is vital to anticipate, prevent and attack the causes of significant reduction or loss of biological diversity at source ... where there is a threat of significant reduction or loss of biological diversity, lack of full scientific certainty should not be used as a reason for postponing measures to avoid or minimize such a threat."

[5] See Message from the President of the United States transmitting the Convention of Biological Diversity to the US Senate, Senate Treaty Doc. 103–20, 20 November 1993, p. 18. In fact, the USA negotiated the Biodiversity Convention to ensure that it did not go beyond existing US law. See M. Chandler, "The Biodiversity Convention: Selected Issues of Interest to the International Lawyer" (1993) 4 *Colorado Journal of International Environmental Law and Policy* 141–175, at p. 155.

[6] As a federal system, responsibility in the United States for the conservation of biological diversity is shared between the federal government and the states. Powers not conveyed to the US government through the Constitution are deemed reserved to the states. Although regulation of wildlife has traditionally been carried out at the state level, the US Constitution in the Treaty Clause, Commerce Clause and Property Clause grants the federal government broad authority to conserve fish, wildlife and plants. The primary source of authority for federal implementation of international wildlife treaties is the Treaty Clause, which was determined by the Supreme Court in *Missouri* v. *Holland* (252 US 416 (1920)) to authorise Congress to enact laws to implement valid treaties, even when such laws may have the effect of overriding states' wildlife laws. See e.g. H. Doremus, "Patching the Ark: Improving Legal Protection of Biological Diversity" (1991) 18 *Ecology Law Quarterly* 265, at pp. 290–294, for a discussion of federal constitutional authority to regulate wildlife conservation.

Biodiversity Legislation in the United States

Species Protection

The Endangered Species Act of 1973[7] (ESA) is the United States' strongest and most comprehensive legislation designed for the conservation and preservation of species diversity.[8] The purpose of the ESA is to preserve and protect threatened and endangered species and the ecosystems upon which they depend.[9]

The ESA provides powerful protection for species that have been pushed to the edge of extinction.[10] It prohibits the taking of endangered species and strictly regulates federal actions that may affect endangered and threatened species and their habitats. The Act authorises the Secretaries of Commerce and Interior (who share responsibility for administration of the Act) to identify endangered or threatened species, designate habitat critical to their survival, establish and conduct programmes for their recovery, enter into agreements with states, and assist other countries to conserve endangered and threatened species. By virtue of the ESA, the condition of 238 threatened and endangered species in the United States is either stable or improving.[11]

Once a species is listed, two general prohibitions become applicable: (a) under section 7 of the Act, federal agencies are prohibited from funding, authorising or carrying out any projects that are likely to jeopardise the existence of threatened or endangered species or result in the destruction or adverse modification of areas designated as critical habitat;[12] (b) under section 9 of the Act, purely private actions are affected by a universal prohibition on the taking of, or trading in, endangered species.[13]

---

[7] 16 USC 1531–1543.

[8] Other federal wildlife protection laws include: Anadromous Fish Conservation Act of 1965, 16 USC 57(b); Fish and Wildlife Coordination Act of 1958, 16 USC 661–667(e); The Lacey Act of 1990, 16 USC 667(e), 701; Bald Eagle Protection Act of 1940, 16 USC 668(a)–(d); Federal Aid in Wildlife Restoration Act, 16 USC 669(a)–(i); Migratory Bird Conservation Act, 16 USC 715(a)–(r); Wetlands Loan Act of 1961, 16 USC 715(k)(3–5); Federal Aid in Fish Restoration Act, 16 USC 777(a)–(k); Wild and Free–Roaming Horses and Burros Act, 16 USC 1331–1340; Marine Mammal Protection Act of 1972, 16 USC 1361–1407; Magnuson Fishery Conservation and Management Act of 1976, 16 USC 1801–1882; Migratory Bird Treaty Act of 1980, 16 USC 2901–2911.

[9] 16 USC 1531(b). Endangered species are those determined to be in imminent danger of extinction throughout all or a significant portion of their range: 16 USC 1532(6). Threatened species are those determined to be likely to become endangered in the foreseeable future: 16 USC 1532(20). Both determinations must be based solely on biological evidence and the best scientific and/or commercial data available: 16 USC 1533(6)(1)(A). The definition of species authorises the Secretary to list vertebrate fish and wildlife at different taxonomic levels – by species, subspecies, or "distinct population segment": 16 USC 1532(16).

[10] See generally J.C. Kilbourne, "The Endangered Species Act under the Microscope: A Closeup Look from a Litigator's Perspective" (1991) 21 *Environmental Law* 499.

[11] W. Robert Irvin, "The Endangered Species Act: Keeping Every Cog and Wheel" (1993) *Natural Resources and the Environment*, at pp. 36–38.

[12] 16 USC 1536(a)(2). All government agencies must consult with the Fish and Wildlife Service of the Interior Department or the National Marine Fisheries Service of the Commerce Department if they have reason to be believe that a project may affect a listed species. No project may go forward unless the relevant agency issues an opinion that such action (or such action as modified by the agency's suggestions) will not jeopardise the species or adversely modify critical habitat: 16 USC 1536(b)(3)(A).

[13] 16 USC 1538(a). Section 9 of ESA makes it unlawful to import, possess, sell, transport or "take" any species listed as endangered under the Act. 16 USC 1538(a)(1). "Take" is defined to include any action that will "harass, harm, pursue, hunt, shoot, wound, kill, trap, capture, or collect, or to attempt to engage in such conduct": 16 USC 1532(19). "Harm" is in turn defined by Secretary of the Interior as any action "which actually kills or injures wildlife. Such an action may include significant habitat modification or degradation where it actually kills or injures wildlife by significantly impairing essential behavioral patterns, including breeding, feeding or sheltering": 50 CFR 17.3.

In addition to the obligation to avoid jeopardy to a species, the ESA in section 7(a)(1) imposes an affirmative obligation on all federal agencies to undertake conservation programmes to facilitate species recovery.[14] In addition to this general obligation, the Secretaries of the Interior and Commerce have the duty to review all programmes administered by them and use such programmes to further the purposes of the ESA.

The ESA provides strong measures for species and habitat protection. However, over the years it has been amended to provide certain exceptions which seek to ameliorate its strictest requirements.[15] The Endangered Species Committee (also known as the "God Squad"), composed of high ranking federal and relevant state officials, is authorised to grant an exemption to the section 7 prohibition on federal agency actions that may jeopardise a species if: (a) there are no reasonable and prudent alternatives; (b) the benefits of such action clearly outweigh the benefits of alternative courses of less harmful action, and such action is in the public interest; and (c) the action is of regional or national significance.[16]

Private parties are also able to obtain an exemption from the takings prohibition imposed by section 9. Under section 10(a),[17] the relevant wildlife agency may permit, under appropriate conditions, any taking of an endangered species if the taking is incidental to carrying out an otherwise lawful activity. To be eligible for such a permit, the applicant must develop a conservation plan which meets a specified set of criteria. Before allowing that action to go forward, the agency must review the applicant's conservation plan, provide opportunity for a public hearing, certify that "the taking will not appreciably reduce the likelihood of the survival and recovery of the species in the wild" and determine that the plan is adequate.[18] The agency may impose monitoring and reporting requirements upon the applicant, and may revoke the permit if the applicant is not in compliance with its terms and conditions.[19]

One of the Act's most valuable provisions is its encouragement of public participation by allowing interested persons to petition the Secretary to list a species or designate critical habitat. In addition to such input, the law grants the interested public standing to sue the federal government to ensure that the rules and regulations are enforced and to prevent any private or government entity from violating the Act.[20]

---

[14] 16 USC 1536(a). Although the scope of the duty is unclear, at a minimum, it provides the basis for federal agencies to take positive steps to implement conservation recommendations of the Fish and Wildlife Service or National Marine Fisheries Service and to actively participate in an interagency recovery plan. See Kilbourne, op. cit., n. 10, at pp. 564–572.

[15] See E.M. Smith, "The Endangered Species Act and Biological Conservation" (1984) 57 Southern California Law Review 361–413.

[16] 16 USC 1536(h). Conditions may be imposed to ensure that the federal agency minimises the adverse effects of the action upon the listed species or affected critical habitat: ibid.

[17] 16 USC 1539(a).

[18] 16 USC 1539(a)(2)(B). Before granting a permit, the agency must determine: (a) that the taking will be truly incidental; (b) that the applicant will mitigate impacts to the maximum extent practicable; (c) that the applicant will insure adequate funding for the plan; and (d) that the plan includes measures required by the agency: ibid.

[19] See Smith, op. cit., n. 15, at pp. 395–400. By incorporating such a planning capability, the ESA provides an incentive for developers to undertake comprehensive ecosystem-oriented planning for those projects most likely to pose the greatest threat to biological conservation. Ibid., pp. 399–400. For a more recent review of habitat conservation plans, see R.D. Thornton, "Searching for Consensus and Predictability: Habitat Conservation Planning under the Endangered Species Act of 1973" (1991) 21 Environmental Law 605.

[20] 16 USC 1540(g). This has allowed non-governmental organisations to both assist the government in collecting relevant information regarding species' status and to ensure that federal wildlife agencies fulfil their legal mandates.

Ecosystem Protection under the ESA

The primary mechanism for protecting the ecosystems upon which endangered and threatened species depend is through the designation of critical habitat and the consequent prohibition of its destruction or adverse modification through federal agency action.[21] Critical habitat is to be designated concurrent to listing, or within a year thereafter.[22] Critical habitat is defined as areas that contain physical or biological features which are "essential to the conservation of the species" and which "may require special management considerations or protection".[23] Examples of physical and biological features which may be considered essential to the conservation of a species include areas important for population growth, food and water resources, shelter, breeding and rearing sites, and habitats that are representative of the historic distribution of the species.[24]

Another important mechanism with potential to protect habitats and ecosystems is the development of species recovery plans. The ESA directs the Secretary to develop and implement recovery plans for the "conservation and survival" of listed species.[25] Priority for development of recovery plans is to be given to species that are most likely to benefit from such plans, particularly those that are, or may be, in conflict with development projects or other forms of economic activity.[26] Recovery plans are technical scientific documents prepared by biological experts from federal, state and local agencies, identifying specific actions required to conserve and allow recovery of particular species. As of 2 January 1991 the Fish and Wildlife Service on behalf of the Secretary of Interior had 276 approved recovery plans in place that covered 363 different domestic species (out of approximately 600 listed species).[27]

Problems of the ESA

Despite its success in keeping certain species from becoming extinct, as the supposed linchpin in federal efforts to conserve biological diversity the Endangered Species Act falls short in at least four major areas.[28] First, the Act precludes early intervention. By the time a species reaches a point where it is known to be endangered, recovery may require extremely expensive measures and the population of an endangered species may lack sufficient genetic diversity to guarantee long-term survival.[29] The ESA thus attempts to save the worst cases while avoiding any attempt at keeping healthy species or ecosystems from becoming endangered.

---

[21] See generally K.S. Yagerman, "Protecting Critical Habitat under the Federal Endangered Species Act" (1990) 20 *Environmental Law* 811–856.

[22] 16 USC 1533(a)(3) and (b)(6)(C). The Secretary is obligated to designate such habitat to "the maximum extent prudent and determinable": 16 USC 1533(a)(3)). An area may be excluded from the critical habitat if the benefits of excluding the area outweigh the benefits of including the area after considering economic impacts, "unless the failure to designate such area as critical habitat will result in the extinction of the species concerned": 16 USC 1533(b)(2) and 50 CFR 424.12. See Yagerman, *ibid.*, pp. 834–838.

[23] 16 USC 1532(5)(A).

[24] 50 CFR 424.12(b).

[25] 16 USC 1533(f)(1). See Kilbourne, *op. cit.*, n. 10, at p. 524.

[26] 16 USC 1533 (f)(1)(A).

[27] Kilbourne, *op. cit.*, n. 10, at p. 525.

[28] See generally Doremus, *op. cit.*, n. 10, at pp. 304–317; Smith, *op. cit.*, n. 15, at pp. 386–407; O.A. Houck, "The Endangered Species Act and its Implementation by the US Departments of Interior and Commerce" (1993) 64 *University of Colorado Law Review* 277, at p. 347.

[29] The average population of animal species listed as threatened or endangered between 1985 and 1991 was fewer than 1000 individuals, while for plants listed during the same period, the average population was fewer than 120 individuals. Irvin, *op. cit.*, n. 11, at p. 40.

Second, the ESA lacks a mechanism for setting priorities among species. Rather than calling for the listing of species based on their taxonomic uniqueness or their importance as indicator or keystone species,[30] the legislative standards for listing and protective measures are vague and open to manipulation. Agency guidelines for listing a species and allocating resources for recovery efforts concentrate primarily on the magnitude and immediacy of threats.[31] As a result, a few, highly visible charismatic species benefit, while other species dwindle. For example, over half of the US$102 million allocated to endangered and threatened species preservation and recovery was spent on only twelve species. Another quarter of the funding was allocated to the second twelve species and only one quarter (US$28 million) was spent on the remaining 570 listed species.[32]

Third, political pressures and administrative interpretations have limited the protection the ESA was meant to provide.[33] For example, only 16% of listed species have had critical habitat designated.[34] Even when designated, agency interpretations have eliminated the requirement that sufficient critical habitat be preserved to allow an endangered species to recover, only to survive.[35] This interpretation essentially deprives the critical habitat provision of any independent meaning, thus allowing habitat destruction to occur so long as at least one breeding population remains.[36]

Fourth, the ESA fails to tackle the larger problem of protecting ecosystems essential to the maintenance of non-endangered species. Identification of critical habitat or other areas to be protected is based on a single species. Modification of critical habitat which does not affect a listed species may go forward, despite adverse effects on other species or other components of the associated ecosystem.[37] The Act provides no basis for a systematic attempt to identify and protect ecosystems critical to a large number of species, or that contribute in other ways to the conservation of biological diversity. Moreover, in some instances, protection of individual species may conflict with protection of ecosystems.[38]

---

[30] See e.g. Doremus, op. cit., n. 6, at p. 307; Smith, op. cit., n. 15.

[31] D.J. Rohlf, "Six Biological Reasons Why the Endangered Species Act Doesn't Work – and What To Do About It" (1991) 5: 3 Conservation Biology 273, at p. 275.

[32] S. Winckler, "Stopgap Measures", Atlantic Monthly, January 1992, at pp. 74–81. However, information necessary to establish priorities may be gathered through the Interior Department's National Biological Survey established in 1993 to examine status and trends of all US wildlife habitats and ecosystems. B. Babbitt, "Protecting Biodiversity", Nature Conservancy, January/February 1994, at pp. 17–21.

[33] See Houck, op. cit., n. 28 for a detailed overview of the way political considerations have influenced agency implementation of the Act.

[34] A study conducted by the Government Accounting Office revealed that as of early 1992, 546 of 641 species had no critical habitat designated or pending. Houck, op. cit., n. 28, at p. 302.

[35] See Houck, op. cit., n. 28, at p. 302. Interior Department regulations issued in 1986 require an "alteration that appreciably diminishes the value of critical habitat for both the survival and recovery of a listed species": 50 CFR 402.02. Although "survival" is not defined, it connotes short-term continuation whereas "recovery" refers to the longer-term, ecologically meaningful continuation of the species generally including an increase in population sufficiently above the minimum viable population level so that the probability of extinction is low. Yagerman, op. cit., n. 21, at p. 842. By requiring that an action jeopardise both survival and recovery, the definition eliminates separate consideration of actions that impinge on – "even ones that could cripple – the chances of a species for recovery": Houck, op. cit., n. 28, at p. 299.

[36] Yagerman, op. cit., n. 21, at p. 842; Houck, op. cit., n. 28, at p. 301.

[37] Doremus, op. cit., n. 6, at p. 308.

[38] Doremus, op. cit., n. 6, at p. 304. Doremus provides an example of potential conflict in the Florida Everglades where lack of water threatens the health of the Everglades ecosystem. Plans to release impounded water to maintain waterflow were threatened because the impounded water supported a population of endangered snail kite. Ibid., p. 309.

For these reasons, the Act neither ensures the long-term recovery of species in their natural surroundings, nor does it offer comprehensive protection to ecosystems.

Terrestrial Habitat Conservation

Despite the limitations of the Endangered Species Act and its single-species focus, the United States has many other laws specifically providing for the protection of habitat.[39] Unlike many countries, the United States is the owner of nearly one-third of the nation's land.[40] It has a system of nature reserves that includes parks and forests, wildlife refuges, wilderness, species protection areas, and recreational, cultural and historic areas. Close to 90 million acres have been set aside for 492 wildlife refuges, 74.2 million acres for national parks, and 200 million acres for national forests, of which 34 million acres are designated pristine wilderness.[41]

The National Forest Management Act[42] (NFMA) actually incorporates the principle of conserving biodiversity into public land management.[43] The National Forest Management Act requires the Forest Service, as part of its forest planning process, to "provide for diversity of plant and animal communities based on the suitability and capability of the specific land area in order to meet overall multiple-use objectives".[44] The implementing regulations provide that biodiversity shall be preserved "so that it is at least as great as that which would be expected in a natural forest".[45]

While not having an explicit mandate to conserve biodiversity, national parks offer the most significant protection to wildlife habitat. The National Park Service Organic Act of 1916[46] establishes a system of national parks dedicated to conservation and recreational use, free from resource extractive activities. Other acts such as the National Wildlife Refuge System Administration Act of 1966,[47] the Wilderness Act,[48] and the Wild and Scenic Rivers Act of 1968,[49] also provide a significant amount of protection by requiring that the designated federal lands be maintained in a relatively primitive state (subject to a limited number of resource extractive activi-

---

[39] Public land management laws include the National Park Service Organic Act, 16 USC 1–8(a); the Refuge Recreation Act of 1962, 16 USC 460(k)–(k)(4); the National Wildlife Refuge System Administration Act of 1966, 16 USC 668(dd)–(ee); the Wilderness Act, 16 USC 1131–1136; the Wild and Scenic Rivers Act of 1968, 16 USC 1271–1287; the Federal Land Policy & Management Act of 1976, 43 USC 1701–1783; the Resource Planning Act of 1974, 16 USC 1601–1613; the National Forest Management Act of 1976, 16 USC 1601–1614.

[40] J.L. Sax, "Nature and Habitat Conservation and Protection in the United States" (1993) 20 *Ecology Law Quarterly* 47–56. Federal ownership is particularly prevalent in the western states, where natural conditions remain relatively unimpaired. For example, the USA owns 82% of Nevada, 64% of Utah, 68% of Alaska and 61% of California. *Ibid.*, p. 48.

[41] Sax, *ibid.*, p. 47.

[42] 16 USC 1601–1614.

[43] J.D. Holst, "The Unforseeability Factor: Federal Lands, Managing for Uncertainty, and the Preservation of Biological Diversity" (1992) 13 *Public Land Law Review* 113, at p. 127.

[44] 16 USC 1604(g)(3)(B).

[45] 36 CFR 219.27(g).

[46] 16 USC 1–8(a). The purpose of the national parks system is "to conserve the scenery and the natural and historic objects and the wildlife therein and to provide for the enjoyment of the same in such manner and by such means as will leave them unimpaired for the enjoyment of future generations".

[47] 16 USC 668(dd)–(ee).

[48] 16 USC 1131–1136.

[49] 16 USC 1271–1287.

ties and recreational uses).[50] The remaining public lands, not previously designated for any specific use, are required to be managed on the basis of multiple use and sustained yield by the Federal Land Policy and Management Act of 1976.[51]

Terrestrial Management Approaches

The traditional method of management for public lands and protected areas has been described by Sax as an "enclave" strategy.[52] The enclave approach sets aside a defined tract of land for a specified purpose, and then manages the land solely for that purpose. For example, national parks are managed to protect their natural flora and fauna and to permit non-destructive human visitation; wildlife refuges are managed primarily for the benefit of specific species (for example, endangered and threatened species, migratory birds, waterfowl); national forests are devoted to multiple uses, including outdoor recreation, range, timber harvesting, watershed protection, and fish and wildlife.[53] According to Sax, "Since there is no general requirement to co-ordinate management across the enclave boundaries, the boundary and not the resource traditionally has determined management strategy".[54]

The United States has only recently recognised the need to adopt a consistent ecological approach to land management problems.[55] In addition to the steps taken by Interior Secretary Babbitt to monitor the status and health of habitats and ecosystems through the National Biological Survey, the National Park Service has injected diversity considerations into its own planning decisions, and is encouraging neighboring federal land managers to integrate diversity principles into their resource management decisions.[56] The Fish and Wildlife Service has adopted an approach called "gap analysis" for its system of wildlife refuges, which is a geographic information system designed to inventory biodiversity on a large scale.[57] The analysis seeks to identify the necessary means of sustaining viable wildlife populations on often isolated segments of wildlife habitat by establishing a network of conservation lands, mixing public and private lands, and developing wildlife transport corridors where necessary.

Problems with Terrestrial Habitat Conservation

In the past several decades forty-two species of native mammals have vanished from fourteen national parks, even though these species were present and completely protected when the parks were established.[58] With increasing population and mounting

---

[50] Holst, *op. cit.*, n. 43, at p. 127, n. 75.
[51] 43 USC 1701–1783. The term "multiple use" is defined to include "the harmonious and coordinated management of the various resources without permanent impairment of the productivity of the land and the quality of the environment": 43 USC 1702(c).
[52] Sax, *op. cit.*, n. 40.
[53] As required by the Multiple Use-Sustained Yield Act of 1960, 16 USC 528.
[54] Sax, *op. cit.*, n. 40, at p. 49.
[55] See Babbitt, *op. cit.*, n. 32.
[56] R.B. Keiter, "NEPA and the Emerging Concept of Ecosystem Management on Public Lands" (1990) 25 *Land and Water Law Review* 43, at 57.
[57] Sax, *op. cit.*, n. 40, at p. 55, quoting The Commission on New Directions for the National Wildlife Refuge System, *Putting Wildlife First: Recommendations for Reforming our Troubled Refuge System* (1992), at p. 21.
[58] Holst, *op. cit.*, n. 43, at p. 128, citing Chadwick, Mission for the 90's: The Biodiversity Challenge, *Defenders Magazine Special Report* (1990), at p. 4.

demand for resources, federally managed and protected areas in the United States face ever greater threats. There is thus little room for a complacent attitude that United States' previous well-intentioned efforts at habitat conservation provide a satisfactory approach to the current requirements for protecting species or ecosystems.

First, as a result of the enclave structure of federal lands and protected areas, the USA administers adjacent parks, forests and wilderness areas according to their distinctly different management mandates, even though they are part of the same ecosystem.[59] Because land managers have no authority to regulate activities outside their boundaries, lack of interagency co-operation or buffer zones means that stringent land use rules within a protected area can fail to protect natural resources if destruction occurring outside the protected area contributes to degradation within.[60]

Second, the impetus behind federal designation of national parks and other protected areas is not always the preservation of living ecosystems; as with other developed nations, selection of areas by the United States for preservation often turns on the scenic beauty or geological uniqueness of the area, as well as its suitability for recreational or educational uses.[61] Areas that have been set aside for "multiple-uses" generally emphasise resource extraction, often at the expense of longer term conservation needs.[62]

Third, areas selected for preservation are generally too small to preserve biodiversity. According to biogeographers, small isolated areas cannot support enough diverse vegetative mass and lower order animal species to provide sufficient energy resources to maintain sustainable populations of large vertebrate species.[63] Smaller vertebrate species may be compromised by restricted gene pools that amplify undesirable traits and cannot accommodate changes in the environment or natural catastrophic events.

Fourth, those areas that have been preserved are not well distributed to maintain a cross-section of biological diversity; almost half of their total area is in the Northern Hemisphere's sub-Arctic and temperate climatological zones.[64] While the future may look bright for high altitude lichens, the USA still lacks a systematic programme for preserving unique or representative ecosystems.[65]

Fifth, without an explicit legal mandate calling for the conservation of biodiversity such as the one governing the Forest Service, current enthusiasm may fade, leaving land management agencies with the discretion to discount diversity concerns to the detriment of existing ecosystems.[66]

---

[59] Sax, *op. cit.*, n. 40, at p. 49. The potential for inconsistencies in such a strategy are illustrated in an example regarding a hypothetical area of bear habitat. The bear habitat that lies within a national park is typically managed for preservation. Yet the park may be surrounded by a national forest devoted to oil and gas development and/or timber harvesting, a military base utilised for weapons testing, and an Indian Reservation, which may lease its land for cattle grazing. Only where an endangered species is involved would these areas be managed with a similar mandate. *Ibid.* See also J.L. Sax and R.B. Keiter, "Glacier National Park and Its Neighbors: A Study of Federal Interagency Relations" (1987) 14 *Ecology Law Quarterly* 207, at pp. 208–9, 215.

[60] Sax, *op. cit.*, n. 40, at p. 49; Holst, *op. cit.*, n. 43, at p. 127.

[61] Smith, *op. cit.*, n. 15, at p. 409.

[62] For example, in National Forests, timber harvesting has long been the dominant use. Sax, *op. cit.*, n. 40, at p. 49.

[63] Holst, *op. cit.*, n. 43, at p. 129.

[64] Smith, *op. cit.*, n. 15, at p. 409.

[65] Holst, *op. cit.*, n. 43, at p. 130.

[66] Moreover, it is critical that such mandate be carefully drafted. According to Holst, the Forest Service has narrowly interpreted its biodiversity mandate, often permitting detrimental activities such as clear cutting because such activities allow species normally absent from dense forests to take advantage of the newly opened areas, thus increasing the literal number or "diversity" of species – to the detriment of more sensitive native species. Holst, *op. cit.*, n. 43, at p. 131, n. 94.

Marine Habitat Conservation

Not shackled by the traditions of land managers and their enclave strategies, the National Marine Sanctuary Program in the United States may come closest to encompassing a programme focused on preserving all aspects of biological diversity. The National Marine Sanctuaries Act[67] was passed in recognition that certain areas of the marine environment possess qualities of special national and international significance which require more comprehensive management than is possible through resource specific legislation. The Act authorises the Secretary of Commerce to designate marine sanctuaries in any area of the marine or Great Lakes environment, including the exclusive economic zone, and to provide for their long term conservation and management. Site-selection criteria emphasise the importance of preservation of all the different aspects of marine biodiversity. These include "the area's natural resource and ecological qualities, including its contribution to biological productivity, maintenance of ecosystem structure, maintenance of ecologically or commercially important or threatened species or species assemblages, maintenance of critical habitat of endangered species, and the biogeographic representation of the site".[68]

The Act is administered by the National Oceanic and Atmospheric Administration (NOAA) of the US Department of Commerce, which is responsible for preparing detailed management plans to ensure long term protection of the Sanctuary and its living resources.[69] The Act gives NOAA extensive powers to regulate any activities that are not compatible with resource protection. Most sanctuaries include prohibitions or more stringent regulations on oil and gas development, harvesting of living marine resources, dumping or discharging of any wastes, vessel traffic, alteration of or construction on the seabed, and other activities that may harm sanctuary resources.[70]

Ecosystem Management of Marine Sanctuaries

Amendments in 1992[71] made four significant changes that improved the Act's ability to protect marine biodiversity through ecosystem-based management:

(a) The scope of the programme was expanded to include explicit recognition of the role of sanctuaries in preserving ecosystems in their natural states[72] and in providing places off-limits to consumptive activities.[73]

---

[67] Title III of the Marine Protection, Research and Sanctuaries Act of 1972, 16 USC 1431–1434.

[68] 16 USC 1433(b)(1)(A).

[69] Although the Act requires NOAA to "facilitate" multiple uses within the Sanctuary to the extent possible, and to engage in broad-based consultations with other federal and state agencies, citizen groups, users, and the general public during preparation of the management plans, management of each area must be guided by the primary objective of resource protection: 16 USC 1431(b)(5). See also B. Thorne-Miller and J. Catena, *The Living Ocean: Understanding and Protecting Marine Biodiversity* (1991), at p. 89.

[70] *Ibid.*, p. 90.

[71] National Marine Sanctuaries Program Amendments Act of 1992, Public Law 102–587, 4 November 1992.

[72] 16 USC 1431(a)(6). This section now recognises that "protection of these special areas can contribute to maintaining a natural assemblage of living resources for future generations".

[73] 16 USC 1431(b)(9). This section adds a new purpose to the Act: "to maintain, restore, and enhance living resources by providing places for species that depend on these marine areas to survive and propagate".

(b) NOAA's authority to control external activities affecting marine sanctuaries was clarified: the Act now prohibits any action that damages sanctuary resources, regardless of whether such action occurs inside or beyond the sanctuary boundaries.[74]

(c) Other federal agencies are now required to consult with NOAA regarding their activities that are likely to injure or destroy any sanctuary resource:[75] this requirement includes private activities authorised by licences, leases or permits issued by any federal agency. NOAA must respond with practical alternatives, which may include conduct of the activity elsewhere.

(d) Opportunities for co-ordination with other federal agencies, as well as with state and local authorities were improved through authorisation of sanctuary advisory councils. Advisory councils are to be composed of relevant federal, state and local government agencies, regional fisheries management councils, and educational, civic, environmental and user organisations and private citizens interested in the long term protection and multiple use management of sanctuary resources.[76]

Problems with the National Marine Sanctuary Program

Notwithstanding substantial improvements in the Sanctuary Act by virtue of these amendments, the Marine Sanctuary Program continues to have problems that impede its ability to protect and conserve marine biodiversity. First, the site selection and designation process has been both slow and non-systematic. In its twenty-two years of existence, only thirteen sanctuaries have been designated. During the eight years of the Reagan administration, only one sanctuary was established – in American Samoa. Thereafter, Congress took the initiative in 1988 and compelled NOAA to designate four specified sanctuaries by 1990.[77] A fifth sanctuary, the Florida Keys, was designated through special legislation, completely bypassing NOAA administrative procedures.[78]

Second, despite the Act's potential to function as a focal point for conserving marine biodiversity, Congress has chosen to downplay this role in favor of designating sites for other reasons (e.g. to stop offshore oil and gas development, to protect popular tourist destinations, to provide protection from ships). Of the twelve recognised biogeographic provinces in US coastal waters, only seven contain sanctuaries. An expert panel's recommendation[79] that the programme should strive for the selection of outstanding marine areas (on a relatively large geographical scale) representative of the biogeographical provinces of the US coast was rejected when the 1992 amendments were adopted.

Third, even with NOAA's new authority to review activities external to the Sanctuary for their impact on Sanctuary health, Sanctuaries remain under siege from

---

[74] 16 USC 1436.

[75] 16 USC 1434.

[76] 16 USC 1446.

[77] Cordell Banks, Flower Garden Banks, Monterey Bay, Western Washington Outer Coast, see the Marine Sanctuaries Amendments 1988, Section 205 of Public Law 100-627, 7 November 1988.

[78] Florida Keys National Marine Sanctuary and Protection Act, Public Law 101–605, 16 November 1990, codified as 16 USC 1433 note.

[79] Marine Sanctuaries Review Team, *National Marine Sanctuaries: Challenge and Opportunity*, A Report to the National Oceanic and Atmospheric Administration, 1991, at p. 22.

unwise land use ashore. The National Marine Sanctuaries Program has virtually no control over deteriorating water quality from sewage, sediment, pesticides, and other harmful substances currently entering the marine environment. An exception may be the Florida Keys National Marine Sanctuary, where Congress mandated that NOAA, EPA and state and local Florida agencies work together to develop a comprehensive water quality protection plan.[80] NOAA otherwise has no way of regulating ongoing polluting activities or coastal development, only new activities that may directly impact sanctuary resources.

Fourth, the programme's mandate for multiple use requires NOAA to "facilitate" many potentially conflicting activities.[81] The problems of reconciling commercial fishing with resource protection are exacerbated due to the fact that fishing regulations must be approved by the appropriate Regional Marine Fisheries Management Council, whose mandate is the promotion of commercial fishing. Moreover, in many sanctuaries, designation has actually served to increase the area's attraction for tourism and recreational activities, leading to problems of overuse and damage due to boat anchors and diver's fins.

Fifth, among the many problems facing the Sanctuary Program, the most significant is insufficient funding. Despite the number of sanctuaries doubling over the last six years, the Sanctuary Program budget has not kept pace. An expert panel convened to review the Sanctuaries Program in 1990 concluded that an annual budget of US$30 million was required for the programme to function effectively.[82] At that time its annual budget was US$4 million. For fiscal year 1994, Congress increased the funding level, but only to US$9 million.

Environmental Impact Assessments

The third legal mechanism in the United States that contributes to biodiversity conservation is the National Environmental Policy Act of 1969[83] (NEPA). Environmental impact assessments required by NEPA ensure that federal agencies carefully consider information concerning significant environmental impacts. Unlike ESA, NEPA's approach also ensures consideration of harms to animal and plant species which are still relatively abundant.

Congress enacted NEPA nearly twenty-five years ago in recognition of "the profound impact of man's activities on the interrelations of all components of the natural environment".[84] NEPA declares it to be national policy, inter alia, to use all practicable means and measures to "preserve important ... natural aspects of our national heritage and maintain, wherever possible, an environment which supports diversity and variety of individual choice". Thus it can be said that the goals of NEPA include the conservation of biological diversity.[85]

NEPA requires federal agencies to assess the impacts of proposed projects and prepare an impact statement for "major federal actions significantly affecting the quality of the human environment".[86] Each detailed statement must assess:

---

[80] The Florida Keys National Marine Sanctuary and Protection Act, Public Law 101-605, sec. 8(a).

[81] 16 USC 1431(b)(5).

[82] Marine Sanctuaries Review Team, op. cit., n. 79, at p. 15.

[83] 42 USC 4370.

[84] 42 USC 4331(a).

[85] R.L. Fischman, "Biological Diversity and Environmental Protection: Authorities to Reduce Risk" (1992) 22 Environmental Law 435, at p. 477, citing Council on Environmental Quality, Environmental Quality: Twenty-First Annual Report (1991), at p. 178.

[86] 42 USC 4332(2)(C).

(i) the environmental impact of the proposed action, (ii) any adverse environmental effects which cannot be avoided should the proposal be implemented, (iii) alternatives to the proposed action, (iv) the relationship between local short-term uses of man's environment and the maintenance and enhancement of long-term productivity, and (v) any irreversible and irretrievable commitments of resources which would be involved in the proposed action should it be implemented.[87]

In addition to direct effects, NEPA mandates federal agency review of the indirect and cumulative effects of proposed federal action on natural resources. The regulations define the concept of "cumulative impact" to mean "the incremental impact of the action when added to other past, present, and reasonably foreseeable future actions".[88] This includes a review of the project's potential effects on the components, structures and functioning of the affected ecosystems.

NEPA has proven to be a powerful tool.[89] The environmental impact statement (EIS) requirement provides the public with a written assessment of potential environmental impacts. Public interest groups can use the findings of an EIS (or its inadequacy, when that is the case) to formally protest a decision permitting an activity. Courts have concluded that NEPA and its implementing regulations impose rigorous procedural requirements on federal agencies, and have used the equitable injunction as a potent remedial device to ensure compliance.[90] NEPA injunctions have halted many an ill-advised project, and the threat of NEPA litigation has frequently sent agency planners back to the drawing board to reexamine their environmental analyses.[91]

Moreover, by encouraging federal agencies and land managers to look beyond narrow jurisdictional boundaries, NEPA facilitates the adoption of an ecological perspective. In addition to requiring review of the indirect and cumulative impacts of a federal agency action, it accomplishes this by facilitating interagency co-ordination among federal agencies and land managers, even those that operate under fundamentally different legal mandates.[92] According to Keiter,[93] NEPA does so by: (a) mandating consultation in the early stages of the environmental review process;[94] (b) requiring federal agencies to identify and evaluate the potential impact of projects at the earliest stages, to notify the affected agencies of the proposed action and afford them an opportunity to comment on the proposal;[95] (c) requiring the environmental impact statement to include a discussion of "possible conflicts between the proposed action and the objective of Federal, regional, State, and local ... land use plans, policies and controls for the area concerned";[96] and (d) authorising the Council on Environmental Quality to mediate interagency disagreements through a referral process that can be initiated by either agency.[97]

---

[87] 42 USC 4332(2)(C).

[88] 40 CFR 1508.7.

[89] Thorne-Miller and Catena, op. cit., n. 69, at p. 104.

[90] Keiter, op. cit., n. 56, at p. 47.

[91] Ibid., p. 45.

[92] Sax, op. cit., n. 40, at p. 53.

[93] Keiter, op. cit., n. 56, at p. 47.

[94] 42 USC 4332(2)(C) provides: "Prior to making any detailed statement, the responsible Federal official shall consult with and obtain the comments of any Federal agency which has jurisdiction by law or special expertise with respect to any environmental impact involved."

[95] 40 CFR 1501.7.

[96] 15 USC 1502.16(c).

[97] 42 USC 1504.

Problems with NEPA

Although the environmental impact assessment process can lead to dramatic improvements in a project's design and planning from both environmental and economic viewpoints, several flaws in the Act make NEPA a less than perfect tool for biodiversity conservation. First, even though NEPA requires all federal agencies to review their authorising statutes, regulations, and policies and procedures to ensure compliance with the *"intent, purposes*, and procedures set forth in this Act",[98] the Act has been interpreted to give emphasis merely to its procedural aspects, and not its intent and purposes. In *Vermont Yankee Nuclear Power Corp.* v. *Natural Resources Defense Council*,[99] the Supreme Court found that NEPA did not mandate a particular result, but merely ensured a fully informed and well-considered decision.[100]

Second, while NEPA requires federal agencies to identify and consider the cumulative environmental impacts accompanying development proposals, it does not ensure that this analysis will encompass the relevant ecosystems.[101] Although the regulations require environmental impact analyses to address the regional implications of a proposal, neither the federal land management agencies nor the courts have been eager to expand NEPA obligations beyond designated enclave boundaries.[102] Other statutes, such as National Forest Management Act and the Endangered Species Act, must be invoked for this to happen.[103]

Third, possible cumulative impacts are only considered to the extent they are reasonably foreseeable and likely to ensue. However, many impacts on biodiversity are not "reasonably foreseeable", though they may be quite possible, particularly should a worst case scenario occur. With the limited biological information currently available, the impacts a particular action will have on the long-term preservation of biodiversity can only be predicted in a limited sense.[104] Therefore, such impacts are ignored. As stated by Holst:

> By limiting our NEPA analysis to only those impacts that are reasonably foreseeable, we have chosen to err on the wrong side of caution when evaluating the biological impacts of potentially destructive activities.[105]

Fourth, in terms of interagency co-operation, NEPA ensures only "process" co-ordination. It does not ensure meaningful substantive co-ordination sensitive to transboundary ecological realities.[106] Because of its procedural focus, federal agencies are not required to adopt or heed the comments or concerns of federal wildlife agencies. Absent a legal obligation to protect shared ecosystem resources, federal land managers can, and (and often do) reject another agency's critical comments, and even ignore its opposition.[107]

Fifth, NEPA only applies to major federal actions, which leave out the universe of state and private actions as well as small-scale federal actions that may have a large

---

[98] 42 USC 4333.
[99] 435 US 519 (1978).
[100] C.A. Cole, "Species Conservation in the United States: The Ultimate Failure of the Endangered Species Act and Other Land Use Laws" (1992) 72 *Boston University Law Review* 343, at p. 360.
[101] Keiter, *op. cit.*, n. 56, at p. 50.
[102] *Ibid.*, p. 52.
[103] *Ibid.*, p. 49.
[104] Holst, *op. cit.*, n. 43, at p. 128.
[105] *Ibid.*, p. 129.
[106] Keiter, *op. cit.*, n. 56, at p. 48.
[107] *Ibid.*

cumulative effect.[108] And, finally, there is no legal requirement that federal agencies review a proposed action's potential impact on biological diversity or that they consider and seek to avoid impacts on biological diversity. Despite statements to the effect that NEPA incorporates biodiversity concerns, federal agencies have yet to consistently incorporate such concerns into their evaluations. At a minimum, incorporation would require consideration of how biodiversity of an area might be affected by habitat destruction, habitat fragmentation, pollution, exotic species, or loss of populations that would result from the proposed action.[109]

Pollution Controls

A fourth important aspect of biodiversity conservation is pollution prevention and control. Although habitat alteration may be clearly the most significant threat to biodiversity, chemical pollution has become a major threat in virtually all parts of the world.[110] Most pollution control laws[111] in the United States have traditionally focused on protecting human health and not on protecting the integrity of ecosystems or vulnerable species. In 1990 the US Environmental Protection Agency recognised the need for a new approach, and committed itself to "attach[ing] as much importance to reducing ecological risk as it does to reducing human health risk".[112]

The environmental law with one of strongest mandates to protect ecological integrity is the Federal Water Pollution Control Act[113] (also known as the "Clean Water Act"). The goal of the Clean Water Act is to "restore and maintain the chemical, physical, and biological integrity of the nation's waters". Originally aimed at eliminating polluting discharges from point sources (i.e. pollutants from industrial and municipal sources discharged via pipeline), the Clean Water Act was amended in 1987 to encompass nonpoint sources of pollution (i.e. land runoff, groundwater contaminants and airborne pollutants).[114] Based on the Clean Water Act's original

---

[108] See Cole, *op. cit.*, n. 100, at pp. 360–361.

[109] See Fischman, *op. cit.*, n. 85, at p. 479.

[110] McNeely *et al.*, *op. cit.*, n. 2, at p. 38. Contaminants in the environment can adversely affect biological diversity at several levels. First, even at concentrations with no measurable health effects on humans, pollution can injure individuals of other species that are more sensitive to particular contaminants. Pollution can either kill outright, or increase the likelihood of death from other stresses, such as drought or habitat fragmentation. Second, pollution affecting a large number of individuals can weaken or destroy whole populations. Third, the effects of a weakened or depleted population can ripple through an entire ecosystem or landscape: habitat alteration may be one result. Fischman, *op. cit.*, n. 85, at p. 443.

[111] Pollution control laws include the Comprehensive Environmental Response, Compensation and Liability Act of 1980, 42 USC 9601-9675; Federal Insecticide, Fungicide, and Rodenticide Act, 7 USC 136(a)-(y); Federal Water Pollution Control Act (33 USC 1251-1376); Clean Air Act, 42 USC 7401-7642; Toxic Substances Control Act, 15 USC 2601-2654; Marine Protection Research and Sanctuaries Act (Ocean Dumping Act), 33 USC 1401-1445; Public Health Service Act (Safe Drinking Water Act), 42 USC 300(f)-300(j)(11); Solid Waste Disposal Act, 42 USC 6901-6991; Prevention of Pollution from Ships Act of 1980, 33 USC 1901-1911. For a detailed analysis of how some of these laws could be more effectively used by the EPA to enhance the conservation of biodiversity, see Fischman, *op. cit.*, n. 85.

[112] Fischman, *op. cit.*, n. 85, at p. 439. This decision was based on a report by EPA's Science Advisory Board, which strongly recommended a reordering of EPA's priorities to incorporate ecological concerns. Science Advisory Board, US EPA, *Reducing Risk: Setting Priority and Strategies for Environmental Protection* (1990).

[113] 33 USC 1251-1376.

[114] Regulation of private land use is traditionally a state or local responsibility (see *op. cit.*, n. 6). However, through the Clean Water Act's authority to control nonpoint pollution, EPA can reach and indirectly regulate land use, and thereby prevent both direct alteration of terrestrial habitat and indirect alteration of aquatic habitat. Fischman, *op. cit.*, n. 85, at p. 455. This is particularly important because such authority can reach the two-thirds of the nation that is not federally owned.

timeline, polluting discharges into navigable waterways were to be eliminated by 1985, with an interim goal of achieving waters with quality sufficient for the propagation of fish, shellfish and wildlife, and for recreation, by mid-1983.[115]

The Environmental Protection Agency administers five major programmes under the Clean Water Act. Two programmes regulate point sources, a third addresses nonpoint sources, a fourth governs dredge and fill activities, and the fifth co-ordinates the National Estuary Program. The first point-source programme, the National Pollution Discharge Elimination System (NPDES), regulates direct discharges of industrial and municipal wastes into navigable waters.[116] Direct discharges permitted under NPDES must comply with EPA established technology-based effluent limitations as well as with state or local water quality standards (based on EPA guidelines).[117] Discharges are generally governed by regulations requiring that a certain level, or type, of technology be used (e.g. "best practicable", "best available"). For particular pollutants, maximum permissible discharge levels may also be set if needed to maintain or attain waters fit for fishing and swimming.

The second point-source programme regulates discharges of industrial wastes that go into municipal sewage treatment facilities, rather than directly into water.[118] Such discharges must meet separate standards established under a National Pretreatment Program. Standards prohibit effluent that can interfere with, pass through or is incompatible with sewage treatment facilities.[119]

Non-point pollution, the source of approximately one-half of national water pollution,[120] is addressed through the third programme, a joint co-operative programme with the states.[121] In areas where water quality falls below certain water quality standards, states must prepare plans to reduce and control all nonpoint sources. EPA provides guidance indicating suggested "best management practices", but implementation is left to the states. Efforts to address non-point pollution in coastal areas were made mandatory through amendments to the Coastal Zone Management Act in 1990.[122] These amendments require EPA and NOAA to elaborate new "guidance measures" for non-point pollution control. States are then compelled to develop and implement a Coastal Nonpoint Pollution Control Program that incorporates

---

[115] 33 USC 1251(a)(1) and (2). Obviously these deadlines were not met.

[116] 33 USC 1342.

[117] Additional requirements based on water quality criteria are imposed on direct discharges into ocean and certain coastal waters. 33 USC 1343(c)(1). Prior to issuing any NPDES permit for discharge into marine waters, EPA must determine that the discharge will not "unreasonably degrade" the marine environment. "Unreasonably degrade" includes significant adverse changes in ecosystem diversity, productivity and stability of the biological community within and surrounding the discharge area. Ocean discharge criteria are the only legislatively mandated criteria based on the sensitivity of the receiving waters, rather than on potential human uses.

[118] 33 USC 1316. See generally, T.A. Gold, "EPA's Pretreatment Program" (1989) 16 *Environmental Affairs* 459–530.

[119] 33 USC 1317(b)(1).

[120] Clean Water Network, *Briefing Papers on the Clean Water Act Reauthorization*, March 1993.

[121] 3 USC 1329.

[122] PL 101–508, Section 6217,16 USC 1455b. The Coastal Zone Management Act of 1972, 16 USC 1451–1464, establishes a voluntary program designed to provide financial encouragement to states to develop comprehensive programs to protect and manage coastal resources. To qualify for federal funding, states must demonstrate that they have programs and enforceable polices to regulate land uses, water uses, and coastal development that give full consideration to "ecological, cultural, historic and aesthetic values as well as to the needs for compatible economic development": 16 USC 1452(2).

the relevant guidance measures.[123] Such programmes are to be developed in conjunction with state activities under the Clean Water Act, but are subject to the approval of both NOAA and EPA.

Habitat destruction in wetlands and other coastal areas due to dredging and filling is the fourth major activity regulated under the Clean Water Act.[124] Permits from the Army Corps of Engineers, based on criteria developed by EPA, are required prior to filling and or discharging dredged material within three miles of shore, including wetlands. Permit criteria require consideration of the effects of the project on fish and wildlife and their habitats, and the effects on recreational values and marine productivity in the vicinity of the project. If a proposed project will have an "unacceptable adverse" effect, the EPA may veto the project.

The National Estuary Program is the fifth major programme authorised by the Clean Water Act.[125] Established in 1987, the National Estuary Program is designed to create comprehensive management plans to protect the "ecological integrity" of nationally significant estuaries threatened by pollution or development. The National Estuary Program brings together all interested parties and the general public to develop a comprehensive plan for selected estuaries. The plans are to address the control of point and nonpoint sources of pollution, implementation of environmentally sound land use practices, the control of freshwater input and removal, and the protection of living marine resources.[126] An outgrowth of the National Estuaries Program, the Watershed Protection Program, seeks to establish a comprehensive approach to water quality protection based on ecological boundaries, as opposed to manmade boundaries.[127] It targets watersheds where pollution poses the greatest risk to human health, ecological resources, desirable uses of the water, or a combination of these. Like the estuaries programme, it also relies on broad based participation and comprehensive solutions, involving both regulatory and non-regulatory mechanisms. Though not directly called on to protect biological diversity, the Watershed Protection Program includes many aspects that enhance its protection. It develops goals based on ecological criteria that include not only the traditional chemical water quality criteria, but also criteria and goals for physical water quality (e.g. temperature, flow, circulation); habitat quality (e.g. channel morphology, composition and health of biotic communities) and biodiversity (e.g. species number, range).

Both these programmes demonstrate that EPA is currently turning towards a more comprehensive, ecologically based approach. By planning for an ecosystem rather that a medium such as water or air, they are better able than conventional

---

[123] 16 USC 1455b(b). The state programmes must also: (a) identify land uses which contribute significantly to waters which fail to meet water quality standards or which are threatened by future increases in pollutant loadings; (b) identify critical coastal areas adjacent to degraded or threatened coastal waters and make them subject to more stringent management measures; (c) provide for the implementation of special management measures; (d) provide for technical and other assistance to local government and the public for implementing the special management measures; (e) provide opportunities for public participation in all aspects of the programmem; and (f) establish mechanisms to improve coordination among state agencies and between state and local officials responsible for land use, water quality, habitat protection, and public health: 16 USC 1455b(b)(1)–(5).

[124] 33 USC 1344.

[125] 33 USC 1330.

[126] Fischman, *op. cit.*, n. 85, at p. 499.

[127] EPA, Office of Water, *The Watershed Protection Approach: An Overview*, EPA/503/9–92/002, Washington DC (1991).

regulatory approaches to respond to the range of threats to biological diversity posed by the broad-range of pollution sources. By establishing partnerships with state and local governments, EPA is better able to address land use issues that contribute to both chemical pollution and habitat alteration.[128] And, most significantly, by developing ecologically based criteria that seek to protect water-dependent wildlife and ecosystem processes, EPA is pursuing its commitment to reducing ecological risks, and not solely human health risks.

Problems with the Clean Water Act

Despite these recent advances, the Clean Water Act has yet to reduce many of the threats to biodiversity. First, most water quality standards under the Act still focus on water chemistry and do not adequately ensure the protection of ecological integrity.[129] The EPA has been slow to issue, and the states slower to adopt and enforce, wildlife, habitat and biological water quality criteria similar to those adopted in the Watershed Protection Program. The Ocean Discharge Criteria are more often waived than enforced. For example, sewage treatment standards are the same nationwide, and are not based on the sensitivity of the receiving ecosystem. Thus, in nutrient sensitive ecosystems such as coral reefs, state sewage treatment requirements (based on EPA guidance) continue to permit the discharge of nutrients which stimulate coral reef-smothering algae.

Second, despite its success in identifying water quality problems in particular estuaries, the National Estuaries Program has failed to move from the identification phase to the implementation phase: the Clean Water Act lacks a specific requirement to implement the comprehensive management plan for an estuary once it has been approved; it also lacks a specific funding mechanism to ensure that the plans will be implemented, monitored and enforced.[130]

Third, EPA and states issue discharge permits to facilities and industry on a case-by-case, outfall-by-outfall basis without regard to: (a) the availability of suitable alternatives (including pollution prevention through cleaner production); or (b) comprehensive monitoring requirements, which would enable immediate action to be taken at the sign of any ecosystem degradation or other adverse impacts.

Fourth, the EPA often fails to examine the effects of proposed regulations on listed species or their critical habitats, as would be consistent with its responsibilities under section 7(a)(2) of the Endangered Species Act.[131] The effects of pollution standards, permits, programme approvals and clean up decisions on all species should be more closely considered. For example, regulations for new pollutants generally appear only after a significant threat to human health or a highly visible species is perceived.[132] A more precautionary approach would trigger immediate action upon notice of any ill effects upon the biological or ecological integrity of a waterbody, regardless of whether established standards are violated or not.

[128] Fischman, *op. cit.*, n. 85, at p. 500.
[129] B. Adler, "Protecting Biological Integrity", in *Briefing Papers, op. cit.*, n. 120, at p. 35.
[130] D. Martin, "Strengthening the National Estuary Program", in *Briefing Papers, op. cit.*, n. 120, at p. 26.
[131] Fischman, *op. cit.*, n. 85, at p. 488.
[132] Holst, *op. cit.*, n. 43, at p. 120.

Conclusion

The United States' policy towards the Biodiversity Convention was to negotiate a treaty that could be implemented through existing federal legislation.[133] While the United States has many useful programmes that contribute to the conservation of biodiversity, its insistence that absolutely no new legislation is required to implement the Biodiversity Convention appears to ignore some of the holes in existing US law. Based on this review of federal legislation in the area, it appears that US laws fall short of several fundamental prerequisites for the comprehensive conservation of biological diversity. *The primary reason is that the United States lacks a legislative mandate to protect ecosystems or to prevent species from becoming endangered.*

The Endangered Species Act cannot provide such a mandate for it cannot address the broader problem of habitat destruction: its species-specific approach allows too little too late. Furthermore, federally owned lands and protected areas – both terrestrial and marine – fail to remedy the ESA's shortcomings because they are not selected or managed with biodiversity goals in mind. Thus they are often too small to sustain species or too few to adequately represent all important ecosystems.

Federal lands, if comprehensively managed, could provide significant ecosystem protection but unfortunately the National Environmental Policy Act does not require federal agencies to co-ordinate their activities. Nor does it require them to avoid or minimise harm to biodiversity. Pollution control laws such as the Clean Water Act that could require the preservation of ecological integrity are making slow progress towards implementing criteria that consider more than human health risks and human uses of resources.

State and local laws and programmes, though not explored here, may serve to fill in some of the gaps, but there is no guarantee of consistency of purpose or effectiveness in action. Moreover, at the state and local level more man-made (political) boundaries exist, increasing the difficulties in planning at the appropriate ecosystem level.[134]

Comprehensive protection of biological diversity would seem to require, at a minimum: (a) a national commitment to preserving biological diversity in all its manifestations and forms: genes, species and ecosystems; (b) a comprehensive national strategy and action plan to address the numerous activities affecting biological diversity and ecosystem health; (c) greater co-operation and co-ordination between and among federal, state and local authorities in implementing management policies for ecosystems straddling jurisdictional boundaries; (d) environmental impact

---

[133] See Chandler, *op. cit.*, n. 5, at pp. 155–158. In the USA's statement to plenary at the close of the treaty's negotiations it sought to explain the reasons for its position:

> In regard to Articles 7–13, the United States has a tightly woven system of state and federal programmes in fish and wildlife management. Our system is undergirded by hundreds of state and federal laws and programmes and an extensive system of federal and state wildlife refuges, wildlife management areas, recreation areas, parks and forests. The United States does not intend to disrupt its existing federal and state authorities. Indeed our Government is committed to expanding and strengthening these relations. Should the United States become a Party to this Convention, its intent would be to meets its conservation obligations through existing federal laws and would look forward to continued cooperation with the various states in this regard.

> Report of the Intergovernmental Negotiating Committee for a Convention on Biological Diversity, United Nations Environment Programme, 7th Negotiating Sess. 5th Sess. of the International Negotiating Committee, at p. 35, UN Doc. UNEP/Bio.Div/N7-INC.5/4 (1992).

[134] See A.D. Tarlock, "Local Government Protection of Biodiversity: What is its Niche?" (1993) 60 *University of Chicago Law Review* 555, at p. 562.

assessments that identify and analyse the full impact, both cumulatively and geographically, of management and development proposals on biological diversity and ecosystem processes; and (e) wildlife and environmental agencies with the legal authority, capacity and obligation to retain and preserve the ecological integrity of terrestrial, aquatic and marine environments.

Recent promising signs that the USA is beginning to take its conservation responsibilities seriously is found in the decision of President Clinton to establish the President's Council on Sustainable Development. The Council was founded in 1993 to "develop and recommend to the President a national sustainable action strategy that will foster economic vitality".[135] Charged with the task of developing bold new approaches to integrate economic and environmental policies, the twenty-five-member Council is composed of high-ranking representatives from industry, government, environmental, labour and civil rights organisations. Its Natural Resources Management and Protection Task Force, one of six task forces, is to "develop an integrated vision of what constitutes sustainability for natural resources considering biodiversity, ecosystems and watersheds, with a focus on issues in the areas of wetlands, fisheries, agriculture, coastal resources and forestry". Ultimately, the Task Force is to develop policies to foster sustainable management and protection of natural resources that will remove existing barriers and catalyze new initiatives. If the Task Force achieves its goals, and the US proceeds to implement them, then perhaps the US once again may be entitled to reclaim its leadership position in global environmental policy.

It is hoped that this review of US legislation will enable both the United States and other governments to improve on the extensive experience of the USA in environmental protection. With over a hundred years of experience in drafting legislation, the US has much to emulate. However, legislation, like the ecosystems it seeks to protect, is dynamic, and thus needs updating to adapt to changing circumstances in an evolving world.

---

[135] President's Council on Sustainable Development: Briefing Packet.

11

The European Community and Preservation of Biological Diversity

Patricia Birnie

Development of EC Environmental Policy on Preservation of Biological Diversity

Introduction

As Kiss and Shelton have pointed out,[1] the origin of European States' interests in protecting what we now perceive as biodiversity lies in their concern to protect their economic resources, which led them to regulate commercial exploitation of such resources as forests, game, fish and fur seals by adopting national laws that date back to at least the eighteenth century. Thus each State gradually struck for itself, in its own way, the balance between the competing economic and environmental interests involved in protection of biological diversity without specifically directing such laws to this aim or taking a broad ecological approach. The fact that the legal systems of European States are based on a variety of legal traditions – most deriving from Roman law, some from German or Scandinavian concepts and others from the English common law system[2] – resulted in further disparity between the laws adopted.

Today the threats to biodiversity, especially from loss of habitat which was not addressed in these early laws, have become grave in many parts of Europe, especially in Central and Eastern Europe. At the same time, both internationally and regionally and especially within the European Community, the need for a more comprehensive and integrated approach has been realised, spurred on by the 1992 United Nations Conference on Environment and Development's (UNCED) adoption of the Convention on Biological Diversity and of Agenda 21.[3] The body best equipped

---

[1] A. Kiss and D. Shelton, *Manual of European Environmental Law* (1993), at p. 12.

[2] *Ibid.*, p 14.

[3] See S. Johnson, *The Earth Summit* (1992) for the texts of these documents with commentary.

International Law and the Conservation of Biological Diversity (C. Redgwell and M. Bowman, eds.: 90 411 0863 7: © Kluwer Law International: pub. Kluwer Law International, 1995: printed in Great Britain), pp. 211–234

with institutions, which can adopt both the necessary policies and laws on an integrated basis, and which is equipped both with a range of instruments for implementing these and with some means of enforcing them on a regional basis, including by recourse to a regional court, is the European Union (EU), formerly known as the European Community (EC).[4]

Despite these advantages, however, the EC suffers from certain disadvantages. First, it was not originally and, despite continuing expansion, is still not comprehensive of the whole continent of Europe; thus neither all relevant species, nor the habitats crucial to their preservation, nor the sources of threats to such habitats, fall exclusively within its competence. Secondly, the EC was established to create a common market for trading purposes; its founding treaty made no direct reference to protection of the environment or natural resources. However, the close relationship of its aim of removing barriers to trade and the economic costs of environmental protection soon became apparent, as did the need to establish a common environment policy to complement the established common policies on, for example, trade and agriculture. None the less, the tension between economic and environmental objectives remains.

The Origin and Relevance of the Five EC Action Programmes on the Environment.

The Declaration of the 1972 United Nations Conference on the Human Environment[5] and its adoption of an Action Plan for the Human Environment provided the catalyst for the adoption of the EC's own series of Action Programmes on the Environment. The first was adopted by the EC Council in 1973.[6] It expressly aimed to balance environmental and economic objectives. The setting and quality of life and surroundings and living conditions of the peoples of the EC were to be improved by an expansion of measures that procured for them an environment that provided the best conditions of life, which reconciled this expansion with the need for protection of the natural environment, and maintained a satisfactory ecological balance that protected the biosphere. It thus aimed to develop an integrated approach to conservation of endangered species (the Convention on Trade in Endangered Species – CITES – was also concluded in 1973)[7] and to protection of their habitats. This led ultimately (in 1979) to the adoption of the directive to

---

[4] The European Community (EC) was established by the Treaty of Rome in 1957; 298 UNTS 11; The European Union (EU) by the Maastricht Treaty on European Union of 1992, 32 ILM 247 (1992); ILM 1693 (1993). In this chapter the term EC is used when the actions concerned arise under the EC Treaty; EU for actions under the latter treaty. For details, see D. Lasok and J.W. Bridge, *Law and Institutions of the European Communities* (5th edn., 1991); T.C. Hartley, *Foundations of European Community Law* (2nd edn., 1991). On its environmental policy see S. Johnson and G. Corcelle, *The Environmental Policies of the European Communities* (1989), N. Haigh, *EEC Environmental Policy and Britain* (2nd rev. edn., 1990), and *ibid.*, "The European Community and International Environmental Policy" in A. Hurrell and B. Kingsbury (eds.), *The International Politics of the Environment* (1992) at p. 228, and L. Krämer, *Focus on European Environmental Law* (1992).

[5] Report of the United Nations Conference on the Human Environment 1972, UN Doc.A/Conf. 48/14 Rev 1 (New York, 1972).

[6] OJ C112, 20.12.73, p. 1 ; this was initiated by the Paris Summit of Heads of State held earlier in 1973 and ran from 1973–76. It was followed by the Second Programme (1977–81), OJ C39 10.10 1977, p.1; Third Programme (1982–86), OJ C46, 17.2.1983, p. 1; Fourth Programme (1987–92), OJ C325, 7.12.1987, p. 1. For Fifth Programme see n. 16.

[7] Convention on International Trade in Endangered Species of Wild Flora and Fauna, 12 ILM 1085 (1991).

conserve wild birds and the undertaking to implement international treaties for protection of wildlife, which are considered in detail below.

The second Action Programme on the Environment in 1977 noted that the Commission was taking part in the development by the Council of Europe of a convention to protect wildlife and biotopes and that it would present, if necessary, proposals to ensure that this was satisfactorily applied in the Community. The treaty in question, the Berne Convention (discussed later), was concluded in 1979. In the 1980s non-governmental organisations (NGOs) began to discuss the possibility of promoting an EC directive to protect habitats and species other than birds. In the UK a strong case began to be made for this, though the EC's Third Action Programme, which ran from 1983-86, still noted in its Preamble that the EC's task was to promote throughout the Community a harmonious development of economic activities and a continuous balanced expansion which, even in the then changed economic circumstances, "was inconceivable without making the most economic use possible of the natural resources offered by the environment and without improving the quality of life and protection of the environment".

By this date the chief form of EC action was the preparation of legislation and other rules, and undertaking such actions as reviewing existing instruments, completing and co-ordinating scientific research and measures to alert public opinion. Much emphasis had been placed in the first and second Action Programmes on pollution, but with the aim of controlling it, rather than preventing degradation of the receiving environment by precautionary measures. Now, however, the case began to be made at Community level for a Community instrument to conserve habitat, whose gradual, irreversible disappearance was seen as the chief threat to survival of species. While recognising that local and regional responses were "decisive", the Programme found that a Community framework was becoming essential in order to give greater cohesion to these efforts. This would ensure that "a network of properly protected biotopes, sufficient in both extent and number, and interlinked in a rational fashion, was set up and maintained", and so designed that the protected habitat would ensure the survival of all species native to the Community. An aim, which, it noted, would be made easier if Community financial resources could be directed to it, possibly including support to voluntary organisations, within a framework of management rules for nature reserves.

By the time of the Fourth Action Programme, which ran from 1987 to 1992,[8] a different note could be and was sounded following the provision in 1986, by the Single European Act (SEA),[9] of a clear legal basis for the adoption of specifically environmental policies and measures, which until then had been controversial, based on interpretation of non-specifically environmental articles.[10] The SEA required that EC policy and action on the environment must preserve, protect and improve the quality of the environment, ensure a prudent and rational utilisation of natural resources, and contribute to the protection of human health.

The SEA favoured preventative action, rectification of damage at source, the inclusion of environment protection as a component of other policies, and that the EC and its Member States co-operate with Third States. These aims, however, were offset by the expressed need to take account of the EC's regional environmental

---

[8] 25 ILM 506 (1987).
[9] 25 ILM 506 (1986); in force 1 July 1987.
[10] For example, the preamble to the Rome Treaty set out the aim of constant improvement of living conditions; see also Article 100 and Article 235.

conditions, the benefits and costs of taking action as compared to non-action, and the economic and sound development of the EC as a whole, as well as the balanced development of its regions. Moreover, the Community was to act only if action was best taken at Community and not at Member State level (i.e. the principle of subsidiarity was acknowledged). None the less, the SEA did state that the EC was to take as the basis of its environmental policy a high level of protection.

By the date of the adoption of the Fourth Action Programme, most EC Member States had become parties to the 1973 CITES, although the Community as such has never been able to become a party thereto as the Convention does not provide for it to do so and its parties have never made the necessary amendment for this purpose. The Community therefore considered that it could take measures to harmonise its Member States laws implementing CITES. The Programme noted that "The time is now ripe for the Community and Member States to make a major new thrust in the field of nature conservation". It thus called for measures that directly or indirectly would ensure preservation of biological diversity, though it did not specifically articulate them in these terms. These included measures to protect and enhance Europe's natural heritage, in particular, by: the implementation of Community instruments in force, such as the Regulation on implementation of CITES in the EC[11] and the Directive on Conservation of Wild Birds;[12] measures for the protection of areas of importance in the Community or particularly sensitive environmentally, such as those referred to it in a 1979 framework directive,[13] and for the revival of areas that had been environmentally impaired. The Programme further included protection of forests against atmospheric pollution and fires on the basis of recent Regulations for encouragement of environmentally beneficial agricultural practices;[14] and overall, integrated protection of the Mediterranean Region. Finally, it encouraged the EC's active participation in relevant international organisations and closer co-operation with developing countries on matters such as protection of natural resources, with particular reference to desertification, water supply, tropical forests, and use of dangerous substances and products, to be accompanied by technological co-operation.

The Fourth Action Programme foresaw the need to improve the scientific basis of environmental policy and to develop efficient economic instruments to ensure application of the polluter pays principle for the remedying of environmental damage. These instruments might include taxes, levies, State aid and rebates.

The proposal for specific EC action on nature conservation met with the approval of all Member States except Denmark,[15] but was not listed among the priorities recognised in the Council Resolution on the Action Programme. When the Commission proposed a Habitat Directive in September 1988 it was thus not received with enthusiasm in all States, including the UK. Its text was not finally agreed until December 1991.

In the run-up to the 1992 UN Conference on Environment and Development (UNCED), the EC, at its Dublin summit in 1990, stressed its special responsibility in the international arena to use more efficiently its position of moral, economic and

---

[11] EC Reg. 3226/82/EEC.
[12] EC Dir. 79/489/EEC.
[13] Dir. 79/409/EEC, OJ L103, 27.4.79.
[14] EC Regs. 352/86/EEC and 3529/86/EEC.
[15] See *Manual of Environmental Policy: The EC and Britain*, Environmental Policy Release 5, at 9.9–7; House of Lords Select Committee on the European Communities, Session 1988–89. Report, Habitat and Species Protection, 1989.

political authority to advance international efforts to solve global problems and promote sustainable development and respect for global commons, and to meet the needs of the developing world, as well as central and Eastern Europe. With the text of the Habitat Directive finalised it could now play an active role at UNCED in concluding the Convention on Biological Diversity.

The Fifth Action Programme, which was entitled "Towards Sustainability",[16] was promulgated on the eve of the UNCED, it referred to the introduction by the Maastricht Treaty on the European Union, concluded on 7th February 1992, of a new principal objective of the EU, viz. promotion of "sustainable growth", not of "sustainable development" (the term used in the UNCED instruments). The Maastricht Treaty, unlike the SEA and the previous Action Programmes, referred for the first time to an "environmental policy", not merely to "actions relating to the environment" and reinforced the adoption of the doctrine of "subsidiarity" by stating in Article 3b that, except in defined areas where the EC has exclusive competence, "the Community shall take action only if and insofar as the objectives of the proposed action cannot be sufficiently achieved by the Member States and can, therefore, by reason of the scale or effects of the proposed action, be better achieved by the Community". It also stressed the need to take decisions "as closely as possible to the citizens" (Article A). Otherwise it reflected much of the rhetoric and terminology of the UNCED instruments, expressed in similarly general terms. "Sustainability" was said to require policies and strategies that enabled economic and social development without detriment to the environment and natural resources on the quality of which continued human activity and further development (including for future generations) depended. This entailed, it concluded, preserving the overall balance and value of *"the natural capital stock"* (emphasis added). In other words, biological diversity was perceived as having an economic value that went to the heart of Community purposes.

For the first time, both the Fifth Action Programme and the EC's accompanying report on the State of the Environment in the EC referred to Biological Diversity as such, a response to the imminent conclusion of the Convention on Biological Diversity under the auspices of the UNCED. The report, in a section headed "Biological Diversity: A Common Heritage to Preserve and Regenerate",[17] noted that several species or populations of animals needing extensive space were in danger of extinction: over 100 species, including the bear, lynx and monk seal, were endangered. Residual forest ecosystems were in a precarious predicament (and in southern regions their continuity could not be guaranteed); the area of open countryside had contracted and was further dramatically contracting under the impact of changes in land use and eutrophication; large aquatic environments had become rare and were at risk. To safeguard these a land management policy was needed that would preserve whole catchment areas. Other types of habitat, such as bogs and reed beds, were equally at risk, following draining of marshes and increasing coastal pollution. Without substantial reinforcement of existing measures and effective implementation thereof these processes would continue; even with prompt and effective action, it would take many years for species and areas affected to regenerate. In the short term it was only in a few regions without major problems that improvements could be

---

[16] COM (92) 23 Final, Vol. II, Brussels, 27 March 1992; an EC Programme of Policy and Action in Relation to the Environment and Sustainable Development; it was accompanied by a *Report on the State of the Environment* evidencing its general deterioration.

[17] *The State of the Environment in the European Community*, section 7, p. 49.

foreseen; in peripheral regions in particular, the danger was that the legitimate need for development would be pursued at the expense of their rich biological diversity to the detriment of sustainable development.

The Report thus concluded that major changes from the earlier Action Programmes on the Environment were needed. There should be more focus on: agents and activities depleting natural resources rather than waiting for damage to occur; changing practices and behaviour patterns detrimental to the environment by involving all States in a shared responsibility, broadening the range of instruments used; setting long term objectives for each main issue and augmenting the flow of information and education. Priorities for action should be set and should include, *inter alia*, depletion of natural resources such as soil, water, natural areas and coastal zones. Amongst the five selected target sectors for action (industry, energy, transport, tourism and agriculture) agriculture is of the most significance for preservation of biological diversity. The Report envisaged reform of the Common Agricultural Policy (which includes the Common Fisheries Policy) to achieve a more sustainable balance between agricultural activity and other forms of rural developmental use of natural resources.

The suggested broadening of the range of instruments used included not only legislative measures but a broad mix of measures. Of course, Community-wide rules and standards would be necessary to preserve the integrity of the internal market, but market based instruments, such as fiscal incentives and disincentives and establishment of civil liability regimes, could, it was proposed, nudge producers and consumers into more responsible use of natural resources, avoiding waste by internalising the external environmental costs and making environmentally friendly products as attractive as others. In addition so-called "horizontal" supporting instruments requiring provision of data, scientific information, use of less polluting technology, better planning, and education programmes were envisaged, as well as financial supporting mechanism such as provision of further structural and other funds.

This ambitious new programme was seen as a turning point for the Community, reconciling environment and development, not just as a programme promoting conservation or environmentalism. It would provide the framework for a new approach to environmental protection by reducing economic and social activity. It looks beyond the short term to the twenty first century and to make it work, calls for the establishment of a wide ranging, representative consultative forum (involving enterprises, consumers, unions, professional organisations, non-governmental organisations, and local authorities) and of an equally wide implementation network under the Commission's supervision, and for monitoring by a high-level Environmental Policy Review Group.

EC Participation in UNCED

The EC participated in the Preparatory Commissions for UNCED and in the conference itself as a full member; Member States often indicated that they were speaking on behalf of the EC as well as themselves.[18] The EC supported the UNCED Declaration on Forest Principles and Agenda 21's programme for the conservation of biodiversity, and favoured the speeding up of work on this by setting a target for the year 2000, not

[18] See Report of the Netherlands for the EC and Member States to the Third Session of the UNCED PREPCOM, WG I, Agenda Item 3a, Geneva, 14 August 1991, and Item 4.

2001. It considered the focus should be on fact finding and clarification of the intrinsic value of various biodiversity functions. *Ex situ* conservation was not so important as conservation of *in situ* genetic resources by the protection of ecosystems, allowing for the necessary interactions between individuals of the same and different species, creating enough protected areas to conserve the maximum possible number of species and ecosystems, and nurturing the relationship between protected areas and core areas, which should be inter-linked to create large ecological networks.

Although this chapter, for reasons of space, cannot include a similarly extended exposition of the EC's actions on protection of biotechnology at UNCED, the EC accepted the potential for the sustainable development of this and the need to provide safeguards against any unintended side-effects. In particular, it expressed the view that its application to food productivity had to be complementary to traditional agriculture and should not replace use of traditional technology and agricultural varieties. The EC sees the need here for an integrated programme of development to preserve varieties best suited to their localities, with biological and integrated pest control being the key issue.

Following UNCED, a meeting of the European Council in Lisbon adopted an eight-point plan on preservation, *inter alia*, of biodiversity, which included publishing national action plans on biodiversity to prepare the way for ratification of the convention adopted at Rio, as well as for implementation of the Forest Principles, the Rio Declaration and Agenda 21. It also proposed provision of financial support to developing countries through official Development Assistance and replenishment of the Global Environmental Facility (GEF). The EC announced that it intended to take a lead in restructuring the GEF to provide a permanent financial mechanism for supporting the Climate Change and Biological Diversity Conventions. In supporting the establishment of the High Level Commission on Sustainable Development, it made clear that it would press for establishment of an international review process of the progress made in the implementing of the UNCED Principles on forests and desertification.

The EC's UNCED commitment to aiding developing countries to promote sustainable development on the basis of Agenda 21, including support for preservation of biological diversity, required a new approach in negotiating its Lomé IV agreement with the ACP States;[19] a specific environment title would have to be included and a more harmonised, preventative approach adopted. Environmental Impact Assessment of co-operation projects would be required and priority accorded to funds allocated to management of forests, soil and prevention of desertification and deforestation.

EC Signature of the Biodiversity Convention

On signing this Convention the EC and its Member States stated that further work on strengthening the Convention should start as soon as possible, and that the EC should take a lead in developing internationally agreed guidelines for biotechnology in relation to safety procedures and risk assessment, as a basis for future binding conventions on these aspects.

It is clear that the Community, though supportive of the Rio Declaration and a party to the Biodiversity Convention, has given priority in its Rio related Fifth Action Programme more to development of policy and non-binding measures, such as provision of assistance, listing of priorities, economic and fiscal instruments, international

---

[19] Fourth ACP EC Convention (Lomé IV), 29 ILM 425 (1990), at p. 783.

co-operation etc., rather than adopting a new full UNCED/Biodiversity specific programme of implementing legislation. Instead it intends to rely, in accordance with the new environmental approach of the Maastricht Treaty, on reviewing and restructuring existing regulations and directives, making them part of a more integrated approach, whilst at the same time, in conformity with the new approach to subsidiarity, leaving more matters to be dealt with at national or, as in the terms used, at local level. Both the Maastricht Treaty and the Fifth Action Programme evidence that the EC is concentrating also on playing a more active role at the international level, by participating actively in relevant international conventions and international organisations.

Much, therefore, now depends on the effectiveness for the purpose of preserving biodiversity, *inter alia,* of the existing framework of measures adopted by the EC that are directly relevant to this specific purpose. Compared to the broad scope and highly ambitious aims of the very general programmes and plans outlined in this part, they are, as will be seen from the account given in Part II below, surprisingly few in number, and have given rise to a number of problems of interpretation and implementation.

Measures Adopted by the EC Relevant to Protection of Biodiversity

Legislative Strategy

In its Fifth Action Programme relating to the Environment and Sustainable Development, the Community not only made the case for protection of nature and biodiversity but indicated that the key to its strategy would be "primarily through sustainable land management in and around habitats of Community and wider importance".[20] An interrelated network of habitats, based on the concept for Natura 2000,[21] should be created through the restoration and maintenance of habitats themselves and of corridors between them. It recognised that the success of this would very much depend on the careful formulation of transport, agricultural and tourist policies. The relationship of this plan to existing legal measures and the strategies referred to above are illustrated in Figure 11.1.

In this "Network" approach it seems that the key legal instrument for conservation of natural habitats is the EC's Habitat Directive,[22] and for protection of endangered species both the Habitat and the Birds Directive[23] and the Regulation on the Implementation of CITES.[24] Curiously, no mention is made of implementation of the 1979 Berne Convention on Conservation of European Wildlife and Natural Habitats[25] (which the Habitats Directive implements) or species other than birds listed on CITES Annexes which are indirectly protected by EC instruments, viz. the EC Regulation on Common Rules for Import of Whales or other Cetacean Products,[26] the EC directive concerning the Importation of Skins of Certain Seal Pups,[27] and the Commission's Regulation on the Prohibition on Importing Raw

---

[20] *Op. cit.,* n. 16, sec. 5.3. Protection of Nature and Biodiversity, at p. 48.

[21] *Ibid.,* p. 47. For a good account of the origins and development of the protected area approach in Europe, see Kiss and Shelton, *op. cit.,* n. 1, at pp. 125–134.

[22] EC Directive on the Conservation of Natural Habitats and of Wild Fauna and Flora, 92/43 EEC, OJ L206, 22 July 1992, p. 116.

[23] EC Directive on the Conservation of Wild Birds, 79/419/EEC, OJ L103 27 April 1979, p. 50.

[24] 3636/82/EEC, OJ L384, 31 December 1982, as modified, OJ L367, 28 December 1983.

[25] *European Treaty Series* 104.

[26] 348/881/EEC, OJ L39, 12 February 1981, p. 123.

[27] 83/129/EEC, OJ L19, 19 April 1983; extended by 85/444EEC, OJ.L259 of 1 October 1985.

Figure 11.1   Diagram on Nature Conservation

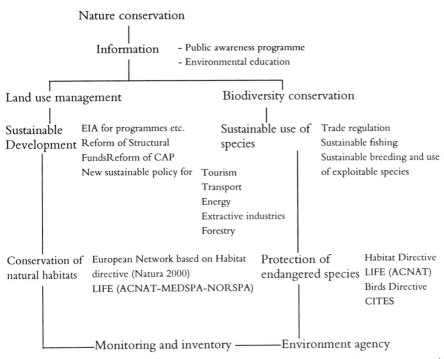

Source: EC Programme of Policy and Action in Relation to the Environment and Sustainable Development, COM (92) 23 Final, Vol. II, Brussels, 29 March 1992.

and Worked Ivory derived from the African elephant into the Community.[28] The Table provided in the Fifth Action Programme detailing the targets and instruments relating to Nature and Biodiversity similarly concentrates on the Habitat and Wild Bird Directives, though it does refer also to regulations concerning internal and international trade in endangered and over exploited species and Regional Agreements concluded under the Bonn Convention (Table 11.1).

The EC is party not only to the Berne Convention, which protects flora and fauna and their habitat, but also to the 1980 Convention on Conservation of Antarctic Marine Living Resources (CCAMLR),[29] which also adopts an ecosystem approach to conservation, and the Bonn Convention on Conservation of Migratory Species of Wild Animals,[30] which provides for conservation agreements to be concluded between States whose involvement in protection of specific migratory routes

---

[28] OJ L240, 17 August 1989, p. 5.
[29] Council Decision 81/691/EEC of 4 September 1981 on conclusion of the CCAMLR, OJ L252, 5 September 1981, at p. 26, as amended.
[30] OJ L210, 19 July 1982, p. 10.

Table 11.1 Nature and Biodiversity

	Targets up to 2000	Instruments	Time frame	Sectors/actors
Maintenance of biodiversity through sustainable development and management in and around natural habitats of European and global value: and through control of use and trade of wild species				Agriculture forestry, fisheries transport Tourism, energy, industry
	1. Maintenance or restoration of natural habitats and species of wild fauna and flora at a favourable conservation status	Habitat Directive	1992 ⇒	EC, MS, LAs, NGOs, farmers
		Updating of Directive 79/409/EEC on wild birds	ongoing	EC, MS, LAs
		Setting of criteria for identification of habitats, buffer zones and migratory corridors	1992–1993	EC, MS, LAs, NGOs, farmers
	2. Creation of a coherent European network of protected sites	Action programmes for the efficient conservation and monitoring of the sites designed for Natura 2000	1991–1993	EC, MS, LAs, NGOs, farmers
	Natura 2000: flagship programmes of carefully selected and managed natural areas within the EC	Inventory, monitoring systems, and recovery plans for endangered and overexploited species	1991–1992	EC, MS, LAs, NGOs, farmers
		regulations concerning internal and international trade of endangered species	1992 ⇒	EC, MS, LAs, NGOs, farmers, UNEP (CITES)
	3. Strict control of abuse and trade of wild species	International Conventions (Biodiversity, Alps, Regional agreements under Bonn Convention)	1992 ⇒	MS + EC + UNEP (CITES + Bonn Convention)
		Reform of CAP (notably zonal programmes for support of environmentally friendly agricultural practices)		ongoing EC, MS, LAs
		Environmental assessment of plans and programmes	1995 ⇒	MS, LAs, EC
		Programmes for promotion of public awareness	1992 ⇒	MS, LAs, EC, NGOs
		Measures to maintain and protect forests	Progressive	EC, MS + forest owners

Source: EC Programme of Policy and Action in Relation to the Environment and Sustainable Development, COM (92) 23 Final, Vol. II, Brussels, 29 March 1992, p. 48.
Note: MS = Member States; LAs = local authorities.

is necessary to afford full protection throughout the whole migratory route of migratory species listed in its Appendix I as endangered (i.e. in danger of extinction throughout all or a significant portion of their range). Formal AGREEMENTS (sic) are to be concluded among Range States of particular species listed on its Appendix II as having "unfavourable conservation status" or one that would benefit from international co-operation.

The Relevant EC Instruments: Their Operation in Practice

As pointed out by Boyle,[31] at the Rio Conference the EC expressed regret at the weakness of the Biological Diversity Convention in contrast to its own directives, but did sign the Convention as did all EC Member States. It has now approved the Convention and 7 Member States have now ratified or acceded to it.[32] The Fifth Action Programme on the Environment, as we have seen, strongly reflects the instruments adopted at Rio. It is thus somewhat surprising that in effect the directives to which the EC was presumably referring, that is those *directly* relating to the Convention's approach to *in situ* and *ex situ* protection of biological diversity, are only 2 in number, namely that on Conservation of Wild Birds adopted in 1979, and that on the Conservation of Natural Habitats and of Wild Fauna and Flora adopted in 1992. Although the other measures referred to at the end of Part I do protect certain animal species to some extent, they do not protect their habitats (though the Bonn Convention protects migratory routes). It is this important aspect of these two directives that distinguishes them for our purposes, the first providing the model for the second.

The Wild Birds Directive[33]

The Birds Directive was adopted to ensure that Member States protected birds in a uniform and constant manner since, as migratory species, they could not effectively be protected throughout their flyways by the taking of stringent measures in one State only, nor was there any economic incentive so to act when others did not. Thus the 1979 measure could be, and was, presented (as required before the SEA or Maastricht Treaty) as one necessary for the harmonious development of economic

---

[31] A. Boyle, "The Convention on Biological Diversity" in L. Campiglio, L. Pineschi, D. Siniscalco and T. Treves (eds.), *The Environment After Rio: International Law and Economics* (1994), at p. 114.

[32] Article 35 of the Convention provides for accession by regional economic integration organisations; in their instrument of accession such organisations must state the extent of their competence with respect to matters governed by the Convention or relevant Protocol. The EC approved the Convention on 21 December 1993; Denmark, the Federal Republic of Germany, Portugal and Spain ratified on that date; Luxembourg ratified on 9 May 1994; Italy on 15 April 1994: the UK acceded on 3 June 1994 (effective 1 September 1994) and extended its instrument to include Jersey, the British Virgin Islands, Cayman Islands, Gibraltar, St Helena and the St Helena Dependencies. At the time of writing, it appears that Belgium, Greece, France, Ireland and the Netherlands have signed but not ratified. The EC and Member States have made a declaration. The UK, on signing the Convention declared first, its understanding that Article 3 of the Convention set out a guiding principle to be taken into account in implementing it, and, secondly, that the decisions to be taken by The Conference of the Parties, under Article 21, para. 1, concern "the amount of resources needed" by the financial mechanism and nothing in it or Article 21 authorised the Conference of the Parties to decide "the amount, nature frequency, or size of the contributions" of the Convention's parties. The latter includes the EC. All EC Member States except Ireland and Luxembourg made similar declarations.

[33] For a good account of its origins and a critique, see Haigh, *op. cit.*, n. 4, at pp. 289–299.

activities throughout the Community and a continuous and balanced expansion, as well as one "improving living conditions".

In fact the demand for conclusion of such a directive arose in the European Parliament in the early 1970s, following petitions presented to it by concerned non-governmental organisations (NGOs). The Commission prepared a draft, largely based on existing UK legislation. The directive's Article 1 imposes a general duty on Member States to "maintain the population of naturally occurring birds (and their eggs, nests and habitat) in the wild state in their European Territory (which term could include territorial waters) at a level that corresponds in particular to ecological, scientific and cultural requirements while taking into account economic and recreational requirements" or " to adapt the population of these species to that level" (Article 2). Member States must additionally preserve, maintain or re-establish a sufficient diversity and area of habitats for all species of birds referred to by means of providing protected areas. The preservation of biotopes and habitats must also include their upkeep in accordance with the ecological needs of habitats both inside and outside the protected zones, re-establishment of destroyed biotopes and creation of biotopes (Article 3). Exceptions can, however, be made for various reasons, including hunting; otherwise the traditional protection techniques of controlling hunting practices are preserved, the oldest and most common form of legislation throughout Europe as a whole.[34]

The directive institutes by a system of Annexes different levels of protection for species listed thereon. There is a prohibition on deliberate killing, capture or destruction of nests and eggs or disturbance of birds in the breeding and rearing season, and the keeping of birds the hunting or capture of which is banned (Article 5). Sale and transport of birds and parts thereof (except for those listed in Annex III) is also banned (Article 6). Species listed in Annex II can, however, be hunted, subject to its specific requirements, as long as this accords with national measures in force, the principles of wise use and ecologically balanced control of the species concerned, and species are kept at the levels required. However, most species (currently 175) are listed on Annex I, and are thus subject, under Article 4(1), to specific measures for conservation of their habitat, designed to ensure their survival and reproduction. Account must be taken of their vulnerability to changes in habitat, their rarity and other special protection requirements. Member States are required to classify the most suitable territories (both land and sea) as special protection areas (SPAs) for these species and similarly protect regularly occurring migratory species that are not listed in this Annex. Special attention is required to be given to protection of wetlands, particularly those of international importance (Article 4(2)). Such measures have to cover the breeding, moulting and wintering areas of these migratory species.

Member States must send information to the Commission concerning the measures taken by them so that the Commission can evaluate them to ensure that they represent a coherent whole protective package. They must in general strive to avoid deterioration of habitats by pollution and other measures, but in the case of the special protection areas they must take appropriate steps to avoid pollution or deterioration of habitats or any disturbance affecting the birds in so far as they are significant to the objectives set out for habitat protection. Most of these measures have now been superseded by the Habitat Directive, based upon a similar system, following its entry into force.

---

[34] Kiss and Shelton, *op. cit.,* n. 1, at pp. 117–118.

Member States are also required to forward a triennial report to the Commission on their national implementing measures. The Commission then prepares a composite report, each Member State's relevant authorities being required to verify the section based on information supplied by their own State. The final version is sent to the Member States but is not published. Member States are also required to encourage research and other work required to provide the basis for protection and management; particularly on subjects listed in its Annex V. A Committee was established to adapt Annexes I and V to technical and scientific progress. Member States are, however, allowed to introduce stricter protection measures than those provided for in the directive. On adopting the directive, the Council also adopted a Resolution calling on Member States to notify the Commission within two years of their classified SPAs and wetlands designated or intended to be designated as those of international importance, and any other areas classified under national legislation for purposes of bird protection. The Commission was to maintain an up-to-date list of these and put forward criteria for determining the SPAs.

Despite the many positive contributions of the directive to protection of habitat and preservation of biodiversity of bird species, the Commission's approach and the directive itself suffered from a number of weaknesses. Many considered that it was undesirable to impose uniform conditions on bird protection throughout the Community; the result was likely to lead to standards based on compromises. The main point of contention in this respect was that the directive provided for rights of derogation for individual Member States in certain circumstances, albeit these were limited (Article 9(1)) both in number and by the specificity of the reasons permitting derogation from the requirement of Articles 5–8. Moreover, derogation was allowed only "where there is no other satisfactory solution". Member States were also specifically required to give various items of detailed information in the derogations (Article 9(2)) and to report annually to the Commission on implementation of the Article. The Commission itself, on the basis of the information supplied, was required to ensure that the consequences of the derogations were never incompatible with the directive and to take somewhat ambiguous "appropriate steps" to this end. Given the enormity of the overviewing task and the limited means of overview, inspection and enforcement available to the Commission, fulfilment of these obligations was likely to prove difficult. Coupled with the fact that agriculturists and hunters resented interference with their traditional rights, this has led to major enforcement problems, particularly in the Southern Mediterranean.

In general, however, conservationists welcomed the fact that the Community as such had proved able to make advances that had been shown to be beyond the internal capacity of certain Member States, however strong the pressure exerted, and expected also that the EC's advance would influence development of measures in neighbouring countries and those constituting the relevant flyways, especially in stimulating action in the Council of Europe and the IUCN. Even in the UK,[35] whose legislation had provided a model for the directive, some new legislation had to be introduced to comply with it; most new measures were included in the Wildlife and Countryside Act 1981, but the UK only belatedly notified the Commission a year later of the steps taken to classify special protection areas (by establishment of national, local and voluntary nature reserves and sites of special scientific interest (SSSIs)), though it noted that not all areas so classified met all the

---

[35] For an account of this and UK law and practice relating to conservation of nature, see S. Ball and S. Bell, *Environmental Law* (2nd edn., 1991), at pp. 403–426.

requirements of the directive and later notified a few derogations. Conservation groups in the UK (notably the Royal Society for the Protection of Birds) have raised some criticism of UK practice[36] (for example, that SSSIs could not adequately be protected below the low water mark under existing legislation) but now have the advantage of referring alleged non-implementation of the directives to the Commission's attention. As UK legislation did not incorporate the directive's general principles on bird and habitat protection, much thus depends on actual practice in implementation. Several complaints of violations have been made.

The Commission can, of course, refer alleged violations, if not resolved through negotiation or issue of Reasoned Opinions, to the European Court of Justice (ECJ). A considerable number of relevant cases have been referred to the ECJ, either on the Commission's own instruction or after prompting by NGOs. Germany, France, Italy and Belgium have been found to be in violation of the directive.[37] Of particular interest was the Court's strict construction of the right to derogation in Case 412/85 *EC Commission v Germany* in which it held that the derogation must be based on at least one of the reasons listed in Article 9(1) of the directive, which it regarded as exhaustive, and that it must also meet the criteria listed in Article 9(2), the aim of which was, in its view, to limit derogations to those strictly necessary and to enable supervision thereof by the Commission. Interestingly, it has since concluded that the need fully to translate Article 9 into national law is especially important when "the management of the common heritage is entrusted" to Member States.[38] It is not clear from whence the ECJ derived this concept, or precisely what legal significance it accords to it. It is not included in the directive, and has no clear or legally significant content outside the UN Law of the Sea Convention, which limits it to deep seabed minerals.[39]

However, even more significant was the decision in the so-called *Leybucht Dyke Case*[40] in which the Commission asked for an injunction to prevent dyke-building works undertaken by Germany and a declaration from the ECJ that by planning or undertaking such works Germany had violated Article 4 of the Birds Directive. The latter was sought on the basis that the works were detrimental to the habitat of birds protected by it in an SPA, and would lead to a reduction in that area. The Commission contended that the provision of the directive that species on Annex I shall be the subject of "special conservation measures" (Article 4(4)) necessitated that Member States take positive measures to avoid deterioration of SPAs and that no exceptions could be made, nor could economic interests be taken into account. The ECJ agreed that Member States could only reduce the extent of SPAs on

---

[36] See D. Pritchard, *Implementation of the EC Directive on the Conservation of Natural Habitats and of Wild Fauna and Flora* (1993) RSPB/Birdlife International.

[37] For example in Case 236/85, *EC Commission v. Netherlands* (1987) ECR 3989; Case 247/85, *EC Commission v. Belgium* (1987) ECR 3029; Case 252/85, *EC Commission v. France* (1988) ECR 2243; Case 262/85, *EC Commission v. Italy* (1987) ECR 3073; *EC Commission v. Germany* Cases 412/85 (1987) ECR 3503 and 288/88, ECR I -2721. But note that the specific requirements in the derogation clauses of the directive (Article 9), and in Articles 5–8, have led the ECJ to reject most derogations made on these Articles; see L. Krämer, *European Environmental Law Casebook* (1993), at pp. 200–203.

[38] Case C-339/87, *EC Commission v. Netherlands* (1993) 2 CMLR 360, at p. 386.

[39] United Nations Convention on the Law of the Sea, United Nations, New York (1983); Part XI.

[40] *Commission of the European Communities v. Federal Republic of Germany* supported by the United Kingdom, Case C-57/89, 28 February 1991, (1991) ECR 883; see case note by D. Baldock, (1992) 4 *JEL* 139–44; and D. Freestone, (1991) 1 *Water Law* 152–156; see also Krämer, *op. cit.*, n. 37, at pp. 219–231.

exceptional grounds that corresponded to a general interest superior to that represented by the directive's ecological objective; the interests set out in Article 2, viz. economic and recreational needs of the State concerned, were not relevant to establishment of a general interest. The need to prevent flooding and ensure coastal protection, however, were relevant and the objectives of the dyke works, including as they did provision of protection of a fishing port and the strengthening of coastal structures in a form that had ecologically beneficial side effects in that they would create water meadows, were sufficiently important to justify derogation from the directive's conservation requirements, though it is doubtful whether the Commission would ever be competent to evaluate the size of such areas, the reasonableness thereof, or the ecological results.[41] The ECJ held, however, that application of the principle of proportionality required that the dyke works should be limited to the strict minimum necessary and should result in the smallest possible reduction in the SPAs.

As Krämer presciently noted, however, the decision was not likely to become a leading case; a legislative response was probable and commentators who welcomed the descision would be disappointed.[42] The judgement left unclear precisely what circumstances would be regarded as sufficiently exceptional and in the general interest to permit damage or disturbance of a protected site and was criticised by many Member States on that account, as most have allowed themselves considerable discretion in considering what constitutes non-essential interests when enacting relevant national legislation. As a result Member States seized the chance presented by the draft Habitat Directive, which would replace Article 4 of the Birds Directive, to amend its text to allow projects of a social and economic nature to be taken into account in considering the management of SPAs, when an overriding public interest is involved (see below). It should be noted, however, that in its later decision in *Commission* v. *Spain*,[43] the ECJ supported the Commission's contention that by failing to take preservation and maintenance measures to fulfil the ecological requirements of habitat protection, Spain had violated Articles 3 and 4 of the Birds Directive and rejected Spain's argument that these requirements were subject to other interests, such as those of a social or economic nature, or at least had to be balanced against these. The Court held that Member States could not arbitrarily rely on such other interests as a justification for derogation from the Birds Directive.

The EC Directive on the Conservation of Natural Habitats of Wild Flora and Fauna

The Habitats Directive, which was introduced to ensure effective implementation of the 1979 Berne Convention on the Conservation of European Wildlife and Natural Habitats,[44] is the key instrument among the EU's measures for protection of biodiversity. It lays down the rules and procedures for instituting the Natura 2000 network, the aim of which is to provide a discrete but coherent European ecological network of special areas of conservation (SACs), which includes the areas classified as SPAs under the Wild Birds Directive as well as the SACs designated under the Habitats Directive.

---

[41] Krämer, *op. cit.*, n. 37, at pp. 229.

[42] *Ibid.*, pp. 224–225.

[43] Case C-355/90, 2 August unreported, but see H. Somsen, (1994) 3 *EEL Rev.* 268–278.

[44] The EC signed the Convention in 1979 but did not ratify it until 1982, following a Council Decision, Decision 82/72/EEC OJ 1982 L 38.

Its main aim is stated to be "to promote the maintenance of biodiversity" but this is qualified by the permissibility of "taking account of economic, social, cultural and regional requirements" and by the pronouncement that "This Directive makes a contribution to the general objective of sustainable development, whereas the maintenance of such biodiversity may in certain cases require the maintenance, or indeed encouragement of human activities"(Preamble). It has, however, the second aim of protecting species as such and certain species and sites that are defined as being in danger of disappearing are identified as priorities.[45] It is innovatory in being the first international attempt comprehensively to protect both habitat and species in a region and in introducing the concept of restoring or maintaining habitats or species "of Community interest" at "a favourable conservation status" (defined in Article 1 in terms the interpretation and application of which require considerable scientific information and knowledge). Its preamble acknowledges the "common responsibility" of EC Member States to conserve habitats and species of Community interest; it also notes that application of the polluter pays principle is only partially relevant to conservation and that implementation of the required measures may be excessively costly for some of the poorer Member States, points which, as we have seen, were developed in the EC Action Programmes on the Environment. Despite the forward looking nature of the directive, the EU's success in achieving its objectives will, therefore, depend on the effectiveness of the implementation of the directive, namely the extent and coherence of the areas actually designated throughout the Community and the stringency with which the required protection measures are enforced, i.e. the zeal with which violations are pursued and punished. Its requirements are demanding, especially the need for a coherent and ecological approach, hitherto not much evidenced in Community practice. In addition, to date existing Community funds used for enforcement protection have been limited[46] and though the directive provides for Community co-financing of measures "essential for the maintenance or re-establishment at a favourable conservation status of the priority natural habitat types and priority species on sites concerned" (Article 8) it is not clear how much financial assistance will be available for this purpose. Caution is thus required in predicting the success of this remarkable initiative.

At the start, the directive having been adopted in 1992, Member States were required to enact provisions to comply with it by May 1994. Such information as has been made informally available to the writer suggests that few, if any, Member States completed this process within this time limit, though some have adopted some piecemeal measures. This is despite the fact that provision of the original draft of the directive which had drawn attention to the need, in order to achieve full protection, to conserve sites *outside* the designated SACs by avoiding pollution or deterioration of habitats therein and to protect distinctive features of the landscape, were amended. There is considerable concern that, particularly in the context of UNCED's promotion of "sustainable development", the EU's and Member States' policy may favour development at the expense of conservation.[47]

---

[45] C. Hatton, *The Habitats Directive: Time for Action* (1992) WWF UK.

[46] About 400 million ECU are available to fund environmental projects over a five year period whereas 157 billion are allocated for structural purposes (structural funds), though these include funds for development of less developed regional rural areas. C. Tydemans and L. Warren, "Legal Mechanisms within the European Community for the Protection of Wildlife and Habitats" (1995) *Journal of Wildlife Management*.

[47] See Hatton, *op. cit.*, n. 45, and Tydemans and Warren, *ibid.*; case studies examined in these papers give credence to those fears, e.g. developments proposed for the Messelongi Wetlands; River Acheloos in Greece; the Burren in Ireland; and in the UK at Cardiff Bay in Wales and Oxleas Wood in England.

A listing system for sites and related species is prescribed. By May 1995 every Member State is required to notify to the Commission a list of sites, indicating the types of natural habitat listed in Annex I[48] of the directive and the species listed in Annex II[49] that are native to its territory which that particular site supports (Article 4 (1)). The lists are amendable in certain circumstances (Article 5). In the case of animals which range over a wide area, Article 4 (1) requires that the sites have to "correspond to the places within their natural range which present the physical or biological factors essential to their life and reproduction". In the case of aquatic species in this category, sites should be proposed for listing only where there is a "clearly identifiable area representing the physical and biological factors essential to their life and reproduction". Annex III lists the criteria for selecting sites of Community importance. There are two stages: first, assessment of the relative importance of sites for the habitat types and species listed in Annexes I and II; secondly, assessment of the importance to the Community of sites included in the national lists. Following this, the Commission must, by May 1995, adopt a list of sites of Community importance drawn from the national lists, identifying sites which have lost one or more priority natural habitat types or priority species.[50] Criteria for establishing the Commission's list are provided under stage II of Annex III within five biogeographical regions and the European territory of Member States to which the EC Treaty applies.

When a site of Community importance has been designated, the Member State concerned must as soon as possible and within six years at the latest (i.e. by 2001) designate the site as a SAC and set the priorities for its maintenance and restoration. When a SAC has been listed by the Commission, the Member States must take special conservation measures (which must include management plans) which meet the ecological needs of the site, an obligation (provided for in Article 6(1)) which replaces those in Article 4(1) of the Wild Birds Directive. Deterioration of natural habitats and habitats of species, as well as disturbance of the latter, must be avoided; an appropriate assessment must be made of the implications for the site of plans or projects which, though not directly connected with or necessary to management of the site, are likely to have significant effects on it. Finally, if a plan or project does proceed despite estimation of its negative implications and if there are no alternatives and "imperative reasons of overriding public interest exist", including social and economic reasons, the Member State must take all the "compensatory measures" that are necessary to establish that the "overall coherence" of Natura 2000 is protected and must inform the Commission of these (Article 6(4)). On the other hand, in the case of sites hosting priority natural habitat types or priority species, or both, the plan or project can proceed only if there are health or safety considerations, or beneficial consequences of primary importance for the environment, or if on the basis of an opinion delivered by the Commission other imperative reasons for overriding the public interest, which are not defined in the directive, exist. Finally, the Commission is required to review periodically Natura 2000's contribution to the

---

[48] Annex I lists nine habitat types the conservation of which require designation as a SAC, namely costal (sic) and halophytic habitats, temperate heath and scrub; sclerophyllus scrub (mattorral); natural and semi-natural grassland formations; raised bogs, mires and fens; rocky habitats and caves; forests.

[49] Annex II lists animal and plant species of Community interest the conservation of which requires designation of special areas.

[50] Articles 4(2) and (3). Article 1(d) and (h) provide definitions, respectively, of both "priority natural habitat types" and "priority species". Some flexibility is permitted in relation to Member States whose sites support habitats of species which represent more than 5% of their national territory.

achievement of the directive's objectives and Member States are required to exercise surveillance over habitats. Member States are left to their own initiatives in being required to try, when *they* consider it necessary, to promote management of landscape features of major importance for wild flora and fauna, e.g. river banks, or traditional systems for demarcating the boundaries of fields, which are vital for irrigation, dispersal and generic exchange.

The directive's provisions (Article 12) concerning the protection of species listed in Annex IV(a) follow the familiar system of requiring hunting controls and measures to prevent disturbance of species, especially during breeding seasons, in terms similar to those laid down in the Wild Birds Directive. Member States must establish systems to monitor incidental capture and killing of species listed. Similar requirements are laid down to protect plant species listed on Annex IV(b). Flora and fauna, for which measures can be taken to ensure that any taking thereof in the wild or exploitation is compatible with their maintenance at a favourable conservation status, are listed in Annex V. Member States are required to prevent use of indiscriminate measures capable of causing local disappearance or disturbance to populations of wild animals listed in Annex V(a) and those listed in Annex VI(a) in relation to which derogations have been allowed under Annex I(b).

Despite criticism of derogations under the Wild Birds Directive, it was inevitable, if only to ensure its acceptance by Member States, that the Habitat Directive would also allow them. Derogation from required protective measures can be granted under Article 16 if there is no satisfactory alternative, the derogation is not detrimental to the maintenance of the populations of the species covered at a favourable conservation status in their natural range, and for other purposes similar, but not identical to those allowed under the Wild Bird Directive.[51] As also under that directive, Member States must report to the Commission biennially (not annually as in the case of the Wild Birds Directive) any derogations which they have made, stating their reasons for doing so, the relevant circumstances and the supervisory measures instituted. The Commission may then give an opinion on the derogation. It does not, however have any express power either to reject the derogation or to subject it to any conditions.[52] Article 17 requires Member States every six years to report also their measures for implementing the directive and to evaluate their impact on the conservation status of listed habitats and species. The Commission must then produce a composite report based on national reports evaluating progress towards achievement of the aims of Natura 2000. After comments by Member States a final report is to be published and sent to Member States, the European Parliament, the Council and the Economic and Social Committee.

Finally, the directive makes provision for the conducting of scientific research,[53] amendment of Annexes,[54] for the institution of a Committee to aid the Commission in performing its tasks under the directive,[55] and requires Member States to control

---

[51] *Viz.*, for purposes of protecting wild flora and fauna in their habitat, to prevent serious damage thereto; in the interests of public health and safety and for other imperative reasons of overriding public interest including those of a social or economic nature and those of beneficial consequences of primary importance for the environment; for research and education; in order to permit limited taking or keeping of species listed on Annex IV (Article 16(1)).

[52] Articles 16(2) and (3).

[53] Article 18.

[54] Article 19.

[55] Article 20.

the deliberate introduction into the wild of any species which is not a native of their territory[56] and to promote education and the dissemination of general information on matters of ecological importance pertaining to the implementation of the directive.[57]

Implementation of the Habitats Directive: UK Proposals and Practice

It is apparent that the success of this innovatory instrument in achieving its conservatory objectives will greatly depend on whether Member States not only adopt the necessary laws and regulations but effectively enforce them by monitoring observance thereof, prosecuting violations, imposing deterrent penalties for those found guilty and making available sufficient funds for restoration of degraded habitats to a "favourable conservation status". As remarked earlier, the Habitats Directive was diluted in scope and vigour during its negotiating process in deference to Member States concerns for the economic and social implications of the cost of the measures originally proposed.[58] It is clear, however, that even in its final diminished form its requirements remain ambitious in relation to the abilities and willingness of EU Member States effectively to implement it. This requires not only adoption of the necessary wide-ranging legislation but its integration and the diversion of considerable financial and administrative resources at a time when the governments of most Member States are reducing public expenditure. As also pointed out, the directive was in large part based on existing UK legislation, which was in advance of that in place in most other Member States. A brief examination of UK legislation thus provides a case study of the likelihood of the realisation of the directive's ambitious aims.

Studies conducted by various reputable UK NGOs do not give grounds for optimism. Not only are there many gaps in existing UK legislation and practice thereunder but there is little of the necessary integration of powers provided under the various instruments and policies[59] concerned that would enable them to be mobilised to protect particular listed sites or species, an aspect that is crucial to the concept of the preservation of the SACs that constitute the Natura 2000 network of protected areas and their linking corridors and surrounding areas. A consultation paper on means of implementation of the Habitats Directive issued by the UK government in 1993[60] left little time for adequate consideration of and response to its proposals for regulations which were laid before Parliament for debate in 1994.[61]

---

[56] Article 22(b).

[57] Article 22(c).

[58] See Tydermans and Warren, *op. cit.*, n. 46.

[59] See, for example, the works cited *op. cit.*, n. 4 and n. 45; also H. Phillips and C. Hatton, "Implementing the Habitats Directive in the UK" (1994) 15 *EUROS* at pp. 17–22; D. Pritchard, *op. cit.*, n. 36 comments on the DoE/Welsh Office consultation paper from the Royal Society for the Protection of Birds; and comments by the World Wide Fund for Nature, United Kingdom (WWF UK) submitted to the Department of the Environment, 11 November 1993.

[60] Implementation of Council Directive 92/43/EEC on the Conservation of Natural Habitats and of Wild Flora and Fauna; see also DoE/WO (draft) Planning Policy Guidance Note on Nature Conservation; DoE Consultation Paper on Coastal Planning Policy Guidance and DoE/WO Consultation Paper on "The Town and Country Planning General Development Order: Permitted Development, Environmental Assessment and Implementation of the Habitats Directive".

[61] The Conservation (Natural Habitats etc.) Regulations 1994, debated in the House of Commons, 19 July 1994, and in the House of Lords 17 October 1994. They were expected to enter into force in November 1994, six months after the due date of 5 June 1994 laid down in the Directive.

These met with considerable criticism from concerned NGOs. The gist of the criticism by conservationists was that the Government's draft regulations adopted a minimalist approach towards implementation of the directive and that the opportunity to redress the weaknesses apparent in practice under the Wildlife and Countryside Act of 1981 and other Acts related to the directive, which would have enabled its effective implementation, had been missed.[62] Though the regulations, which ran to seventy-three pages, did focus their attention on protection of SACs, the measures proposed would not achieve the requisite "favourable conservation status" either for marine or terrestrial habitats or for the species listed in the directive's Annexes, since they did not apply outside the SACs; a review of all existing measures and schemes for protecting the countryside in general was required in order to direct them to protection of areas outside the SACs, as required by the directive. The provisions concerning protection of marine habitats and species were regarded as particularly deficient since no powers were provided for nature conservation agencies to introduce bye-laws to maintain the integrity of marine sites both inside and outside the SAC areas; there was a particular lack of clarity in prescription of the protective responsibilities outside marine SACs, conflicts of interest between concerned marine and land-based Government Departments and a lack of any mandatory positive management scheme for SACs in general, both marine and land-based. The National Audit Office had confirmed that since 1987 there had been over 800 recorded cases of damage to the SSIs protected under the Wildlife and Countryside Act 1981[63] and a marked decrease in biodiversity in the decade from 1984-1994.[64] Though various government agencies had appropriate powers they had singularly failed to use them because of administrative complexities and resource implications.[65]

Amongst particular examples cited was the case of Westhay Moor, a hay meadow which qualified as a Ramsar Wetlands Convention site and Special Protection Area under the Wild Birds Directive[66] and had accordingly been subjected to a Nature Conservation Order (NCO) in 1987 because its owner refused to enter into the required managements agreements. The owner objected to the NCO and thus a public inquiry was held. In the meantime, a Compulsory Purchase Order (CPO), the only one ever to have been issued for an SSI) was served in 1985 as, unmanaged, the site's nature conservation value declined. The owner appealed against the CPO and thus another public inquiry had to be held which supported the CPO. The Nature Conservancy Council (then the appropriate body) offered to lease the moor in order to undertake its management, but this offer was refused. The owner then signed a management agreement, which automatically negated the CPO, but resorted to arbitration of the amount of compensation offered; this had not been concluded by mid-1994. In addition to such problems, referred to by two commentators as a "bureaucratic nightmare",[67] it has been pointed out that to resort to use of

---

[62] See the works cited *op. cit.*, n. 59.

[63] National Audit Office, *Protecting and Managing Site of Special Scientific Interest in England* (1994).

[64] Statistics released by the Institute of Terrestrial Ecology *et al.* for the Department of the Environment Countryside Survey (1994).

[65] See briefs prepared by Wildlife and Countryside Link, London, 19 July 1994; see also Wildlife Link, *SSIs – A Health Check*, Wildlife and Countryside Link, London, 1991; and detailed comments by WWF UK on the UK Consultation Paper on the Habitats Directive, dated 11 November 1993.

[66] See Wildlife Link briefs, *ibid.*; for other examples see Hatton, *op. cit.*, n. 45, at pp. 6–15.

[67] Phillips and Hatton, *op. cit.*, n. 59.

a CPO to purchase large estates which qualify as SACs, for example Mar Lodge in Scotland, would absorb Scottish Natural Heritage's entire budget. Even a revised form of Management Agreement to encourage owners to manage their land, as the above example illustrates, will be ineffective where an owner proves recalcitrant. Similar problems will no doubt emerge in other Member States.

There was also particular concern in relation to the inequality of measures for marine as compared to terrestrial systems under existing legislation and the failure of the draft regulations to remedy this. Though powers existed under current legislation to enact bye-laws to maintain the integrity of Special Areas of conservation and the relevant natural conservation agencies can use them to protect areas both inside and outside territorial SACs, they cannot do so for marine SACs. The UK government has preferred to rely on a voluntary approach to restricting activities or interests which damage the marine environment. The WWF UK had called for enactment of legislation to protect the marine environment and had even produced a draft of a Marine Areas Bill, but the draft regulations place on local authorities and public bodies which can exercise jurisdiction over the marine environment only a general duty "to have regard to the requirements of the Directive" (Regulation 3). Though this will apply to the Department of the Environment and the Ministry of Fisheries it is unclear how efficient it will be, divided as it is between departments, in offering integral protection inside and outside SACs, given that pollution and other damaging disturbances do not recognise boundaries.

It is not yet known whether any amendments will be made to the draft Regulations during their passage through Parliament but in any event, given that the UK is regarded as a State whose legislation in these matters is more advanced than that of most EU Member States, the gaps therein and the difficulties encountered in implementation do not auger well for the success of Natura 2000. It must be recalled, however, that there are other mechanisms available within the EU system which offer other means of environmental protection. Some of these are discussed below.

Other European Community Measures Impinging on Habitat Protection

The European Environment Agency (EEA)

The EC adopted in 1991 a Regulation[68] establishing a European Environment Agency (EEA). Prolonged wrangling about the location of the EEA initially delayed its entry into force, but it is now operative and situated in Copenhagen. It has no powers to initiate or overview EU legislation but has the task of setting up a European environment information and observation network that will provide objective information that can be relied on for promulgation of new and revision of old measures throughout the EU. It is an autonomous body with its own Director, Management Board and scientific committee. It will gather information on the state of the environment on a Community-wide basis and collect and assess it, with the aim of ensuring its comparability. It will report triennially on the State of the Environment. It is also charged with the task of developing forecasting techniques and developing methods of assessing environmental costs. Its assessment studies will

---

[68] Council Regulation 1210/90/EEC of 7 May 1990 on the establishment of the European Environment Agency and the European Environment information and observation network, OJ L120, 11.5.1990, p. 1.

focus on the pressures imposed on the environment and its quality and sensitivity. As it provides in this context for the inclusion of "transfrontier, plurinational and global phenomenon", it is not difficult to see how useful its services could be to enhance the effective implementation of the Natura 2000 network within the EU in relation to its transboundary threats and effects. This is especially so as it may, subject to certain restrictions, publish information and make it generally available and it is open to participation by non-Member States of the EU.

Environmental Impact Assessment (EIA)

The EEC Council Directive on the Assessment of Certain Public and Private Projects on the Environment,[69] though it has many inherent limitations because of its exceptions and the ambiguity of its terms, does also offer indirect protection to habitats and species since it requires environmental impact assessment of "public and private projects" which are likely to have significant effects on the environment. Though it excludes various "national interest" projects, including defence, it requires Member States to adopt all the measures necessary to ensure, before consent is given, that projects that are likely by virtue *inter alia* of their nature, size or location to have such signficant effects, are made subject to an assessment regarding those effects. The determination of both "significant" and "effects" is left to the State's discretion. Projects are subdivided into those for which an EIA is obligatory and others which need not be assessed unless they are likely to have the above effects (Annex I and II projects respectively). The information to be provided is detailed in a further Annex III.

Though its possibilities for enlarging habitat protection are obvious, its application has in fact given rise to numerous interpretation difficulties and wrangles between both the Commission and Member States and private individuals and their governments and the directive is one that has resulted in more complaints to the EC Commission than any other Directive relating to the environment. The Commission has sent many formal complaint letters to Member States and has also threatened to commence proceedings before the ECJ but as yet there is little case law on the subject, though a recent case in which the National Trust for Ireland and WWF UK complained that the EC Commission had violated the terms of the directive by permitting the use of structural funds for a project in Ireland[70] and cases which have arisen in national courts, indicate ways in which NGOs could act as "watchdogs" of the Habitats Directive. On the other hand a report on the directive's Implementation in Member States[71] recorded non-compliance with various requirements in most Member States as well as disparity in means and extent of implementation. This also indicates the problems facing the Habitats Directive. Proposals by the Commission for a new directive extending EIA to government programmes and policies excluded from the 1985 Directive have met strong resistance from some Member States, although the EC's Fifth Environmental Action Programme called for this. The new EEA is, however, now charged with the task of developing criteria for EIA with a view both to applying and possibly revising the 1985 Directive. It has to be recalled that the EU is a dynamic organisation, exceptional in its institutional and legislative and judicial structures, which can, by pro-

---

[69] Council Directive 85/337/EEC of 27 June 1985, OJ L175/1, 5.7.85.
[70] Case G 407/92, OJ C27, 30.1 1993, p. 4.
[71] COM (93) 28 Final, 13 vols. (1993); see (1993) 221 *ENDS Report* 20.

posal and interpretation, build upon the framework powers laid down in directives, which is, of course, a main reason why some Member States have recently pressed for more subsidiarity.

Right of Access to Enviromental Information

The EC was the first international body to adopt an instrument providing for such a right.[72] It aims to ensure that all persons in EC Member States have free access to environmental information held by public authorities throughout the EC and that such information is widely disseminated. It also sets out the terms and conditions on which this information should be made available, with some exceptions. The directive is cast in broad terms, which give rise to interpretation problems, as Governments are more likely to adopt restrictive interpretations than are individuals or companies. Again, however, the directive should sufficiently assist determined NGOs to obtain the kind of information that would be necessary for monitoring effective implementation of the Habitats Directive. As this chapter illustrates they have already begun to act in this role.

Other Relevant EC Measures and Policies

Reference has already been made throughout this chapter to various EC regulations and directives and common policies that contribute to protection of species and habitat. Space does not permit consideration of the numerous ways in which they do so. Relevant measures include the Regulation implementing CITES; those on import of whale products, and of ivory from the African Elephant; and some of those adopted under the Common Agricultural[73] and Common Fisheries Policies.[74] Its numerous measures on control of pollution from such sources as air, water, chemicals etc. also indirectly protect habitat. Its institution of Structural Funds that can be used, *inter alia*, for environmental protection[75] and new funds such as LIFE[76] and Cohesion Funds[77] are also developments that can be used to protect habitat.

Conclusion

The EC, as a party to the Berne Convention and having signed and approved the Biodiversity Convention, participated in the UNCED processes that led to the negotiation of the latter, and having committed itself in its Fifth Action Programme to preservation of biodiversity, has made a serious and encouragingly innovatory attempt to provide for this in its Directive on Conservation of Natural Habitats and Wild Fauna and Flora. However, as can be seen from this account of its develop-

---

[72] Council Directive 90/313/EEC of 7 June 1990 on the freedom of access to information on the environment, OJ L158, 23.6.1990, at p. 56.

[73] For application to nature protection of directives made under this policy, see Haigh, *op. cit.*, n. 4, at pp. 315–319.

[74] For details, see R.R. Churchill, *EEC Fisheries Law* (1987); see also the report on Monitoring Implementation of the Common Fisheries Policy, Commission of the EC, SEC (92) 394 final Brussels, 6 March 1992; Council Regulation (EEC) no. 3760/92 of 20 December 1992 establishing a Community system for fisheries and aquaculture, OJ L389, 31.12.92.

[75] Council Regulation EEC/1210/90 of 24 June 1988 on the tasks of structural funds.

[76] Council Regulation EEC/1973/92 of 21 May 1992 establishing a Financial Instrument for the Environment (LIFE), OJ L206/1 22.7.92.

[77] Council Regulation EEC/792/93 establishing a Cohesion Financial Instrument, OJ L79/74, 1.4.93.

ment and the first attempts to implement it, and especially in the context of problems arising in the implementation of related regulations and directives, this ambitious instrument has many deficiencies and ambiguities. It remains to be seen whether in time, in the light of international developments and pressures in the international fora in which its Fifth Action Programme commits the EC to participate, these difficulties will be overcome. Much will depend on the opportunities provided to the ECJ to interpret these directives and to the vigilance of NGOs in monitoring and reporting violations and maintaining a high level of public concern and awareness since the EC has never been provided with an environmental inspectorate of its own.

As Krämer remarks, the Habitats Directive "provides for a comprehensive system of habitat conservation which is to be set up within the next fifteen years. It is to be hoped that nature in Western Europe will survive until then"[78] and also one must add the integrity of the habitats within it that conserve and restore biodiversity.

---

[78] Krämer, *op. cit.*, n. 4, at p. 231.

12

Developing Countries, 'Development' and the Conservation of Biological Diversity

R. Jayakumar Nayar and David Mohan Ong

Introduction

One of the most significant facts which has come to light in dealing with the conservation of biological diversity is that by far the largest number of plant and animal species are present in tropical lands and seas – in rainforests and marine ecosystems in particular.[1] These species-rich areas of the world are, in turn, nearly all present within the territorial jurisdictions of developing countries. Consequently, attempts at the international regulation of the conservation of biodiversity must take into account (and perhaps more importantly, be seen to take into account) developing countries' perspectives on this matter. The attitudes of these countries to the international law in respect of biodiversity are themselves shaped by the wider stance adopted by developing countries towards the development of international environmental law, and public international law generally.

Our aim in this chapter is to provide an analysis of international attempts to regulate the conservation and sustainable utilisation of biological resources, as reflected in the 1992 Convention on Biological Diversity, which take into account the "development" imperatives that impact on biodiversity. The point of departure for the approach adopted in our analysis, however, is the conceptualisation of "development" that informs our discussion. The confrontation between developing and

---

[1] The number of new species catalogued is increasing almost daily, along with the estimated total number of species present. There are huge variations in these estimated figures. Gary Hartshorn for example, quotes 1.5 million known species and notes that the estimate of the total number of species has increased three-fold, from 2 to 3 million in the early 1970s to 10 million in 1980: see G. Hartshorn, "Key Environmental Issues for Developing Countries" (1991) 44 *Journal of International Affairs* 393, at p. 394. Another view puts the known number of species at 1.4 million but "guestimates" that the total number of species on earth may be anywhere from 10 million to 100 million: see Meadows, Meadows and Randers, *Beyond the Limits: Global Collapse or A Sustainable Future* (1992), at p. 64. All these writers are united, however, in noting that tropical forests, coral reefs and wetlands have the greatest variety of plant and animal species.

International Law and the Conservation of Biological Diversity (C. Redgwell and M. Bowman, eds.: 90 411 0863 7: © Kluwer Law International: pub. Kluwer Law International, 1995: printed in Great Britain), pp. 235–253

developed countries with regard to the apportioning of rights and responsibilities is not unfamiliar in negotiations relating to international environmental policy. Although there is no denying the importance of appreciating the specific concerns of developing countries as expressed by governments at the international level, a development perspective on the issues relating to biodiversity importantly also extends the focus of analysis to incorporate a human-centred approach to conservation. What is significant in this respect is that it requires a consideration of the dynamics of the development process from a grassroots perspective; shifting the balance away from the macro to the micro level of development discourse.

A two-pronged approach to the question of development and the conservation of biodiversity will be adopted here. The first part of the chapter will deal with what can be regarded as the official stance of developing countries with respect to biodiversity issues. Special attention will be paid to intergovernmental declarations and other governmental policy statements on the biodiversity issue. The focus will then shift to a comparison between the official developing countries' governmental policy on biodiversity and the alternative views held by concerned non-governmental organisations (NGOs) and other grassroots organisations on the success (or otherwise) of this officially endorsed policy. Central to the alternative view is the questioning of the assumptions which underpin much of the official policy response, both at the national and international levels, and the analysis of biodiversity conservation from a grassroots perspective of development.

The Developing Countries' Approach to the Conservation of Biological Diversity

Insofar as it is possible to generalise a developing country approach to the problem of conservation of biological diversity, a couple of observations may be made. First, it may be noted that many developing countries were initially of the view that concern for threats to biological diversity voiced mainly by so-called specialists from developed/industrialised western countries constituted yet another form of disguised imperialism,[2] and that to bow before such concern would merely bring about an unwelcome neo-colonial style intrusion upon their own national development policies.[3] In this sense, the reaction of most developing countries to concern over biological diversity is akin to their reaction to the more progressive notions of the principle of self-determination amongst peoples, and the protection of individual human rights in democracies. The extent of the applicability of these principles of international law within newly independent, and often politically and economically unstable, developing nations, even those which embrace democracy, has always

---

[2] Vandana Shiva has labelled these perceived threats to Third World biodiversity as "First World Bio-Imperialism", accusing the Northern countries and their industries not only of exploiting the biodiversity present in the South but also of attempting to replace this biodiversity with monocultures of crops which are more susceptible to disease. See V. Shiva, *Biodiversity: A Third World Perspective*, at pp. 14–18.

[3] For example, the Prime Minister of Malaysia, Dr. Mahathir Mohamad is reported to take the following stance in response to concern expressed mainly in Western developed countries over the large-scale felling of tropical forests: "Now that the developed countries have sacrificed their own forests in the race for higher standards of living, they want to preserve other countries' rain forests – citing a global heritage – which would indirectly keep countries like Malaysia from achieving the same levels of development." This statement is reported in the *Far Eastern Economic Review*, 1 August 1991, at p. 20 and also in the Chronicle Section (1991) 1 *Asian Yearbook of International Law* 295. See also, S.H. Bragdon, "National Sovereignty and Global Environmental Responsibility: Can the Tension be Reconciled for the Conservation of Biological Diversity" (1992) 33 *Harvard International Law Journal* 381, esp. at p. 387.

been a major bone of contention between the ruling elites of these fledgling states and Western, developed countries in the international arena.

Second, and not unrelated to the above point, it is the widely held perception amongst developing countries that their possession of the mainly untapped resource potential of species biodiversity within their territories presents them with an unrivalled opportunity finally to gain what may euphemistically be called lost development ground on the developed nations in the new, biotechnology-based, industries that have been widely touted as the next wave in the historical progression of development. Access to these resources should therefore be jealously guarded, especially from would-be competitors who lack such species biodiversity within their own jurisdictions but have both the capital and technological capacities to utilise them. In other words, the geographical occurrence of this biodiversity of species allows developing countries an hitherto almost unprecedented "say" in the legal construction of the system for their exploitation, and these countries are keen to exert the maximum amount of leverage from this fact.

The legal principles upon which they ultimately rely for this leverage over the developed/industrialised states are those of the permanent sovereignty of all states over their natural resources and the establishment of a New International Economic Order. Developing countries are also keen on asserting their rights over species extracts by relying on the principle of extraterritorial jurisdiction, based on an extended appreciation of their interests in the exploitation of species biodiversity that initially originates from their own countries.

It may be argued that developing countries wish to receive the maximum possible returns for the use of the plant and animal species extracts that are initially found within their territory.[4] It is the full value of what may be extracted from these species that they want in return, rather than their initial raw material market value.[5] This enhanced market value would need to be reflected in the total cost of these extracts, thus forcing the pharmaceutical and other companies that are keen to exploit these resources either to lower their profit margins for the products they develop from them, or increase their retail market prices for these products in order to pay the increased economic rent for the extracts in question. This method of assigning real values to the species extracts used in the new biotech industries of the future would entail a complete restructuring of the present world market system for pricing raw materials used in industrial production.

It was for precisely this reason that the former American president, George Bush, refused to sign the 1992 UNEP Convention on Biological Diversity in Rio de Janeiro. Bush's arguments at the time centred upon the apparent unwillingness of

---

[4] In this sense, the substance of the developing countries' position with respect to their ownership of the wealth of species biodiversity is similar to current attempts by economists to "internalise" the costs of environmental pollution by industrial processes. Such environmental pollution is at the moment an "externality" in the sense that the costs of nullifying such pollution presently borne by society in general and therefore not accounted for in any decision to allow such pollution to occur. This internalisation of previously external environmental costs is deemed necessary in order to provide the environment with an intrinsic economic value. Such environmental costs were previously external factors in the economic equation of industrial production. Their internalisation thus allows for a full accounting of the costs incurred when using natural resources for the production of industrial goods and services. See for example, D. Pearce et al., *Blueprint 3: Measuring Sustainable Development* (1993), at p. 81.

[5] For example, this has resulted in a plan by the Group of 15 (G-15) non-aligned, developing countries to set up a gene bank for herbs and medicinal plants in order to be able to assign property rights more effectively on them. See K. Peterson, "Recent Intellectual Property Trends in Developing Countries", in Recent Developments (1992) *Harvard International Law Journal* 277, at p. 286.

the United States of America to countenance any significant changes to the present system of assigning market values to raw materials and other primary commodities, which do not reflect their value-added nature in the end product. Although President Clinton reversed this US position and signed the Convention soon after coming into power, it is arguable that he was able to do so only because the relevant provisions in Articles 15, 16 and 19 had been significantly watered down in order to allow for a variety of possible interpretations. Indeed the Clinton administration had initially hoped to attach an interpretative statement to the signature which would have attempted to restrict the scope and ambit of the area to which the provisions of the Convention apply. Although this interpretative statement did not officially see the light of day, the US position on the so-called Intellectual Property Rights to derivatives from raw materials obtained only in developing countries but held by transnational corporations remains the same.[6]

In the run-up towards the Earth Summit in Rio de Janeiro, the developing countries' position on the conservation of biological diversity, as well as their position on all the other major environmental concerns that were due to be addressed at the UN Conference on Environment and Development, was developed in a series of intergovernmental meetings convened for the express purpose of articulating a common, united stance on the various environment and development issues that formed the purpose of the Rio Conference on Environment and Development. In general, therefore, the developing countries' (or Group of 77's) policy towards the biodiversity issue can be analysed within the context of the various declarations that were agreed upon at these intergovernmental meetings immediately prior to Rio. The object of the exercise is to examine how far the resulting provisions in the 1992 Convention reflect the developing countries' perspective on this issue.

The evolution of a common stand by the developing countries in the run-up to Rio was spearheaded by several emerging, newly industrialised, developing countries which believed that the concern expressed at the loss of species biodiversity represented a direct threat to their socio-economic national development programmes through its emphasis on *in situ* conservation of species, ostensibly in order to safeguard their future well-being. Furthermore, the idea that the natural biological diversity of the planet's plant and animal species might be treated as the common heritage of mankind caused much consternation amongst these countries, the overwhelming majority of which have only very recently embraced independence. The prospective threat of the common heritage of mankind principle being applied to their newly acquired permanent sovereignty over natural resources within their own territorial jurisdictions was therefore to be avoided at all costs. Thus, it was hardly surprising to find that the initial pronouncements by developing countries in various fora on this issue of conservation of biological diversity took on similar overtones to the continuing North–South debate over trade, aid, development and the environment generally.

In the 1989 Declaration of Brasilia by environmental ministers of Latin American and Caribbean countries, for example, paragraph 1 identifies the

> imperative need to strike a balance between socio-economic development and environmental protection and conservation, through the proper management of natural resources and control of environmental impacts as a source of common concern to the countries of the region having the utmost priority. Such a recognition is a statement

[6] G.S. Nijar and Chee Yoke Ling, "Intellectual Property Rights: The Threat to Farmers and Biodiversity", in *Third World Resurgence*, Issue No. 39, November 1993, p. 35, at pp. 39–40.

about the indissoluble relationship that exists between environmental affairs and socio-economic development and about the obligation to ensure the rational exploitation of resources for the benefit of present and future generations.[7]

Paragraph 2, however, emphasises the sovereign rights of these countries freely to administer their natural resources, as well as the need for the establishment of a new, fair and equitable International Economic Order. Paragraph 3 further cites the improvement of economic and social conditions as the key to preventing the defacement of the environment of these countries. This prevailing rhetoric of the developing countries was replicated in their responses to all the main environmental problems facing the world, including the threat of the extinction of numerous animal and plant species.

The Langkawi Declaration on the Environment, formulated at the 1989 Meeting of the Commonwealth Heads of Government, the majority of whom are from developing countries in Latin America, Africa and Asia, also states that the need to protect the environment from such threats should nevertheless be viewed in a balanced perspective with due emphasis being accorded to promoting economic growth and sustainable development.[8] In paragraph 6, the link between economic growth and environmental concern is described as necessary for the sustainable development of these countries. Developed countries were therefore warned not to use global environmental concerns as a pretext for introducing conditionalities in aid and development financing or for creating trade barriers. Having established the strict parameters for addressing such environmental concerns, the developing countries of the Commonwealth then felt able to agree with the rest of the developed Commonwealth countries (namely the UK, Canada, Australia and New Zealand) on a programme of action which included, *inter alia*, "support (for) activities related to the conservation of biological diversity and genetic resources, including the conservation of significant areas of virgin forest and other protected natural habitats".[9]

By the time of the first Ministerial Conference of Developing Countries on Environment and Development in Beijing, the developing countries or Group of 77, spurred on by a handful of prominent Asian countries such as Malaysia, had advanced the economic arguments of their position. The prominence of the Asian representation within the developing countries' group, particularly those members of the Association of South-east Asian Nations (ASEAN), has been attributed to a number of factors: African countries are traditionally too poorly equipped to lobby effectively in international fora, whilst Latin American countries have shied away from confrontation with Western developed countries which continue to service a large proportion of their enormous collective debts. ASEAN and other East Asian countries on the other hand have long experience in aid and development issues and the complexities of multilateral negotiations.[10] The emergence of Japan as a key player in the financing of several of the agreed mechanisms for the protection of

---

[7] Declaration of Brasilia emanating from the Sixth Ministerial Meeting on the Environment in Latin America and the Caribbean, held in Brasilia from 31–31 March 1989. See H. Hohmann (ed.), *Basic Documents of International Environmental Law* (1992), Vol. 3, Document 71c, at p. 1580.

[8] Para. 5 of the Langkawi Declaration on the Environment by the Commonwealth Heads of Government, done in Langkawi island, Malaysia on 21 October, 1989. See R.R. Churchill and D.A.C. Freestone (eds.), *International Law and Global Climate Change* (1991), at p. 331.

[9] Following the programme of action agreed under para. 8 of the Langkawi Declaration, *ibid.*, at p. 332.

[10] M. Vatikiotis, "Priming for Rio: Malaysia sets tone for Earth Summit agenda", *Far Eastern Economic Review*, 14 May 1992, at p. 22.

developing country environments also enhanced Asia's overall profile at the Earth Summit in Rio.[11]

Thus, after reasserting their understanding of the principles of international law which ought to govern the relationship between the developed and developing nations on the subject of global environmental protection,[12] the participating ministers from forty-one developing countries then addressed the various "sectoral issues" identified in the Beijing Declaration including, *inter alia*, biodiversity. Paragraph 15 of this Declaration stated that the biodiversity convention then being negotiated should clearly recognise "the linkage between access to genetic material and transfer of bio-technology, research and development in the country of origin, sharing the fruits of scientific research and the commercial profits".[13] In relation to these issues, the question of intellectual property rights was to be satisfactorily resolved so that such rights did not become an obstacle to the transfer of biotechnology which could be used by the countries of origin of biological resources for their own purposes. Significantly, the Biodiversity Convention was also required to recognise and reward the innovative work done by rural populations in developing countries for the protection and utilisation of biodiversity. Indeed, paragraph 17 provides for management plans to integrate living resource conservation and development priorities and goals, taking into account the needs of the local communities, including their habitats.

The former point concerning access to genetic material and the transfer of biotechnology was further elaborated upon in the Second Ministerial Conference of Developing Countries on Environment and Development, which was convened in Kuala Lumpur in late April 1992, attracting the participation of fifty-five states. In the Kuala Lumpur Declaration issued at the end of this meeting, paragraph 26 underlined "the need for the Convention to establish mechanisms to give effect to the rights of countries which possess biological and genetic resources in *in situ* conditions".[14] It was reiterated that "the Convention on Biological Diversity must include legally binding commitments to ensure the link between the access to the genetic material of developing countries and the transfer of biotechnology and research capabilities from developed countries, as well as sharing of commercial profits and products derived from the genetic material". Within this context, it is extremely interesting to note that, unlike the Beijing Declaration, the Kuala Lumpur Declaration did not include a proposal for the recognition of the importance of indigenous populations or local communities in the discovery, initial utilisation and conservation of species biodiversity. Neither did the Kuala Lumpur Declaration confirm the right of community participation in the protection of such resources. It

---

[11] *Ibid.*

[12] Amongst these "General Principles" are the sovereign equality of states, the sovereign rights of states to use their own natural resources in keeping with their developmental and environmental objectives and priorities, the principle of non-interference in the internal affairs of developing countries, the non-imposition of conditionalities in trade, aid or development issues and the establishment of a new and equitable international economic order. See paras. 3–9, esp. paras. 5 and 6, in the Beijing Ministerial Declaration on the Environment and Development, adopted unanimously at the Ministerial Conference of Developing Countries on Environment and Development in Beijing on 19 June 1991. Text in Churchill and Freestone, *op. cit.*, n. 8, at p. 363.

[13] *Ibid.*, at p. 364.

[14] Kuala Lumpur Declaration on Environment and Development, issued at the end of the Second Ministerial Conference of Developing Countries on Environment and Development, 29 April 1992. Text in S.P. Johnson, *The Earth Summit: The United Nations Conference on Environment and Development (UNCED)* (1993), at pp. 38–39.

is arguable that this subtle yet very definite shift in emphasis of the Group of 77's position away from recognition of the not inconsiderable input provided by local, indigenous populations in the cultivation of biodiversity represents an unjustifiably narrow interpretation of the economic dimension of biodiversity protection. For example, the UNEP Governing Council's Decision on the preparation of an International Legal Instrument on Biological Diversity notes that the economic dimension of biodiversity should include, *inter alia*, financial transfers and transfer of biotechnology to the owners and managers of biological resources, i.e. the people concerned, and not necessarily only to the states where these resources may be situated.[15]

As may be seen from the above discussion, the developing countries' policy position mainly revolves around the need for an equitable return for the utilisation of their resources in plant and animal species biodiversity, including the transfer of technology in order to be able fully to exploit their resources in this area. This position, as mentioned earlier, is largely based on an extended interpretation of the doctrine of permanent sovereignty over natural resources which is coupled with the desire (if not the ability) to recover most of the full value of the genetic resource originating in their territorial jurisdiction by virtue of proposals that are not dissimilar to those which have been recognised as an assertion of extraterritorial jurisdiction.[16] In this respect, it is important to note that under Article 4(b) of the 1992 Convention on Biological Diversity, the jurisdictional scope of the provisions of the Convention apply to processes and activities under a Contracting Party's jurisdiction or control, within the area of its national jurisdiction, and even beyond the limits of national jurisdiction, but significantly not within the territorial jurisdiction of another state, whether or not that state is a party to the Convention.

Alternative Views on the Conservation of Biodiversity in Developing Countries

Having considered the developing countries' official stance on the conservation of biodiversity, this can now be usefully compared and contrasted with the alternative non-governmental organisation (NGO) views on biodiversity. Perhaps unsurprisingly, such alternative views do not sit easily with the largely accepted paradigm governing the biodiversity debate between the developed and developing countries, which focuses on the economic benefits to be reaped when plant and animal extracts are utilised for the manufacture of new products. These alternative interest groups tend instead to stress the importance of biodiversity to the indigenous local communities that depend for their very survival on the many different varieties of plant and animal life that surround them.[17] Biodiversity is therefore deemed to be important for the continuation of an indigenous lifestyle, rather than for crude financial gain.

---

[15] UNEP Governing Council Decision 15/34, Preparation of an International Legal Instrument on the Biological Diversity of the Planet, 25 May 1989, para. 4. Text in M. Molitor, *International Environmental Law – Primary Materials* (1989), at pp. 47–48.

[16] See, however, "*Lujan* v. *Defenders of Wildlife*: The Need for a Uniform Approach to Extraterritoriality", a Comment which notes that the principle of extraterritoriality has not been judged, at least in the US, as applying to environmental regulation (1993) 19 *Brooklyn Journal of International Law* 1009.

[17] H. Schücking and P. Anderson, "Voices Unheard and Unheeded", in V. Shiva, P. Anderson, H. Schücking, A. Gray, L. Lohman and D. Cooper, *Biodiversity: Social and Ecological Perspectives* (1991), especially at pp. 31–33.

The conservation of biodiversity is championed at the alternative level for at least two other non-development oriented reasons. First, for the purely aesthetic value that may be attached to the propagation of variety in all types of life forms on this planet.[18] Second, for the smooth functioning of the land and marine ecosystems on this earth, recognising that the preservation of the biodiversity of species that inhabit these ecosystems enhances their survivability, notwithstanding the fact that these species may not by themselves hold any anthropocentric values. This argument follows from the fact that not all biodiversity can be turned into biological resources with an attached commercial value.[19] All programmes for the conservation of biological diversity must therefore give priority to the preservation of ecological functions and subsistence uses over and above new schemes of commercialisation of biological diversity for the international market. In other words, the "bottom line" in the conservation of biodiversity must always be defined in ecological, rather than economic terms.[20]

In essence, therefore, the conservation of biological diversity is regarded as an important contributor to the preservation of the cultural diversity of peoples that depend on the biodiversity of their surrounding environments for their lives.[21] Cultural diversity may in turn be seen to contribute to the continuation and even the enhancement of biological diversity. If cultural diversity is seen to be another imperative of an overall strategy towards global biodiversity then such a strategy must perforce incorporate an element of European Union-style "subsidiarity" within its structure for implementation. In the words of one expert, "the relationship between global and local policies should be fluid and democratic, i.e. grassroots experiences should inform international law-making through various consultation mechanisms".[22] Indeed, it has been argued that such a global strategy should also take into account the cultural, social, political and economic problems affecting each region rich in biodiversity.[23] However, a coherent and co-ordinated policy for the incorporation of national and local policies for biodiversity within a global strategy for the preservation of biodiversity is obviously not yet in place.[24] This is especially pertinent when we consider the fact that local species biodiversity forms a natural resource of medicinal, edible and other useful extracts. Thus, the destruction of species biodiversity through the loss of habitat for allegedly "development" purposes often has the effect of forcing hitherto independent local communities to

---

[18] See Ehrenfield, for example, who argues strongly against assigning merely economic value to biodiversity on the basis that such a simplistic process of valuation would do more harm than good for the conservation of global biodiversity, in D. Ehrenfield, "Why Put a Value on Biodiversity?", in E.O. Wilson (ed.), *Biodiversity* (1988), at pp. 213–214.

[19] See also B. Norton, "Commodity, Amenity, and Morality, The Limits of Quantification in Valuing Biodiversity", in Wilson (ed.), *ibid.*, at pp. 200–205.

[20] Shiva *et al.*, *op. cit.*, n. 17, at pp. 28, 30.

[21] S. Bilderbeek (ed.), *Biodiversity and International Law* (1992), at pp. 9–11. See Recommendation 5(c) of the Hague Recommendations on International Environmental Law, 16 August, 1991, in Bilderbeek, *ibid.*, pp. 194–202, at p. 199.

[22] Reply by Stephanie d'Orey to Questionnaire of the Global Consultation on the Development and Enforcement of International Environmental Law, with a Special Focus on the Preservation of Biological Diversity, in Bilderbeek, *ibid.*, at p. 12.

[23] Reply by Juan Mayr to the same Questionnaire, *ibid.*, at p. 14.

[24] Article 8(j) of the Convention merely provides that "the Contracting Parties ... shall as far as possible and as appropriate subject to national legislation, respect, preserve and maintain indigenous lifestyles". See however paras. 15.5, 15.6 and 15.7 of Chapter 15 of Agenda 21, which attempt to provide a plan of action for the protection of biological and cultural diversity at both the international and national levels, in Johnson *op. cit.*, n. 14, at pp. 289–291.

participate in the commercially oriented, market-based economy, which they are ill equipped to do.[25] Consequently, these communities become poorer than they were before the advent of development.

The most obvious threat to the cultural diversity amongst human communities which natural biodiversity both begets and nurtures is manifested in the consequences of natural biodiversity degradation upon the indigenous peoples who actually live amongst this biodiversity. The estimated total of more than 5,000 different cultures in the world today, when compared with the 180 or so recognised national state cultures, means that indigenous peoples constitute about 90-95% of the known cultural diversity of the world.[26] The connection between natural and cultural biodiversity becomes clear when it is realised that the land areas with the highest natural biodiversity are also the homelands of many of these indigenous peoples. However, the connection between the protection or conservation of the natural biodiversity in these areas and the protection of the indigenous peoples who traditionally live there has not been adequately transposed to the 1992 Convention on Protection of Biological Diversity, an anomaly which may be critical to the continued survival of both types of biodiversity.

In the Declaration of San Francisco de Quito of 8 March 1989 by the Member States of the Treaty for Amazonian Co-operation, the importance of the biodiversity to be found in Amazonian ecosystems was recognised. Significantly, this Declaration also made the connection between genetic and biotic conservation and the cultural identity of the human populations in Amazonia, although it qualifies this relationship by stating that the protection of both biological and cultural diversity is to be in accordance with the policies decided by each Amazonian country concerned.[27] The need to protect the diversity of the local Amerindian populations in the Amazon region is reiterated in the same Declaration, where the effective participation of these communities is included through the promotion of indigenous knowledge in regional development programmes.[28]

The Quito Declaration was followed on 6 May 1989 by the Amazonian Declaration adopted at Manaus, Brazil by the Presidents of the self-same states parties to the Treaty for Amazonian Co-operation.[29] Paragraph 3 of the Declaration affirms their support for the two then recently created Amazonian Special Commissions on Environmental and Indigenous Affairs respectively. The lack of evidence to suggest that these commitments have been met and that the measures implemented have been effective in prioritising grassroots-led initiatives is alluded to below in the discussion of the grassroots perspectives on the conservation of biodiversity.

In this respect, the 1991 Hague Recommendations on International Environmental Law, which were adopted by a group of experts for consideration by the United Nations Conference on the Environment and Development (UNCED), provide a more complete catalogue of rights for the local and indigenous communities in relation to the genetic resources that may be found in their surroundings.

---

[25] Bilderbeek, p. 18.
[26] A. Gray, "The Impact of Biodiversity Conservation on Indigenous Peoples", in V. Shiva *et al.*, *op. cit.*, n. 17, at pp. 61–62, citing a personal communication from G. Hendricksen, University of Bergen.
[27] Fourth paragraph of Section II on Environment Policy in the 1989 Quito Declaration, see Hohmann, *op. cit.*, n. 7, Vol. 3, Document 71a, at p. 1572.
[28] Second paragraph of Section III on Co-operation for Amazon Indian Affairs, see Hohmann *ibid.*, at p. 1573.
[29] *Ibid.*, Document 71b, at pp. 1578–1579.

These Recommendations call for a duty upon states to ensure local access and control over genetic resources, as well as their receipt of any benefits from the exploitation of these resources.[30] These objectives are to be met by strengthening local property rights and formulating new intellectual property rights, as well as incorporating such genetic resource protection and benefits into other international agreements and negotiations such as GATT.[31] Furthermore, it was proposed that the principle of equitable sharing of costs and benefits from the conservation and sustainable use of biodiversity be applied, with commercial users paying for such use in a manner which adequately reflects the cost to the natural resource base.[32]

Biodiversity: A Grassroots Perspective

Thus far, the discussion has focused on the politically and economically motivated "tug-of-war" between developing and developed countries with respect to the questions of rights and responsibilities over biological resources. The central issue in the debate is one of control, both over access to the resources themselves, and over the benefits that potentially derive from their use. In this section, the question of control will be considered not by reference to the political and economic jockeying of States *per se*, but rather by reference to the very "philosophy" which underpins the importance of, and the need to conserve, biodiversity. What is intended here is an analysis of the Rio Convention which goes beyond the assessment of the operational viability of the Convention as an international legal instrument, to encompass a holistic perspective of biodiversity within the broader context of human welfare.

Biodiversity and Development: The Concept of Sustainable Use

Article 1 of the Rio Convention on Biological Diversity sets out as the objectives of international action the conservation of biological diversity and the sustainable use of its components, with fair and equitable sharing of the benefits arising from the utilisation of genetic resources. Although Article 1 necessitates that "conservation" be approached in conjunction with "sustainable use", it is not obviously clear how conservation and sustainable use are to be interlinked within the framework of international policy. The conservation/sustainable use duality presents from the onset crucial interpretational questions which determine the subsequent thrust of international policy with respect to biodiversity and development:

(a) is the Convention essentially concerned with international rights and responsibilities with respect to the economic exploitation of natural resources (in this case biological resources), conservation being a "management" principle for resource exploitation based on the "economic growth" development paradigm?
(b) or, alternatively, is the Convention concerned with the conservation of global biodiversity within a framework which incorporates sustainable human use of biological resources, economic resource exploitation being part of the broader context of human development?

---

[30] Recommendation 5a, in Bilderbeek, *op. cit.*, n. 21, at p. 198.
[31] *Ibid.*
[32] *Ibid.* A similar though less specific provision for such equitable sharing is now in Article 15(7) of the Convention.

It is submitted that what is critical in determining the fundamental philosophical thrust of international policy with respect to biodiversity is the understanding of what is meant by "sustainable use" in the context of the Convention.

"Sustainable use" is defined in the Convention as follows:

> the use of components of biological diversity in a way and at a rate that does not lead to the long-term decline of biological diversity, thereby maintaining its potential to meet the needs and aspirations of present and future generations.[33]

By adopting language reminiscent of the Brundtland Commission's definition of sustainable development,[34] the Convention leaves open the same questions regarding the meaning of sustainable use as those which have plagued that concept: the use by whom and sustainability for what purpose and for whose needs and aspirations? It is upon these questions that the philosophical crux of the international legal response to the issues of global biodiversity is dependent. This can be demonstrated by suggesting two possible interpretations of the motivation underlying Article 1 of the Convention:

(a) *Interpretation 1*
Motivation: financial gain.
Focus: economic exploitation, and management of biological resources.
Threat: denial of access to biological resources, reduction of potential for profit, and exhaustion of resource supply.
Aspiration: guaranteed continuous and unimpeded access to resources, including safeguarding resource stock-base.

(b) *Interpretation 2*
Motivation: sustainable livelihood.
Focus: conservation of biological resources and its diversity.
Threat: expropriation of control over resources, destruction of biodiversity and the degradation of life-support systems and livelihoods.
Aspiration: protection of biodiversity, retention of control over resources.

The two possible readings of the philosophical thrust underpinning the Convention, it must be pointed out, do not necessarily preclude the incorporation of both motivations into a successful formula for international action. What is revealed here, however, is the ambiguity of priorities within the broad definition of "sustainable use" contained in the Convention which must be clarified if a meaningful assessment of the Convention is to be made. The question that needs to be addressed is essentially a simple one; what is the purpose of the conservation of biological diversity? A recourse to rhetoric will be useful to resolve the definitional gap with respect to the central concept of sustainable use.

The fundamental reasons for the conservation of biodiversity, and thus the basis of international action, are set out in the Preamble to the Convention as being:

> – the intrinsic value of biological diversity and ... the ecological, genetic, social, economic, scientific, educational, cultural, recreational and aesthetic values of biological diversity and its components;[35] and;

---

[33] Article 2 of the Convention on Biological Diversity, signed by over 150 countries in Rio de Janeiro, Brazil on 5 June 1992, text in Johnson, *op. cit.*, n. 14, at p. 82.

[34] See WCED, *Our Common Future* (1987), at p. 43.

[35] Preamble to the Convention on Biological Diversity, para. 1.

– the importance of biological diversity for evolution and for maintaining life sustaining systems of the biosphere.[36]

It is worth noting that the economic dimension of biological resource utilisation is merely one aspect within a whole spectrum of the values of biodiversity that directs what is to be understood as "sustainable use". The containment of economic exploitation of resources within the broader context of human welfare, implied by the pronouncements in the Preamble, provides us with the first analytical pointer for the assessment of the strategies put forward in the Convention.

The emphasis on the human, as opposed to the purely economic, dimension of biodiversity conservation, although seemingly a restatement of the obvious, is significant in two respects; first, because it directs attention to the proper under-standing of the threats to biodiversity from a grassroots-based, livelihoods perspec-tive; and second, because it necessitates conservation strategies which are consistent with the fundamental livelihoods requirements of resource utilisation. When viewed from a policy perspective, therefore, this translates into the prioritisation of grass-roots-led initiatives to first identify problems in relation to resource management, and subsequently to preserve *in situ* the sustainable utilisation of biological resources.[37] Based on this scheme of priority allocation, therefore, market-oriented resource exploitation measures must be taken as surplus potential for sustainable use, subsequent and secondary to the primary relationships between humanity and nature. The following discussion will seek to return the debate to this fundamental level of analysis.

Threats to Diversity

"Concerned that biological diversity is being significantly reduced by certain human activities…"[38] This oblique statement provides the only reference in the Convention to recognition that there exist threats to biodiversity. Failure to identify with any degree of specificity the nature of the human activities which lead to the biodiversity loss represents a major weakness of the Convention; one which leads to the questionability of the strategies subsequently put forward. The omission is unlikely to be the result of a simple oversight.

The issues that have taken centre stage in discussions relating to the conservation of biodiversity, particularly in the context of development, have tended to be:

(a) the nature of differentiated responsibility based on levels of 'development';[39]

---

[36] Preamble, para. 2.

[37] Recognition of the priority of *in situ* conservation is evident in the Convention; see Article 8. The issue that needs to be considered, however, is whether the provisions of Article 8 reflect the 'bottom-up' approach to policy formulation which is imperative for sustainable livelihoods-oriented conserva-tion strategy. Critical to this line of enquiry is an understanding of the dynamics of the "development" process particularly with respect to the question of control over access to, and the decision-making process in relation to biological resources. This point will be addressed below.

[38] Preamble, para. 6.

[39] See, for example, Article 20 (4) of the Convention:

> The extent to which developing country Parties will effectively implement their commitments under this Convention will depend on the effective implementation by developed country Parties of their commitments under this Convention related to financial resources and transfer of technology and will take fully into account the fact that economic and social development and eradication of poverty are the first and overriding priorities of the developing country Parties.

(b) the terms and conditions of technology transfers, which are seen as central to the ability of developing countries to implement policies aimed at the sustainable use of their biological resources;[40]
(c) the provision of "new and additional" financial resources, which is taken as a prerequisite to enable developing countries to undertake conservation measures;[41] and
(d) the terms of equitable sharing of the benefits resulting from the sustainable use of biodiversity.[42]

The underlying assumptions which inform the prioritisation of these issues can be discerned by reference to the Preamble of the Convention, notably in paragraphs 7, 15, 18 and 19. In short, these statements provide that "development" through the increase of financial investments, modern technologies and greater governmental and international involvement in general, is the way forward to the long-term conservation of biodiversity. It is worth emphasising that the validity of these assumptions is in no way substantiated, in the Convention, by an analysis of the problems of biodiversity loss. The intellectual leap from the concern that is expressed regarding the loss of biodiversity through certain human activities to the confidence with which the strategies for action are prescribed appear to have been achieved less from an understanding of the issues concerned than from a stubborn faith that old medicines will eventually provide the remedy. Whether or not the strategies emphasised are appropriate and relevant to the problems arising from biodiversity loss is a question which needs to be addressed rather than being taken for granted. It is necessary, therefore, first to redress the lack of attention paid to understanding the primary causes of biodiversity loss – revelations which possibly sit uncomfortably with the preferred diagnosis – and then to consider the subsequent solutions that are commonly prescribed at international fora.

Expressing the views of the periphery, of grassroots subsistence communities and the multitude of NGOs, which represent populations directly and are intimately affected by biodiversity loss, Vandana Shiva distinguishes between primary and secondary causes of the destruction of biodiversity.[43] In contrast to the dominant wisdom of national and international policy makers, which ascribe much of biodiversity loss to such factors as local population pressures, Shiva argues that it is essential to understand the primary causes which lead to such pressures resulting in over-exploitation of resources:

> stable communities, in harmony with their ecosystem, always protect biodiversity. It is only when populations are displaced by dams, mines, factories, and commercial agriculture that their relationship to biodiversity becomes antagonistic rather than co-operative. The displacement of people and displacement of diversity goes hand in hand, and displaced people further destroying biodiversity is a second order effect of the primary causes of destruction ...[44]

[40] Articles 16–19.
[41] Articles 20 and 21.
[42] Article 15 (7).
[43] See V. Shiva, "Introduction" in Shiva et al., op. cit., n. 17, at pp. 4–9.
[44] Shiva, ibid., at p. 9.

Two primary causes for the large-scale destruction of biodiversity are identified. First, the destruction of biodiversity due to development projects in forest areas, and secondly, the displacement of diversity through the introduction and perpetuation of monocultures in commercial export-oriented agriculture. The destructive ecological and social impact of "development" projects in forest areas has long been recognised and documented.[45] Diverse governmental and international initiatives ranging from the construction of large scale dams[46] such as the Narmada[47] and Tehri[48] in India, the Nam Choan in Thailand[49] and the Batang Ai and Bakun hydroelectric projects in Sarawak, Malaysia,[50] road building schemes such as the trans-Amazonian highway in Brazil[51] and under the Plan Pacifico scheme in Colombia,[52] and logging and forest resource development schemes such as those in Thailand[53] and Papua New Guinea,[54] are a few examples of the many ongoing development projects that have been identified as ecologically and socially destructive and yet defended defiantly as the solutions to the problems of "underdevelopment" and as the means of achieving future environmental stability.

In conjunction with the ever increasing encroachment upon forest areas, the advent and increasing dominance of industrial agriculture, with its emphasis on large-scale production of economically viable monocultures, further diminishes the prospects for a perpetuation of diversity, both of resources and of forms of production. Two striking examples of the detrimental impact of monoculture production on biodiversity can be seen in the areas of food production and agroforestry, the result of international policy expressed for example in the UN Food and Agricultural Organisation's policy documents, *World Agriculture: Toward 2000, An FAO Study*,[55] and the *Tropical Forest Action Plan* (TFAP),[56] a joint initiative of the FAO, the World Resources Institute, the World Bank, and the United

---

[45] For a general introduction, see World Rainforest Movement, *Rainforest Destruction: Causes, Effects and False Solutions* (1990). For a more personalised account of the social dimension of the problems relating to the destruction of livelihoods through ecological degradation and dispossession, see C. Caufield, *In the Rainforest* (1985).

[46] For a most comprehensive study on the impact of internationally funded, mega-projects involving the construction of large dams, particularly for the purposes of hydro-electric generation, see E. Goldsmith and N. Hildyard, *The Social and Environmental Effects of Large Dams*, Volume 1: *Overview* (1984).

[47] The Narmada dam project is perhaps the most publicised case of the conflict between "development" and the environment. For a critical analysis of the project, the detrimental effects of which were recognised from the onset, see C. Alvares and R. Billorey, *Damming the Narmada: India's Greatest Planned Environmental Disaster* (1988).

[48] See F. Pearce, "Building a Disaster: The Monumental Folly of India's Tehri Dam" (1991) 21/3 *The Ecologist* 123.

[49] See B.S. Cox, "Thailand's Nam Choan Dam: A Disaster in the Making"; and P. Hirsch, "Nam Choan: Benefits for Whom" (1987) 17/6 *The Ecologist* 212–223.

[50] See E. Hong, *Natives of Sarawak: Survival in Borneo's Vanishing Forest* (1987).

[51] See A. Shankland, "Brazil's BR-364 Highway: A Road to Nowhere?" (1993) 23/4 *The Ecologist* 141.

[52] See J. Barnes, "Driving Roads through Land Rights: The Columbian Plan Pacifico", *ibid.*, at p. 135.

[53] See P. Hirsh, "The State in the Village: The Case of Ban Mai", in (1993) 23/6 *The Ecologist* 205.

[54] See G. Marshall, "The Political Economy of Logging: The Barnett Inquiry into Corruption in the Papua New Guinea Timber Industry" (1990) 20/5 *The Ecologist* 174.

[55] Alexandratos (ed.) (1988).

[56] FAO, Rome (1985).

Nations Development Programme.[57] The combined effect of the dominant culture of production practised in industrial agriculture is succinctly summarised by Shiva:

> Plant improvement in agriculture has been based on the enhancement of the yield of a desired product at the expense of unwanted plant parts. The "desired" product is however not the same for agri-business and a Third World peasant. Which parts of a farming system will be treated as "unwanted" depends on one's class and gender. What is unwanted for agribusiness may be wanted by the poor, and when it squeezes out those aspects of biodiversity, agriculture "development" fosters poverty and ecological decline.[58]

What the examples above demonstrate is the inextricable link between the problems related to the loss of biodiversity and the broader issue of what is understood as "development". The dominant assumption that underpins international law in general and international environmental law in particular is that many of the problems facing humankind are attributable to the evils of some predetermined notion of "underdevelopment" or "poverty";[59] that "development" is a "good" to be attained, both through robust national action, and greater international involvement. A more sensitive analysis of the problems of biodiversity loss and the human dimensions behind the economic and "pure" environmental calculations of species diversity reveals, however, fundamental challenges to these basic assumptions. When strategies to "conserve" biodiversity are considered, therefore, it is imperative that the true causes of biodiversity loss are retained as the primary focus of analysis despite the obfuscation of issues that is the result of governmental expropriation of the debate.

Conservation: A Question of Control

What is apparent from the preceding discussion is the need to shift the focus away from technocratic solutions to the problems of biodiversity loss, and on to the more

---

[57] For a damning critique of the policies pursued by the FAO in relation to global agriculture and sustainable resource utilisation, see "The UN Food and Agriculture Organisation: Promoting World Hunger" (1991) 21/2 *The Ecologist*, and especially in relation to food production, E. Goldsmith and N. Hildyard, "World Agriculture: Toward 2000, FAO's Plan to Feed the World", at p. 81. On the TFAP, see also M. Colchester and L. Lohmann, *Tropical Forestry Action Plan: What Progress?* (1990). The FAO's approach has since been revised with the release of *Agenda for Action for Sustainable Agriculture and Rural Development (SARD)*, jointly initiated with the Dutch Ministry of Agriculture, Nature Management and Fisheries, subsequent to a conference in 's-Hertogenbosch, the Netherlands, in 1991. However, whether or not this apparent shift in policy represents a true reorientation of priorities from the past emphasis on capital intensive industrial agriculture as the dominant mode of production, is yet to be seen; for an updated analysis of the FAO approach, see N. Hildyard, "Sustaining the Hunger Machine: A Critique of FAO's Sustainable Agriculture and Rural Development Strategy" (1991) 21/6 *The Ecologist* 239.

[58] Shiva, *op. cit.*, n. 43, at p. 8.

[59] These definitional assumptions, so long the linchpin of development and environmental policy, are now being challenged with great potency both by development thinkers and activists alike. The resulting analyses of human deprivation and environmental degradation that have emerged from this "alternative" dialogue presents directions for action that emphasises the vernacular knowledge and the empowerment of "the poor" as active "participators" as opposed to "participants" in their own development. For examples of this grassroots-led dialogue of "development", see W. Sachs (ed.), *The Development Dictionary: A Guide to Knowledge as Power* (1992) and especially the contributions by G. Esteva, "Development", at p. 6, W. Sachs, "Environment", at p. 26, and M. Rahnema, "Poverty", at p. 158. See also, "The Discovery of Poverty", in *The New Internationalist: Development*, No. 232, June 1992, at p. 7.

fundamental questions relating to the control and, thus, the sustainable utilisation of biological resources.

"Conservation" and "sustainable use" are ultimately determined by those who control both access to the resources in question, as well as to the decision-making process in relation to their use. All the evidence thus far suggests that responsibility towards the natural environment is best placed in the hands of those who are intimately connected with nature in an on-going life cycle of production and conservation. In the context of an international approach to the conservation of biodiversity, with all its implications on resource exploitation and economic development, and, therefore, the potential for conflicting claims on biological resources, what is essential for an effective conservation strategy is a clear expression of priorities relating to the utilisation of these resources. This, it is submitted, can best be achieved by adopting a "rights"-based approach.

The issue of control over resources is not neglected in the Convention. This is evident in Article 3, which lays down as the main foundation of conservation and sustainable use the principle that States have, in accordance to the Charter of the UN and the principles of international law, the sovereign right (subject to their international responsibilities) to exploit their natural resources. The deficiency of the Convention's approach to the question of control, however, is the failure to appreciate the true nature of the internal dynamics of resource exploitation within States. It can be discerned from the preambular paragraphs, when read together with Article 3, that the Convention is based on the precept that States necessarily act, first and foremost, in conformity with the needs of the most vulnerable sections of the population. If history has taught us anything, however, it must surely have taught us the folly of such an assumption.

Related to the above point is the obvious lack of emphasis on the rights over biological resources of grassroots communities who depend on biodiversity for their continued survival. This is all the more astonishing given the prevalence of worldwide struggles being waged by subsistence communities to protect the viability of both cultural and biological diversity. No doubt, comforting references to "promote and encourage participation" of local communities have not been omitted,[60] and form part of the obligation undertaken by States parties under the Convention. Despite the reassurance that these references to grassroots participation may bring to some, however, the danger that they are nothing more than conscience-alleviating means of expressing "business as usual" must not be overlooked.

The lack of priority that is accorded to the so-called "participation" of resource-dependent, subsistence populations, is revealed in the extreme form in preambular paragraph 12:

> Recognising the close and traditional dependence of many indigenous and local communities ... and the *desirability* of sharing equitably benefits arising from the use of traditional knowledge, innovations and practices ... [our emphasis]

The extent to which past practice has deemed it *desirable* to share equitably benefits that have resulted from the direct expropriation of traditional knowledge is only too apparent in the case of what has been called the "intellectual piracy" of the neem tree,[61] i.e. the patenting of the method of chemical extraction of azadirachtin

---

[60] See Preamble, para. 13, Article 8(j) and Article 10(c)(d).
[61] See V. Shiva and R. Holla-Bhar, "Intellectual Piracy and the Neem Tree" (1993) 23/6 *The Ecologist* 223.

from the seed of the neem. The following justification was provided by W.R. Grace, owner of at least four US companies with neem-based patents:

> Although traditional knowledge inspired the research and development that led to these patented compositions and processes, they were considered sufficiently novel and different from the original product of nature and the traditional method of use to be patentable.[62]

The possible implications of this current trend on the livelihoods of farmers, as pointed out by Shiva and Holla-Bhar, are that

> (their) seed-stock, animal breeding-stock and natural pesticides may gradually become the intellectual property of national or multinational companies; they will lose their independence and be forced to pay high prices for products that they could formerly provide for themselves.[63]

The example of the battle over the neem tree, outlined above, presents us with an illustration of the complex nature of conflicting interests which impact on biodiversity. Assuming that international action, through the Convention on Biological Diversity, is truly concerned with the long-term viability of biodiversity in all its richness, as opposed to merely being an exercise of deception for what is essentially a Convention on the Exploitation of Biological Resources, the real prospects for the conservation of biodiversity lie in the interpretation and application of a neglected and inconspicuous provision within Article 1 of the Convention providing for the objectives of this Convention, which are, *inter alia,*

> the conservation of biological diversity, the sustainable use of its components and the fair and equitable sharing of the benefits arising out of the utilisation of genetic resources ... *taking into account all rights over those resources* ... [our emphasis]

Setting aside all the political haggling that has dominated international discussions relating to the issue of biodiversity, it is the commitment with which *all the rights* over biological resources are respected, and the prioritisation of those rights, which will ultimately determine whether biological diversity will be conserved and sustainably used for the benefit of present and future generations.[64]

The Way Forward? Grassroots Perspectives

> Often making use of what James Scott calls the "weapons of the weak", groups, communities and individuals the world over are successfully resisting the web of enclosure and reclaiming a political and cultural space for the commons. Whilst UNCED has been mainly interested in "solutions" that will permit industrial growth to continue, those who rely on the commons are carving out a very different path. Their demands centre not on formulating treaties, but on reappropriating the land, forests, streams and fishing grounds that have been taken from them; on re-establishing control over decision making; and on limiting the scope of the market. They seek to rejuvenate what works and to keep alive or elaborate strategies that meet local needs.[65]

Two examples of grassroots-led initiatives that are directly relevant to the conservation of biodiversity are provided in the Charter of Farmers' Rights[66] and the

---

[62] *Ibid.*, p. 225.
[63] *Ibid.*, p. 227.
[64] Preamble, para. 23.
[65] "Reclaiming the Commons" (1992) 22/4 *The Ecologist: Whose Common Future?* 195. Emphasis in the original.
[66] "A Charter of Farmers' Rights", in *Third World Resurgence*, No. 39 (November 1993), at pp. 28–29.

Declaration of the International Movement for Ecological Agriculture. The different perspective that is adopted in the analysis of the problem and the alternative approach to conservation strategies are self-evident. In the Charter of Farmers' Rights, the rights to land, control over natural resources and the conservation, reproduction and modification of seed and plant genetic material are to be placed firmly in the hands of the farmers and communities that utilise them directly, as opposed to the governments of the states in which these resources are located. Paragaphs 1–9 of the Declaration of the International Movement for Ecological Agriculture[67] chart the problems of the present system of global agriculture and, in paras. 10–12, offer alternative, grassroots-based ecological agricultural systems in order to redress these problems. The two documents, when read together, provide the common strands of the general strategy that is required for a holistic approach to the problem of achieving sustainable development, including the conservation of biodiversity. Inherent to this approach is the idea that the twin objectives of environmental protection and development are not necessarily in opposition to each other in the developing country situation. Instead, both can be taken together in a system which prioritises local needs in terms of development. In this respect, it is interesting to consider whether this claim to local control over the conservation and utilisation of biodiversity can be accommodated by broader interpretations of the concept of sovereignty over natural resources than is currently included within Article 3 of the Convention.

Conclusions

The relevant international law on biological diversity, as provided in the 1992 Convention, tends to view the problem in terms of the need to provide a legal framework for the sustainable exploitation of a resource. However, this paradigm of international utilisation of resources that developing country governments have also taken on board in terms of their national development plans does not leave much room for the preservation and protection of the social and cultural diversity which is present in local communities and which depends for its very survival on the ability of these local communities to continue with their own methods of species utilisation. Even more important, as Vandana Shiva and other writers point out, the wholesale take-over of the traditional methods of utilisation of species biodiversity by First World intergovernmental institutions like the World Bank and the transnational companies involved in the development of value-added pharmaceutical and other products represents a real threat to the continued propagation of plant and animal species biodiversity.[68]

As demonstrated in our earlier discussion of the developing countries' views on biodiversity conservation, much of the emphasis during the debate has focused on negotiating compromises with respect to the economic exploitation of biological resources. Such prioritisation of the economic dimension of biodiversity can hardly be taken as surprising given the intergovernmental nature of the international negotiation process. Despite the tendency, however, in both political and legal discus-

---

[67] "From Global Crisis Towards Ecological Agriculture", Declaration of the International Movement for Ecological Agriculture (1990) 21/2 *The Ecologist* 107.

[68] See Shiva *et al.*, *op. cit.*, n. 17; also see V. Shiva, *Biotechnology and the Environment* and the same author: *Biodiversity: A Third World Perspective*.

sions relating to issues of biodiversity conservation to focus on the economic and financial aspects of international co-operation, it is important that attention is not deflected from the central concern which, at least in rhetoric, lies at the heart of the debate; that the ultimate objective of the conservation of biological diversity as a "common concern of humankind"[69] is the protection and enhancement of human welfare.

The shortcomings that have in many instances led to the destruction of biodiversity and to human deprivation are not the result of the want of "development" but rather the perpetuation of ever more "maldevelopment", i.e. "development" that is designed to nurture and perpetuate the interests of groups that see the relationship between humans and our natural environment not as an intimate, subsistence-defined symbiosis, but rather as one of domination, whereby biodiversity is a commodity to be exploited for profit.

[69] Preamble, para. 3.

13

Biodiversity and Indigenous Peoples

John Woodliffe

The belated realisation that the developed world has much to learn from indigenous peoples about the "holistic traditional scientific knowledge of their lands, natural resources and environment",[1] comes at a time when such peoples are faced with the threat of "being civilised to extinction".[2] There are, however, recent encouraging signs of the international community's determination not only to halt but also to reverse this trend.[3] This tergiversation acknowledges "the interrelationship between the natural environment and its sustainable development and the cultural, economic and physical well-being of indigenous people".[4] A comprehensive framework of human rights law (including the principles of non-discrimination and self-determination) would thus appear to provide the appropriate point of departure for tackling these fundamental issues of group identity and survival.[5] There are those, however, who call for a more radical, anti-statist approach – one that challenges the "sense of settled rights of existing sovereign states" *vis-à-vis* indigenous people by legitimising

---

[1] Agenda 21, Chapter 26.1.

[2] N. Seufert-Barr, "Seeking a New Partnership" (1993) 30/2 *UN Chronicle* 40.

[3] The Working Group on Indigenous Populations of the UN Sub-Commission on Prevention of Discrimination and Protection of Minorities was set up in 1982. See P. Thornberry, *International Law and the Rights of Minorities* (1991), p. 377. At its eleventh session in August 1993, the Working Group agreed the text of a draft declaration on the rights of indigenous peoples: UN Doc E/CN.4/Sub.2/AC.4/1993/CRP.4. The draft was to be considered by the Sub-Commission at its forty-sixth session in 1994: Res. 1993/46. The year 1993 was proclaimed as the International Year of the World's Indigenous People: UNGA Res.45/164; 47/75. Several references to indigenous people (sic) appear in The Vienna Declaration and Programme of Action, adopted on 25 June 1993 at the UN World Conference on Human Rights: text in (1993) 32 ILM 1661. At the end of 1993 the UN General Assembly launched the International Decade of the World's Indigenous People. See generally, E. Stamatopolous, "Indigenous Peoples and the United Nations: Human Rights as a Developing Dynamic" (1994) 16 *Human Rights Quarterly* 58.

[4] Supra, n. 1.

[5] See I. Brownlie, *Treaties and Indigenous Peoples* (1992), Chapter 3.

International Law and the Conservation of Biological Diversity (C. Redgwell and M. Bowman, eds.: 90 411 0863 7: © Kluwer Law International: pub. Kluwer Law International, 1995: printed in Great Britain), pp. 255–269

in certain circumstances the dismemberment of such States.[6] The present chapter is concerned only indirectly with this broader political canvas. Its aims are more narrowly focused, namely to describe, in ecological terms, the connections between biodiversity and indigenous people; to analyse existing national legislation and international instruments that purport to accord legal recognition to these relations; and, more generally, to reflect on whether the obstacles to integrating protection of the values and interests of indigenous peoples with measures to conserve biodiversity can successfully be overcome.

For present purposes, the defining characteristics of indigenous people will be based on the now widely accepted description advanced by Martinez-Cobo.[7] First, there must be an historical continuity with pre-invasion and pre-colonial societies that developed on their territories – a continuity reaching into the present and taking the form, *inter alia*, of the occupation of ancestral lands, common ancestry with the original occupants of these lands, distinctive culture, language and residence in certain parts of a country or region of the world. Second, indigenous peoples regard themselves not only as distinct from other sectors of the societies now prevailing in the above mentioned territories but also as constituting a non-dominant sector of those societies. The third characteristic is a determination on the part of such peoples to preserve their ancestral territories and ethnic identity in accordance with their culture and institutions.

This explanatory definition is no more than that; it should not disguise the wide differences between peoples and communities that are classed as indigenous. Present estimates are that among the 250 million indigenous peoples spread over seventy – mainly developing – countries there are some 5,000 peoples distinguishable by culture, language and geographical separation.[8] Only in a few of these countries are indigenous peoples numerically in a majority. What all these communities have in common is "a profound relationship with the land ... and respect for nature."[9]

The Nature and Importance of the Links between Biodiversity and Indigenous Peoples

In their seminal study of the Amazon forests, Hecht and Cockburn observed:

> Overlooked in virtually all the accounts of the distribution of species and the structure of forests is the role of humanity. There is in fact a growing body of knowledge on how indigenous and local populations manage their natural resources and sustain them over time.[10]

---

[6] R. Falk, "The Rights of Peoples (In Particular Indigenous Peoples)" in J. Crawford (ed.), *The Rights of Peoples* (1988), at pp. 33–34; B. Kingsbury, "Self-Determination and 'Indigenous Peoples'" (1992) PASIL 383.

[7] Final Report of the Sub-Commission of the UN Human Rights Commission on the problem of discrimination against indigenous populations, UN Doc.E/CN.4/Sub.2/1986/7Add.4, paras. 378–80. For discussion of the definition see Brownlie, *op. cit.* n. 5, at p. 60. A definition of indigenous populations is employed in the International Labour Organisation's Convention No. 107 adopted in 1957 and Convention No. 169 adopted in 1989; for discussion see Thornberry, *op. cit.* n. 3, Chapters 38–41.

[8] J. Burger, *The Gaia Atlas of First Peoples: A Future for the Indigenous World* (1990), pp. 18–19, 180–85 (Index of peoples).

[9] *Ibid.*, p. 17

[10] S. Hecht and A. Cockburn, *The Fate of the Forest: Developers, Destroyers and Defenders of the Amazon* (1990), at p. 33.

Indeed, the notion of these forests as "the outcome of human as well as biological history"[11] is one that applies equally to other fragile habitats such as the Arctic, tundra, boreal forest, riverine and coastal areas.

Geographical Distribution of Biodiversity

It has been estimated that 85% of all known plant species are situated in areas that are the traditional homelands of indigenous peoples.[12] At the same time, between 50 and 80% of global species diversity are found in just twelve countries.[13] In addition, tropical rain forests, which account for only 7% of the earth's land surface and provide the habitat for 50 million indigenous peoples, are thought to contain well over half of the species in the entire world biota.[14] It is for these reasons that much of the ensuing discussion is concerned with tropical forest regions.

Sustainable Land Use and Resource Management Developed by Indigenous Peoples

There is now a considerable body of interdisciplinary research which lends strong support for the thesis that principles of sustainability rather than simple subsistence are the bedrock of indigenous economic and cultural life.[15] The systems of land use and resource management that have been developed down the centuries encompass nomadic pastoralism (involving animal husbandry in the harshest of climates),[16] shifting cultivation (often referred to as "slash and burn"),[17] agro-forestry, terrace agriculture, hunting, herding and fishing.[18] In addition, indigenous knowledge of plants, soils, animals, plants and climate is used to achieve a balanced ecosystem.[19]

One of the most extensively studied communities are the Kayapo Indians of central Brazil.[20] In common with many other Amazonian Indian communities, the Kayapo practice shifting cultivation, gather many species of wild fruit, tubers, nuts, leaves and medicinal plants and, in ways that are "astonishingly subtle and complex",[21] manage and manipulate these forest resources to ensure their own nutritional needs and the regeneration of the forest. Thus, selected native plants are grown in "resource islands" and agricultural plots, not only to provide food but also to attract and maintain populations of wild animals, an important food source.[22] The Kayapo also move medicinal and ritual plants to locations beside their trails in the forest.[23]

---

[11] Ibid.
[12] Burger, op. cit., n. 8, at p. 32.
[13] C. Shine and P. Kohona, "The Convention on Biological Diversity: Bridging the Gap between Conservation and Development" (1992) I RECIEL 278.
[14] E.O. Wilson, "The current state of biological diversity" in Wilson (ed.), Biodiversity (1988), at p. 8; UNEP, The State of the World Environment 1991 (1991), Chapter 3.
[15] D. Posey and W. Balée (eds.), Resource Management in Amazonia: Indigenous and Folk Strategies, Advances in Economic Botany, Vol. 7 (1989).
[16] E.g. the Tuareg of West Africa; Burger, op. cit., n. 8, at p. 48.
[17] E.g. the Karen people of Thailand; ibid., p. 44.
[18] E.g. the Inuit in the Arctic; ibid., p. 28.
[19] Burger, op. cit., n. 8, p. 40.
[20] See in particular the work of Darrell Posey, op. cit., n. 15.
[21] K. Taylor, "Deforestation and Indians in Brazilian Amazonia" in E. Wilson (ed.), Biodiversity (1988), at p. 140.
[22] Ibid.
[23] Hecht and Cockburn, op. cit., n. 10, at p. 35.

The Local and Global Benefits of Biodiversity in the Fields of Medicine and Agriculture

Wild genetic resources, that is, species of plants and animals and the variations within them, are now recognised as constituting the "raw materials for future medicines, food and fuels".[24] Approximately 75% of the 119 plant-derived prescription drugs currently in use worldwide have been developed from indigenous medicine.[25] This illustrates the global benefits of conserving natural habitats that contain plant species. What is easily overlooked, however, is the vital importance of these medicinal plants to the health care systems of developing countries.[26] It is estimated that 80% of the population of these countries worldwide relies on plant-derived medicine for its primary health care needs.[27]

Twenty species of plant supply 90% of the world's food and over 50% derives from wheat, maize and rice alone.[28] It therefore becomes imperative to prevent the loss of the wild ancestors of these plants so that cultivar diversity and crop germplasm are maintained.[29] This is well illustrated by the discovery in 1977 of a wild species of disease-resistant perennial maize in Mexico. When cross-bred with modern commercial varieties of maize, the estimated annual savings to farmers worldwide could be as much as US$4.4 billion.[30]

The Impact of Loss of Diversity on Indigenous People

It was noted above that much of the world's biological richness is found in those areas of the globe that form the traditional homelands of indigenous peoples. There is a direct correlation between the rate of diversity losses in terms of ecosystems, species and genes and the destruction of the habitat of indigenous people. A host of factors contribute to the enforced departure of indigenous people from their homelands, which in turn are left in conditions of increasingly irreversible environmental degradation. Among the factors are,[31] first and foremost, deforestation, new human settlements prompted by population growth, mining of mineral resources, logging, dam-building, cattle ranching and the use of homelands for carrying out weapons testing and siting military installations. In addition, further general loss of biodiversity is occurring as a result of pollution, global climate change, over-exploitation of resources and the introduction of exotic species.[32]

---

[24] J. Nations, "Deep ecology meets the developing world" in E. Wilson (ed.), *Biodiversity* (1988), at p. 82.

[25] Burger, *op. cit.*, n. 8, at p. 32. The 119 plant-derived drugs come from fewer than ninety species of plants: N. Farnsworth, "Screening plants for new medicines", in E. Wilson (ed.), *Biodiversity* (1988), at p. 93. The 119 plants are listed in Table 9.1, *ibid.*

[26] K. Brown, *Medicinal Plants, Indigenous Medicine and Conservation of Biodiversity in Ghana*, Centre for Social and Economic Research on the Global Environment, Working paper GEC 92-36.

[27] Farnsworth, *op. cit.*, n. 25.

[28] UNEP, *The State of the World's Environment 1991*, (1991), at p. 20.

[29] Hecht and Cockburn, *op. cit.*, n. 10, pp. 58–9.

[30] *Our Common Future* (1987), World Commission on Environment and Development, p. 155.

[31] See Burger, *op. cit.*, n. 8, pp. 84–121; S. Davis, *Victims of the Miracle: Development and the Indians of Brazil* (1977), Part III.

[32] *Op. cit.*, n. 28, pp. 22–23. For an account of the ecological damage caused by the introduction in 1935 of the cane toad in Australia to combat infestations of beetles in sugar crops, see S. Lewis, Cane Toads: *An Unnatural History* (1993).

Thus, conservation of the biological diversity of tropical forests and the survival of the indigenous people who live there are indivisible. Without the land and its resources that are the central element in their physical and spiritual existence, indigenous communities will suffer a systems failure consequent upon the loss of language, knowledge, institutions and sacred beliefs.[33]

Attention will now be turned to a consideration of the legal frame of reference disclosed by the above discussion.

The Legal Dimension

A United Nations report published in 1991 rightly observed that indigenous people are at one and the same time "victims of environmental degradation and protectors of vulnerable ecosystems".[34] The same report asserts, however, the absence in both areas of any tangible legal regulation:

> The special relationship indigenous people have to the land and the environment has yet to be recognised by a human rights instrument of the United Nations. Nor have ... their special knowledge and successful practices in protecting the environment found a place in the emerging international law concerning the environment.[35]

Whether the Rio Conference on Environment and Development has succeeded in closing this gap will be examined in a later section. First, an attempt will be made to test the above assertions in the light of existing international and national legal practice.

Status of Indigenous Peoples and Their Lands in International Law

Historically, international law, after some initial hesitation,[36] imposed no obstacles to the conquest and appropriation of hitherto undiscovered areas of the globe. The rules governing the acquisition of territory – powerfully symbolised in the doctrine of *terra nullius* – and the absence of constraints on the conduct of sovereign states *vis-à-vis* native populations,[37] together served to deny indigenous peoples a modicum of legal personality.[38] The expansion of colonial empires was thus allowed to proceed unhindered.

A recent study of the legal status of indigenous groups in Latin America, prompted by the quincentenary of the discovery of that continent by Columbus in 1492, demonstrates how little the legal position has changed in the era of the UN Charter.[39] The author traces the history of UN involvement with matters concern-

---

[33] Burger, *op. cit.*, pp. 122–123.

[34] *Human Rights and the Environment*, Preliminary Report of the Special Rapporteur, Mrs Ksentini, submitted to the Sub-Commission on Prevention of Discrimination and Protection of Minorities, 2 August 1991. UN Doc E/CN.4/Sub.2/1991/8, para 23.

[35] *Ibid.*, para. 30.

[36] See Franciscus de Vitoria, *De Indis et de Jure Belli Relectiones* translated in Scott (ed.), The Classics of International Law (1964); "the aborigines undoubtedly had true dominion in both public and private matters ... and neither their princes nor private persons could be bespoiled of their property on the ground of their not being true owners" (p. 128).

[37] See E. Williamson, *The Penguin History of Latin America* (1992), Part I.

[38] See generally, G. Schwarzenberger, *International Law as applied by International Courts and Tribunals*, Vol. 1 (3rd edn., 1957), Chapter 4.

[39] A. Stuyt, "The UN Year of Indigenous Peoples 1993–Some Latin American Perspectives," (1993) 40 *Netherlands International Law Review* 449.

ing indigenous people back to a little-known Resolution adopted by the General Assembly in 1949.[40] While in terms restricted to Amerindians the principles set out in the Resolution apply to all indigenous groups. Essentially, any assistance to or study of these groups by organs or specialised agencies of the UN is conditional on the territorial States concerned "requesting such help". Significantly, the Resolution conferred no equivalent *locus standi* on the indigenous peoples who live in the territories of these States. This remains the position today.[41]

The International Labour Organisation is the author of the only legally binding international instruments specifically to protect indigenous populations. The first of these instruments, ILO Convention No. 107, was adopted in 1957; this was revised in part by ILO Convention No. 169 adopted in 1989.[42] Both Conventions address the question of land rights. For present purposes only the text of the later Convention will be referred to. Article 13 lays down as a general principle of conduct by governments respect for the special importance of the relationship of indigenous peoples with their lands, in particular, its collective aspects. Governments are also enjoined to recognise "the rights of ownership and possession of the peoples concerned over the lands which they traditionally occupy" and to "guarantee effective protection of such rights" including "adequate legal procedures for resolving land claims by the peoples concerned".[43] There must be special safeguards for the rights of such peoples to the natural resources pertaining to their lands, including rights to participate in the use, management and conservation of those resources.[44] It is thus clear from the language of the convention that title to these lands is not based on any grant by the state but on immemorial possession.[45] In addition, by emphasising "possession" as an alternative to ownership, Convention No. 169 goes a considerable way to meeting the criticism that formal concepts of private property and land ownership are alien to groups practising shifting cultivation or nomadism.[46]

However, one reason why the effectiveness of the two ILO Conventions has been castigated as "shocking",[47] is the small number of States that are parties to them.[48] Another, more significant reason, lies in the title of the two ILO Conventions which refer to "indigenous peoples … in Independent Countries". Accordingly, it is the governments of these countries that are given "the responsibility for developing, with the participation of the peoples concerned, co-ordinated and systematic action to protect the rights of these peoples".[49] Admittedly, the ILO has a well-developed system for supervising observance by Member States of their obligations under ILO Conventions that embraces both non-contentious and con-

---

[40] UNGA Res 275 (III), 11 May 1949.

[41] Stuyt, *op. cit.*, n. 39, at p. 452.

[42] Text in (1989) 28 ILM 1382. The preamble to Convention No. 169 calls attention to the distinctive contribution of indigenous people to "the ecological harmony of humankind".

[43] Article 14.

[44] Article 15.1.

[45] Thornberry, *op. cit.* n. 3, at p. 357.

[46] L. Swepston and R. Plant, "International standards and the protection of the land rights of indigenous and tribal populations" (1985) 124 *International Labour Review* 91, at p. 97.

[47] Stuyt. *op. cit.*, p. 468.

[48] There are twenty-seven States parties (including twelve from Latin America) to Convention No. 107 and four (including three from Latin America) to Convention No. 169.

[49] ILO Convention 169, Article 2.1. Indigenous peoples must be accorded the power to institute legal proceedings to protect their rights: Article 12.

tentious procedures.[50] The institutional structure of the ILO reflects, however, a preoccupation with the economic and social concerns of industrialised countries far removed from those of indigenous groups.[51]

Thus, indigenous people, lacking any power of independent action on the international plane, are *prima facie* not subjects of the international legal order and their status and treatment are determined *au fond* by national legal orders.[52] The next section deals briefly with trends in national legal practices. Whether recent developments herald a change in the current overall legal position will be considered shortly.

Status and Treatment of Indigenous People and Their Lands under National Law

Mention has already been made of the threats posed to the very survival of many indigenous populations and to the conservation of the natural resources found in their traditional lands. These facts alone are telling evidence of the ineffectiveness, indifference or even absence of national legal regulation. Amongst the best sources of information on national legal practices in this field are the regular reports[53] filed with and scrutinised by the ILO Committee of Experts on measures taken by State parties to give effect to the provisions of ILO Convention 107[54] concerning the protection and integration of indigenous populations. One detailed study of reports by ten Latin American countries concluded that legislation concerning autochthonous groups reflected, especially in respect of citizenship and status, "attitudes dating from the nineteenth century and earlier".[55] Another problem concerns central government agencies and administrative systems set up specifically to protect indigenous people and their lands which in practice work against the interests of such people.[56] Above all, the principal of integration and assimilation of indigenous people that informs Convention No. 107 has tended to accelerate the loss of ancestral lands. Whilst this approach is now generally disowned by policy makers in favour of encouraging States to extend greater recognition to the separate identity of indigenous people,[57] the damage has been done and may be irreversible. Chapter VIII of the Brazilian Constitution promulgated in 1988[58] offers a ray of hope for the future. While lands traditionally occupied by the Indians are "the property of the Union" such lands are "intended for their permanent possession (and) they have exclusive usufruct rights to the riches of the soil, the rivers and lakes existing therein".[59] By Article 232, Indian organisations and communities are granted *locus standi* to protect their rights and interests. The effectiveness of these guarantees has yet to be demonstrated.

---

[50] See F. Wolf, "Human Rights and the International Labour Organisation" in T. Meron (ed.), *Human Rights in International Law: Legal and Policy Issues* (1984), Chapter 7.

[51] Thornberry, *op. cit.*, p. 367.

[52] Stuyt, *op. cit.*, p. 468.

[53] These are required by Article 22 of the Constitution of the International Labour Organisation.

[54] 328 UNTS 247.

[55] L. Swepston, "Latin American approaches to the Indian Problem," (1978/2) 117 *International Labour Review* 179, at p. 194. For examples of "protective" legislation see *ibid.*, p. 182.

[56] *Ibid.*, pp. 187ff. The Brazilian agency FUNEI, has in the past acquired a reputation for corruption and brutality; see Hecht and Cockburn, *op. cit.* n. 10, at p. 154.

[57] See ILO Convention No. 169, *loc. cit.* n. 42.

[58] Text in Blaustein and Flanz, *Constitutions of the Countries of the World*, Binder III.

[59] Article 231.2.

Limited Recognition of Rights of Indigenous Peoples in Existing International Environmental Treaties and Instruments

The indigenous skills in resource management and sustainability identified earlier might be expected to have gained some recognition in treaties and other instruments concerning environmental law. This, however, as will be seen, is not the case.

The first sustained phase of global treaties concerned specifically with conservation of biological diversity took place in the 1970s.[60] As the harm or threat thereto which these treaties sought to prevent, limit or undo emanated chiefly from the activities and policies of the industrial countries of the developed world, it is not surprising to find the absence of provisions recognising the role in conservation played by indigenous people. This position scarcely changed in the 1980s with the next wave of global and regional agreements that explicitly embraced, in full or in part, the goals of conservation and sustainable use of natural resources. What is surprising is the absence of any human dimension even in the most advanced conservation treaties of that period: the ASEAN Agreement on the Conservation of Nature and Natural Resources 1985[61] and the International Tropical Timber Agreement 1983.[62] Other important international instruments produced at that time reveal a similar picture. They include the UN General Assembly Resolution on the World Charter for Nature;[63] the general principles concerning natural resources and environmental interferences adopted by the Experts Group on Environmental Law of the World Commission on Environment and Development;[64] and the UN Declaration on the Right to Development.[65] The markedly different approach taken by the Rio Conference towards the place of indigenous values in conserving biological diversity is considered below.

To complete the picture, however, reference must be made to three conservation conventions concerned with specific species located in defined geographical areas in respect of which a derogation from stipulated conservation norms in favour of indigenous rights is expressly sanctioned. The earliest of these conventions is the International Convention for the Regulation of Whaling 1946.[66] The International Whaling Commission (IWC) established by the contracting parties is empowered to amend the provisions of the Schedule attached to the Convention by adopting regulations with respect to conservation and utilisation of whale resources.[67] These powers, which are exercised at annual sessions of the IWC, include the allocation and revision of quotas.[68] By the 1980s, with the backing of the United States, the policy of the IWC had begun to shift towards an outright ban on commercial whaling.[69] However,

---

[60] See the Ramsar Convention on Wetlands of International Importance especially as Waterfowl Habitat 1971; Paris Convention concerning the Protection of the World Cultural and Natural Heritage 1972; Washington Convention on International Trade in Endangered Species of Wild Fauna and Flora 1973.

[61] Text in H. Hohmann, *Basic Documents of International Environmental Law* (1992), Vol. 3, p. 1550.

[62] Text in I. Rummel-Bulska and S. Osafo (eds.), *Selected Multilateral Treaties in the Field of the Environment*, Vol. II (1991), at p. 271. One of the stated objectives of the agreement is the development of national policies aimed at sustainable utilisation and conservation of tropical forests and their genetic resources. A new agreement was concluded in 1994, text in (1994) 33 ILM 1014.

[63] UNGA Res 37/7, adopted 28 October 1982.

[64] R. Munro and J. Lammers (eds.), *Environmental Protection and Sustainable Development: Legal Principles and Recommendations* (1987).

[65] UNGA Res. 41/128.

[66] 161 UNTS 72; Hohmann, *op. cit.* n. 62, at p. 1291.

[67] Article V.

[68] The amendments to the Schedule adopted in 1989 are reproduced in Hohmann, *op. cit.*

[69] See generally, P. Birnie, *International Regulation of Whaling* (1985) 2 Vols; A. D'Amato and S.K. Chopra, "Whales: Their Emerging Right to Life" (1991) 85 AJIL 21, at p. 43.

after some initial hesitation, the IWC decided to permit the continuation of aboriginal whaling of the Arctic bowhead whale, a highly endangered species.[70] The Schedule describes this as "aboriginal subsistence whaling" and fixes catch limits for the bowhead, gray, minke and humpback species of whales. As the purpose of these exceptions to the general moratorium is to "satisfy aboriginal subsistence need" at any given period, there is a standard requirement that the "meat and products" of the whales are used "exclusively for local consumption by the aborigines".[71] Those opposed to the continuation of the derogations contend that the subsistence rights of indigenous groups involved should be relinquished in the higher interest of the whales' entitlement to the right to life; in return, compensation would be paid to the indigenous groups by those countries that have historically benefited from commercial whaling.[72]

The Agreement on Conservation of Polar Bears 1973,[73] allows a departure from the general prohibition on the taking of polar bears when carried out by local people using traditional methods in the exercise of their traditional rights and in accordance with the laws of a Contracting Party.[74] Last, the ban on sealing imposed by the Convention on Conservation of North Pacific Fur Seals 1976,[75] expressly does not apply to named groups dwelling on the coast of certain designated waters of the Pacific Ocean who

> ... carry on pelagic sealing in canoes not transported by or used in connection with other vessels, and propelled entirely by oars, paddle or sails, and manned by not more than five persons each, in the way hitherto practiced and without the use of firearms; provided that such hunters are not in the employment of other persons or under contract to deliver the skins to any person.[76]

As is evident from the whaling example above, there is a danger that traditional indigenous values become invested with an absolutist status.[77] The Rio Convention on Biological Diversity 1992, touches on this problem in defining the relationship between the provisions of that Convention and those of any existing international agreement. Rights and obligations deriving from the latter remain unaffected except where their exercise "would cause a serious damage or threat to biological diversity".[78] This loosely worded proviso is a potential source of acrimonious debate.

The Ownership of Wild Genetic Resources and Biochemical Resources

The proven economic benefits of biodiversity, both locally and globally, in the areas of medicine and agriculture have been noted; *a fortiori* it is imperative to demonstrate the economic value of biodiversity conservation and to identify mechanisms for the appropriation and capture of this value.[79] There must also be a resolution of the

---

[70] D'Amato and Chopra, *ibid.*, at p. 42, n. 136.

[71] 1989 amendments, para. 13.

[72] D'Amato and Chopra, *op. cit.*, pp. 57–61.

[73] Text in A.C. Kiss (ed.), *Selected Multilateral Treaties in the Field of the Environment*, Vol. I (1983), at p. 401.

[74] Article III.d.

[75] Text in Kiss, *op. cit.*, n. 73, at p. 460.

[76] Article VIII.

[77] See Brownlie, *op. cit.*, n. 5, p. 73.

[78] Article 22.

[79] D. Pearce and S. Puroshothaman, *Protecting Biological Diversity: The Economic Value of Pharmaceutical Plants*, CSERGE Discussion Paper GEC92-27; T. Swanson, "Conserving Biological Diversity" in D. Pearce (ed.), *Blueprint Two: Greening the World Economy* (1991).

antecedent issue of access rights to prospect for wild genetic resources. In implementing all of these goals, account must be taken of the vital part played by indigenous peoples. Yet to imbue these facts, values and benefits with legal significance is, even at the simplest level, an extraordinarily complex matter whether at national or international level.

Many of the uses of wild genetic resources employed by indigenous peoples constitute, in economic terms, a "store of knowledge and therefore a public good".[80] Over generations, indigenous peoples have helped informally to conserve, nurture and improve species by using methods of cultivation and husbandry calculated to halt the erosion of genetic diversity.[81] Whatever the form of legal ownership over the habitat that provides the genetic resource, whether the benefits from that resource are subject to exclusive ownership rights is a matter for national law to determine. Yet, as a recent study on the subject demonstrates, wild land biodiversity prospecting is currently operating in a "policy vacuum".[82] Historically, unimproved genetic material such as wild species and crop and livestock varieties developed by traditional farmers have been regarded as "ownerless, open-access resources"[83] and, for the limited purpose of free exchange of information, have been described as the "common heritage of mankind".[84] Developing countries have long questioned "a system that labels their resources as 'open access' but then establishes private property (in the form of intellectual property regimes) for improved products based on these resources".[85]

The position under customary international law concerning access to wild genetic resources and their exploitation is clear cut. While in economic terms these resources are "collective goods" in the sense of their potential benefit to humanity worldwide, in legal terms they are part of the natural resources of the State on whose territory they are located, and thus in no way constitute the common heritage of mankind as that notion is presently understood in international law. In spite of the pressure of environmental concerns, the principle of State sovereignty over natural resources still remains a central tenet of the international legal order that has developed since 1945. The firmly entrenched nature of the principle is evident from the Declaration of San Francisco de Quito[86] agreed in 1989 by the Foreign Ministers of the member states of the Treaty of Amazon Co-operation.[87] The Declaration acknowledges the "great importance of the Amazonian ecosystems from the point of view of their biodiversity" and describes them as "one of the most important natural patrimonies" appertaining to each country. However, use and protection of that patrimony, while "respecting the rights of the populations which live there" shall be "in accordance with the policies established by each Amazonian Country ... exercising the right inherent to sovereignty over its Amazonian areas".[88] Again, the language is indicative of the vulnerable status of indigenous people discussed earlier.

---

[80] R. Sedjo, "Property Rights, Genetic Resources, and Biotechnological Change" (1992) 35 *Journal of Law and Economics* 199, at p. 204.

[81] C. Correa, "Biological Resources and Intellectual Property Rights" (1992) 4 EIPR No. 5, 154.

[82] World Resources Institute, *Biodiversity Prospecting* (1993), at p. 2.

[83] *Ibid.*, at p. 19.

[84] Sedjo, *op. cit.* n. 80, at p. 202.

[85] *Op. cit.*, n. 82, at pp. 19, 32.

[86] Text in Hohmann, *op. cit.*, n. 61, Vol. III, at p. 157.

[87] Concluded 3 July 1978; text in (1978) 17 ILM 1045.

[88] Part II: Environmental Policy.

Another factor that complicates the task of building a legal framework is that many of the wild genetic resources have no more than a "potential payoff".[89] Thus, it is estimated that there are upwards of 250,000 species of plants – mostly in developing countries – that have yet to be tested for their possible pharmaceutical properties.[90] Realisation of any "pay-off" in the form of a viable commercial product is, however, ultimately dependent on the application of biological tools in the hands of developed countries.[91] Moreover, it will be only in a minority of cases that the discovery of new medicines or genes suitable for agricultural breeding will be attributable to indigenous knowledge; most will derive from screening methods developed by modern science.[92]

This overview may be crudely described as a conflict between the "gene-poor but technology-rich North and the gene-rich but technology-poor South".[93] The conflict is reflected in the competing legal regimes that have been proposed.[94] What is called for is a legally just accommodation between the interests of the two groups of countries; one that avoids allocating exclusive rights of ownership of products to either side, preserves access to wild genetic resources and the technology that makes use of them[95] and, above all, ensures that indigenous peoples are rewarded for their contribution to the development of a new drug or crop.

Whether the UN Conference on Environment and Development (UNCED) held in Rio de Janeiro in June 1992, has done anything to clarify these myriad issues will now be considered.

The United Nations Conference on Environment and Development

Indigenous groups mounted a vigorous lobby in the run up to UNCED. If the incidence of the use of the term indigenous people in the treaties and related instruments issuing from the Rio Conference is taken as a measure of success, then at first glance, their campaign can be counted as such. Closer inspection of the texts, however, indicates that very little has changed. The centrepiece is without doubt the Convention on Biological Diversity, analysed in depth elsewhere in this volume. For present purposes the following general points are to be noted. First, the Convention uses the term "indigenous and local communities" rather than indigenous people. Early negotiating sessions on the fifth revised draft of the Convention had incorporated this term in the text. By the seventh session this had been watered down to "indigenous populations" which was in turn expunged following a recommendation by a sub-working group on definitions and use of terms.[96] Somewhat inconsistently, other Rio Conference instruments – admittedly of a non-legally binding character – do use the term indigenous

---

[89] Sedjo, *op. cit.* n. 80, at p. 204.
[90] Farnsworth, *op. cit.*, n. 27, pp. 92–93.
[91] M. Tolba and O. El-Kholy (eds.), *The World Environment 1972–1992* (1992), at pp. 200–202.
[92] *Op. cit.*, n. 82, at p. 35.
[93] *Op. cit.*, n. 80, at p. 202.
[94] See Correa, *op. cit.*, n. 81. Other chapters in the present volume explore these legal regimes in detail.
[95] *Op. cit.*, n. 91, at p. 202.
[96] UNEP/Bio Div./N7–INC.5/2, Appendix I, p. 38.

people.[97] Second, the *leitmotif* of the Convention is the legally incomplete bargain struck between the developed and developing States, couched in language that attempts to moderate the principle of State sovereignty by reference to global communitarian values, to which end access to and sharing of genetic resources and technologies are indispensable.[98] The detailed working out of this bargain is discussed in several of the contributions to this volume. Here, it will simply be noted that the Convention, after reaffirming the sovereign rights of States over their biological resources[99] leaves to national legislation of Contracting Parties the determination of questions of access to genetic resources, access to and transfer of technology and the distribution of benefits arising from biotechnologies based on genetic resources.[100]

The third and most significant issue is the explicit recognition given to indigenous communities by the Convention. Building on the broader formulation contained in the Preamble,[101] Article 8, which is headed "In-Situ Conservation", states:

> Each Contracting Party shall, as far as possible and as appropriate ... (j) Subject to its national legislation, respect, preserve and maintain knowledge, innovations and practices of indigenous and local communities embodying traditional lifestyles relevant for the conservation and sustainable use of biological diversity and promote their wider application with the approval and involvement of the holders of such knowledge, innovations and practices and encourage the equitable sharing of the benefits arising from the utilisation of such knowledge innovations and practices.[102]

The weak or "soft" nature of the obligations undertaken by Contracting Parties is evident from the qualified language used. The greater part of paragraph (j) does no more than restate the already familiar agenda of issues concerning indigenous people. In one respect, however, it breaks new ground, by stipulating that promotion of the wider application of indigenous knowledge relating to the conservation and sustainable development of biological diversity shall take into account several desiderata. First, the "approval and involvement" of the holders of such knowledge; second, encouragement of the "equitable sharing of the benefits" arising from that knowledge. The beneficiaries, however, are left unspecified, as are the methods for quantifying an equitable share.[103]

---

[97] See Authoritative Statement of Principles for a Global Consensus on the Management, Conservation and Sustainable Development of all Types of Forests, 13 June 1992, paras. 2(d), 5(a), 12(d): text in (1992) 31 ILM 881; also, Agenda 21, Chapter 26, "Recognising and Strengthening the Role of Indigenous People and their Communities": text in S. Johnson, *The Earth Summit: United Nations Conference on Environment and Development* (1993), at p. 415.

[98] Second preambular paragraph and Article 1.

[99] Fourth preambular paragraph and Article 3.

[100] Articles 15, 16 and 19. Significantly, the Convention does not address the issues of ownership present in these provisions: see The Convention on Biological Diversity: An Explanatory Guide, IUCN Environmental Law Centre, English version, October 1993; p. 67. The Convention does, however, define "domesticated or cultivated species" as species "in which the evolutionary process has been influenced by humans to meet their needs". This would exclude wild species used in their wild state by humans: *ibid.*, p. 21.

[101] Twelfth paragraph.

[102] A broadly similar formulation appears in Principle 22 of the Rio Declaration on Environment and Development.

[103] Compare Statement of Forest Principles, *loc. cit.*, n. 97, para. 12(d): "benefits arising from the utilisation of indigenous knowledge should ... be equitably shared with such (indigenous) people".

It is apparent, therefore, that the Biodiversity Convention foresees progress in tackling both issues as dependent on action at national level.[104] This underscores the marginalised position of indigenous people under the Convention which is concerned first and foremost with reconciling the interests of developing and developed States. Any legal solution must, therefore, get to grips with the multilayered nature of the relationships and interests involved; these arise not only between governments but also between governments and indigenous people as well as between the latter and non-governmental organisations and private commercial firms. In addition, if the claims and aspirations of indigenous people, along with the goals of conservation and sustainable use of genetic resources, are to be dealt with effectively and fairly then certain matters have to be subject to clearly defined legal regulation. The initial questions are: who determines access to resources and on what terms? In this respect, the principle of informed consent is of paramount importance. Next, the modalities for calculating what is fair and equitable compensation for the contribution of all those – including indigenous communities, researchers, collectors, producing companies and source countries – who have invested in the discovery, use and continued existence of genetic resources, must be addressed.[105]

A Way Forward?

There is a body of expert opinion that holds that the existing legal regime of intellectual property rights is of limited and uncertain value in promoting the policy aims outlined above.[106] More promising in the near term is an approach that not only "closes the loop between studying, saving and using biodiversity" but also captures greater economic benefits for indigenous people.[107] Typically, at the hub of this approach is a contractual relationship between the collector responsible for making an inventory of wild genetic resources and a public agency or private pharmaceutical company that screens the resources for their commercial development potential. The most widely published "biodiversity prospecting" contract is that entered into by the US pharmaceutical firm, Merck and the National Biodiversity Institute of Costa Rica (INBio), a private non-profit organisation, in September 1991.[108] In essence, INBio will provide Merck with samples of plants, insects and micro-organisms from designated wildland areas in Costa Rica. Merck is given exclusive rights for two years to screen these samples and to keep the patents to any commercial products that follow from screening. In return, Merck will provide INBio with US$1 million as well as royalties from sales of commercial products.[109] Costa Rica intends to earmark half of any royalties for the conservation of biological diversity in its national parks.[110] Three further points about this agreement need to

---

[104] Articles 20 and 21 of the Convention envisage the establishment of a mechanism for transferring financial resources to developing countries but make no mention of indigenous people.

[105] *Op. cit.*, n. 82, at pp. 2, 38.

[106] See M. Gollin, "An Intellectual Property Rights Framework for Biodiversity Prospecting" in *Biodiversity Prospecting, op. cit.*, n. 82.

[107] *Ibid.*, pp. 32, 33.

[108] *Ibid.*, pp. 1, 2. For examples of other similar arrangements see pp. 8–13, *ibid*. For a comprehensive analysis see M.D. Coughlin, "Using the Merck-INBio Agreement to clarify the Convention on Biological Diversity" (1993) 31 Col JTL 337.

[109] *Ibid.*, p. 356.

[110] *Ibid.* On the work of INBio see Gamez *et al.*, Costa Rica's Conservation Program and National Biodiversity Institute", in *Biodiversity Prospecting, op. cit.*, n. 82.

be stressed. First, it is underpinned by national legislation regulating access by collectors to Costa Rica's "national patrimony".[111] Second, the agreement does not appear to have any indigenous dimension.[112] Last, the terms of the agreement are not open to public inspection.[113]

More recently, attempts have been made to build on the Merck–INBio agreement by formulating model terms for contracts between collectors and companies that make provision for, *inter alia*, the recognition of the ethnobotanical knowledge of indigenous people, the return of benefits to them, and collectors' obligations towards such people.[114] One such draft contract[115] stipulates that the collector must not obtain samples or information about the ethnobotanical use of the samples from indigenous peoples without their informed consent.[116] Where that sample or information leads to the identification of a sample from which is ultimately derived a product for a use similar to that specified by the indigenous people, such people are to receive each quarter a fixed percentage royalty in respect of net sales worldwide.[117] An appendix to the draft contract imposes on collectors the highest standards in respect of the conservation of biologically diverse ecosystems. In particular, the collector must warrant that an environmental assessment of the cumulative effects of collection activities has been carried out by an independent and experienced specialist in environmental auditing. The audit will include interviews with indigenous people and determine violations of applicable laws appertaining to environmental standards or to relations with or rights of such people.[118] In addition, the collector warrants that it has obtained the informed consent necessary to create the agreement and to carry out activities under it, from appropriate representatives of indigenous people that traditionally reside in or use an area in which those activities are to be conducted.[119] At all stages there must be the fullest consultations with local communities and any conservation measures must be compatible with and build upon indigenous cultures.[120] Last, provision is made for indigenous people – who, significantly, are not parties to the contract – to enforce obligations created by the supplementary agreement.[121] The authors of the draft contract envisage that this optional agreement could also be concluded directly between the collector and the indigenous people; in any event the terms of the agreement "may merely prefigure future national regulatory standards".[122] It must be emphasised, however, that a private contract along these lines remains subject to any national legal requirements that may be imposed on the private contracting parties.[123] Thus, the local state may make access to genetic resources conditional on payment of a licence or user fee or

---

[111] Wildlife Protection Law, 12 October 1992, summarised *op. cit.*, n. 82, p. 39.
[112] The indigenous population of Costa Rica is 20,000, just 1% of the total: Burger, *op. cit.*, n. 8, at p.181.
[113] *Op. cit.*, n. 82, vi.
[114] Laird, "Contracts for Biodiversity Prospecting," in *Biodiversity Prospecting*, *op. cit.*, n. 82.
[115] *Op. cit.*, Annex 2.
[116] Article 3D.
[117] Article 6B. In addition, the indigenous person or people can insist that its contribution is acknowledged in any publication: Article 7C.
[118] *Op. cit.*, Appendix C, section 1.
[119] *Ibid.*
[120] *Ibid.*, sections 1.b,4.
[121] Section 8.
[122] *Ibid.*, pp. 266, 272.
[123] *Ibid.*, p. 281.

on compliance with other rules.[124] Moreover, there is nothing to prevent the State from refusing to recognise the *locus standi* of indigenous people in legal proceedings.

The Outlook

What emerges clearly from this study is the extent to which the prospects for the conservation and management of biodiversity are bound up with the survival of indigenous people; in a very real sense, both stand or fall together. On the positive side, there have been some recent notable successes in restoring the land base and self-government of indigenous people.[125] The problems of indigenous people are now firmly on the international agenda, exemplified by the burgeoning activities of the United Nations Economic and Social Council's Working Group on Indigenous Populations.[126] However, there are several powerful countervailing forces at work. Above all, governments of developing countries – in which the greater proportion of biological diversity is located – are mostly wedded to a Western industrialised model of development that in the past has been destructive of the values of indigenous people. International development agencies have often been unwitting allies of those promoting such policies. At the same time, population growth and poverty in developing countries are resulting in either the overexploitation of natural habitat of greatest diversity or its conversion to agricultural use.[127] Moreover, successful management and sustainable development strategies for this diminishing habitat demand systems of land tenure that are inimical to the collectivist systems favoured by indigenous people.[128] Finally, nothing can prevent the absorption of indigenous people into a wider inclusive society with the consequent loss of cultural identity and internalisation of the values of that society.[129] As the anthropologist, Darrell Posey, has warned, we must avoid the patronising myth that indigenous people "live in perfect harmony with nature and therefore should stay there and never change".[130]

---

[124] *Ibid.*, pp. 41, 180.

[125] N. Seufert-Barr, "Preserving the Past, Providing a Future: Land Claims and Self-Rule", (1993) 30/2 *UN Chronicle* 46; see also *Mabo* v. *Queensland* (1992) 66 *Australian Law Journal* 408.

[126] See Report of the Working Group on its eleventh session: UN Doc E/CN.4/Sub2/1993/29; also, the Matuutua Declaration on Cultural and Intellectual Property Rights of Indigenous Peoples, June 1993: UN Doc E/CN.4/Sub2/AC.4/1993/CRP5.

[127] Swanson, *op. cit.*, n. 79, p. 186.

[128] *Ibid.*, p. 202.

[129] Leaders of the Kayapo tribe in Brazil are reported to have protested against a court ruling declaring illegal the building of roads through Indian lands to extract timber from their homeland: *The Guardian*, 10 April 1993.

[130] Hecht and Cockburn, *op. cit.*, n. 10, Appendix B, p. 250.

14

Financial Aid, Biodiversity and International Law

Sam Johnston

Introduction

International aid or overseas development assistance ("ODA") has an important impact upon efforts to conserve biodiversity. For example, many important biodiversity conservation projects in developing countries, where the majority of the world's biodiversity is found, are entirely dependent upon international financial assistance. Indeed, the obligations of the Convention on Biological Diversity ("CBD") are for developing countries explicitly conditional upon the provision of "new and additional funding"[1] which initially must largely be made up of increased ODA. Agenda 21 estimated the cost of implementing measures necessary to conserve biodiversity to be approximately US$3.5 billion per annum of which US$1.75 billion should be "from the international community on grant or concessional terms".[2] ODA not only has the potential to help conserve biodiversity, but also can be a significant cause of its loss with projects funded by ODA contributing to the unnecessary loss of important natural habitats in many countries. For instance, a significant cause of the deforestation in the Amazon has been projects which without ODA funding would never have been viable.

This chapter examines ODA and its sources, detailing amounts and methods employed to ensure conservation of biodiversity. It also briefly discusses some of the more important legal issues which have arisen as a result of efforts to make ODA more effective in conserving biodiversity.

Overseas Development Assistance

ODA is a technical term which refers to concessional aid provided by governments. ODA is delivered either directly by donor countries – bilateral aid – or via

---

[1] Article 20(4).
[2] Agenda 21, Chapter 15.8.

International Law and the Conservation of Biological Diversity (C. Redgwell and M. Bowman, eds.: 90 411 0863 7: © Kluwer Law International: pub. Kluwer Law International, 1995: printed in Great Britain), pp. 271–288

multilateral institutions such as the United Nations and the multilateral development banks – multilateral aid. Almost all ODA comes from the OECD countries.[3] Concessional aid includes grants and loans made at less than market interest rates, but not other types of official financial transfers such as export credits, grants by private voluntary agencies or private flows at commercial rates. In 1992, developing countries received a total of US$55.3 billion worth of ODA which accounted for approximately 35% of the US$159.1 billion in financial resources transferred to developing countries. Non-concessionary bilateral and multilateral disbursements accounted for 8% and private investment made up the remaining 57%.[4]

In 1992, the average OECD ODA contribution was 0.33% of GNP. This average masks considerable discrepancy between the performance of the USA (0.20%) and Japan (0.30%), on the one hand, and the Scandinavian countries (all above 1.0%) and the Netherlands (0.86%) on the other. The OECD target of 0.7% of GNP, first proposed by the Pearson Commission in 1969 and reconfirmed in Agenda 21, is still not meet by the majority of OECD countries, with only the Scandinavian countries and the Netherlands achieving this goal. Nine other countries were above the OECD average of 0.33%. This group was led by France (0.63%). In terms of the amount of aid donated, the United States at US$11.7 billion and Japan at US$11.2 billion are the largest donars.[5]

The Obligation to Provide ODA

Traditionally, ODA has been provided on a voluntary basis and legal obligations only arose in the context of specific treaties. Support for the notion that there now exists some customary international legal obligation to provide ODA was significantly boosted by UNCED, where the provision of assistance was clearly an important condition for developing countries' participation in the UNCED process.

Donors have clearly accepted the legal duty to provide new and additional financial resources to enable developing countries to meet the "agreed full incremental costs" of implementing measures which conserve biodiversity which has global benefits.[6] The exact meaning of this commitment has, however, remained enigmatic. Indeed, its ambiguous nature was the very reason that the term was adopted in the CBD and the FCCC when it became obvious that the traditional split along North/South lines could not be resolved within the time frame of the negotiations.

A key task of UNCED was to clarify the meaning of this ambiguous commitment. The UN General Assembly Resolution establishing UNCED called upon the Conference to "Identify ways and means of providing new and additional financial resources, particularly to developing countries, for environmentally sound development programmes and projects ... and for measures directed towards solving major environmental problems of global concern".[7] Yet even though calls for additional funding are numerous in the UNCED documentation and the issue is the subject of

---

[3] Although non-OECD countries such as the Arab countries and countries from the Soviet Bloc used to contribute as much as a third of ODA, this has largely disappeared in recent years with the OECD countries providing 98.5% of ODA in 1992: see A.V. Lowe, *Development Co-operation: Aid in Transition*, 1993 Report of the OECD's Development Assistance Committee, at p. 101.

[4] *Ibid.*, p. 65.

[5] *Ibid.*, pp. 81–82.

[6] GEF Instrument para. 2, and the CBD, Article 20.2.

[7] Resolution 44/228 of 22 December 1989.

a chapter in Agenda 21, there was little in the way of genuine clarification. Rather, there were simply renewed calls for new and additional funding and more efficient use of existing aid. Even where there appeared to be progress, it was often illusory. For instance, paragraph 33.13 noted with regard to the target of 0.7% of GNP for ODA: "Developed countries reaffirm their commitments to reach the accepted United Nations target of 0.7 per cent of GNP for ODA and, to the extent that they have not yet achieved that target, agree to augment their aid programmes in order to reach that target as soon as possible." Yet by only reaffirming preexisting commitments, it in no way affected the position of those countries, such as the USA, that have never accepted the 0.7% target. Furthermore, Chapter 33 of Agenda 21 deviated from the widely accepted phraseology and rather than calling for "incremental cost", instead called for "adequate and predictable" funding. Although not substantively different, recasting the phrase rather than clarifying the obligation merely adds to the uncertainty surrounding it.

An identical commitment to provide ODA can be found in the FCCC.[8] Yet despite the fact that the meaning of the phrase has received considerable attention in the INC process, clarification of the term has been just as elusive and little, if any, help can be gained from an examination of these deliberations.

Similar commitments have been made under the Global Environment Facility[9] ("GEF") and the Montreal Protocol,[10] which represent the only examples of attempts to apply the term in practice. The GEF interpreted the term as referring to "the extra costs incurred in the process of redesigning an activity *vis-à-vis* a baseline plan – which is focused on achieving national benefits -in order to address global environmental problems". This incremental funding "thus covers that part of the expenditure that is not offset by nationally appropriated benefits."

There are many problems with this interpretation of "incremental". An independent evaluation of the GEF,[11] called for by the Participants in 1993, noted that determining the difference between domestic and global benefits has been difficult in practice. Also, the awareness that poorer countries would not otherwise possess the resources necessary to effect conservation has generally led to the conclusion that all biodiversity benefits from conservation projects accrue globally. Another problem which the report noted with the GEF's approach was the determination of the "baselines" necessary to ascertain the flow of incremental benefits from a project. Although this concept may not be so problematic in the context of ozone depleting substances, due to the existence of fairly accurate and well defined parameters, it tends to become complex when applied in contexts less well defined and with many alternative possibilities. In the case of biodiversity, the concept becomes almost devoid of meaning or even relevance, since the alternative to a certain conservation project would be the absence of such a project, which leads to the conclusion that the entire cost of the project is "incremental".

From a legal perspective, the most important problem noted by the report was that the pragmatic approach which the GEF took to the term "incremental" has meant that management of the GEF portfolio has been conducted on a project-by-

---

[8] Article 11.
[9] Para. 2 of the Instrument for the Establishment of the Restructured Global Environment Fund, 31 March 1994.
[10] Article 10, as amended.
[11] *Independent Evaluation of the Global Environment Facility–Pilot Phase*

project basis. This has meant that little attention has been paid to strategic policy development. The only formal attempt to develop such strategic criteria was found to be "not very useful in practice since they were numerous and not ranked in terms of importance". There has been virtually no attempt to "extract lessons from the totality of GEF experience in order to guide Participants in their deliberations". In the end, the concept was found to have been applied in a highly subjective manner, thus providing little in the way of clarification as to the legal content of the term. The findings of the Independent Panel were largely supported by the GEF's own internal review of its Biodiversity Portfolio. It found that the criteria were "only partially applied" and in any event they were "to vague to provide guidance to the design and selection of GEF Biodiversity Projects". It concluded that "far more work needs to be done in developing the methodology".

The concept of incremental costs thus remains nebulous, with many of its fundamental features such as whether it should be calculated on a gross or net basis; what discount rates should be applied for future benefits or even which developed countries are subject to the duty, remaining undecided and very little of any substance having been "agreed". As a result, the commitment cannot really be considered as creating any substantive obligations. At best, it can only be described as a guiding principle. Consequently, ODA in practice remains a voluntary act of donor countries.

Bilateral Aid

In 1992, direct bilateral aid accounted for some US$38.5 billion or nearly 70% of all ODA.[12] Determining the extent that this bilateral aid is specifically applied to biodiversity conservation is, however, fraught with difficulty. The two principal problems are that, firstly, few donors actually account for biodiversity conservation as a distinct type of spending and second, the cross-sectoral nature of the issue means that many projects which do not appear to have any relevance to biodiversity conservation will in practice have a significant impact upon biodiversity.

For example, the perception that biodiversity pertains only to wild species, *ex situ* gene storage or protected areas will often divert attention from the role of species or genetic diversity in agriculture, aquaculture, forestry, fishery and other rural development projects. Projects aimed at developing natural resource management may involve some loss of biodiversity. Often the direct impact of donor allocations for biodiversity will be negligible compared to the real effects of much larger sectoral expenditure on forests, agriculture, transport, etc. For instance, whilst BMZ (the German Government aid agency) earmarked DM3.5 million for biodiversity in 1991, they also reported forest sector expenditure of DM325 million in 1990.[13] Even though setting aside money explicitly conveys BMZ's concern for biological diversity, the positive impact of the amount set aside will be small relative to the forestry expenditures if a reasonable percentage of the forestry funds goes towards preservation or sustainable use of the forests. On the other hand, if a large portion of forest sector expenditure supports unsustainable logging activities, the overall

---

[12] *Op. cit.*, n. 3, at p. 65.
[13] World Conservation Monitoring Centre, *Global Biodiversity: Status of the Earth's Living Resources*, Chapman & Hall, 1992 ("WCMC"), Table 32.2, at p. 506.

negative impacts of this expenditure would overwhelm the potential benefits gained from the biodiversity funds.

Given the cross-sectoral nature of biodiversity conservation, perhaps the most important feature of any ODA programme is not how much is specifically targeted for biodiversity conservation, rather that there is an effective project appraisal procedure which will ensure that biodiversity conservation is taken into consideration at the design stage for all projects. If projects are screened for negative impacts on biodiversity then this may do much more to conserve existing biodiversity than channelling trivial amounts to specific biodiversity conservation measures. It will also mean that funding allocated towards biodiversity conservation will assume greater importance instead of generating suspicion that it is simply compensation for the negative effects of the remaining development portfolio. The principal legal technique for ensuring this occurs is environmental impact assessments ("EIAs") of projects and loan proposals.

All aid agencies currently subject their projects to some form of EIA under which impacts to the "environment" are assessed and minimised. A number of countries include evaluation of a project's impacts on "biodiversity" into the appraisal process (such as Canada and the USA). Article 14.1(a) and (b) of the CBD obliges governments to start evaluating such effects and, as a result, all ODA should be screened for negative impacts on biodiversity in the future. The extent to which this will minimise adverse impacts of ODA upon biodiversity in practice is, however, uncertain.

Experience of EIAs generally has been that rarely do they fulfil their potential[14] and no more is this so than in relation to biodiversity.[15] This because of factors such as inexperience, lack of resources, and methodological deficiencies. For example, there is a lack of objective criteria to assess the potential impacts on biodiversity. Such impacts are usually assessed on a purely subjective basis with even semi-objective methods, such as cost-benefit analysis, rarely used. Other common deficiencies include a lack of follow-up monitoring, little or no consideration of alternative proposals, and predictive techniques which often owe more to guess work than to any scientific methodology. Finally, as the procedure is often employed at the end of the project evaluation cycle, it loses its most important feature; the synergies that planning and forethought can produce to avoid unnecessary adverse effects.

Bilateral aid is not always simply new funding for a particular project allocated to the general budget of the project. Often the aid may be given to a recipient country on certain conditions, typically, that the aid applied to the purchase of goods and services from the donor country, in which case it is known as tied aid.[16] This practice inflates the price of the project and may result in aid funds being substituting for commercially available finance rather than being additional to such finance.[17] Recently, as many donor countries have been under acute budgetary pressures, some rather inventive methods have begun to appear in order to inflate ODA budgets. One technique which has had particular significance for biodiversity conservation is the use of debt forgiveness programmes to finance projects. Two

---

[14] See M. Prieur, *Les Etudes D'Impact et le Droit Comparé de L'Environnement: Pour Grande Bretagne* (1993) and W. Wood and C.E. Jones, *Monitoring Environmental Assessment and Planning* (1991).

[15] See A.F. Krattiger (ed.), *Widening Perspectives on Biodiversity* (1994), Section 6.

[16] An estimated 26% of bilateral aid was tied during 1989–91: *op. cit.*, n. 3, at p. 94.

[17] For instance, it is claimed that it inflates costs by 15% in *ibid.*, p. 94 or by 10–20% in L. de Silva, *Development Aid: A Guide to Facts and Issues* (1982).

particularly noteworthy projects in this regard are the US's Enterprise for Americas Initiative ("EAI") and the Canadian Latin American Debt Conversion Initiative.

Under the EAI Scheme, certain Latin American countries and Caribbean countries are able to have part of their US debt forgiven if they have in place certain measures such as; an IMF standby arrangement or other IMF restructuring programme; structural adjustment loans from the World Bank; major investment reforms in conjunction with an Inter-American Development Bank loan or other investment reforms; agreement with commercial lenders on debt restructuring; and a democratically elected government. Later amendments imposed certain political criteria such as an acceptable record on: human rights; narcotics control matters; and counter-terrorism. Once the US President agrees that a country meets these criteria, the President may apply to Congress for an appropriation to cancel the old debts and issue new loans with reduced principal and interest rates. The USA and the beneficiary country may then enter into a sustainable development framework agreement. Once this is in place, the beneficiary country may start to make the interest payments in local currency to a trust fund to support the goals of the framework agreement rather than continuing to make the interest payments to the US government in US dollars.

Under this programme, US$1,626 million has received approval for reduction.[18] Agreement has been reached with Colombia, Bolivia, Chile, Uruguay, El Salvador, Argentina and Jamaica. Debt reductions have ranged from 10% to 80% of the face value of the eligible loans, with an average reduction of 54%. Concessional interest rates are between 2 and 3%. The funds generated over the life time of the new loans will be US$134 million and these are to be applied largely to biodiversity conservation.[19]

The Canadian Latin American Debt Conversion Initiative takes the EAI concept one step further and hypothecates principal as well as interest to the environment funds established under its auspices, which also have a heavy emphasis on biodiversity conservation. Only concessional ODA debt, amounting to a potential CUS$145 million, is eligible for the programme. So far under this scheme, some CUS$100 million of debt has been converted, at discounts ranging from 75% to 0%. Countries which have participated include; Colombia, Peru, El Salvador, Nicaragua and Honduras. Other eligible countries are; Brazil, Cuba, the Dominican Republic and Guatemala.

The biodiversity emphasis of these programmes is not only due to the fact that most of these countries have large amounts of biodiversity and are considered megadiverse countries but also that the structure adopted for the EAI and Canadian initiative is particularly suitable for biodiversity conservation given the long-term nature of the obligations which they establish. These debt reductions are, however,

---

[18] Although a total of US$5.2 billion worth of debt is potentially eligible for reduction, further reductions seem unlikely because EAI debt reductions are treated for US accounting purposes as a gift or donation to the debtor, as opposed to being written off (which is the normal accounting treatment that they would receive in a commercial bank). This means that, before the debt reduction may occur, the US Congress has to make an appropriation equal to the amount that is being forgiven. Even though all that is happening is a fictional appropriation, Congress has been reluctant to come up with the total amount necessary for full reductions and further appropriations under this scheme are unlikely to occur. The 1994 budget request has been totally refused and the President has not even made a request for a 1995 budget allocation.

[19] For further details see, the World Bank, *World Debt Tables 1993–94: External Finance for Developing Countries*, Box 2.1, at pp. 35–36.

treated as a gift or donation to the debtor and the value of the reduction in debt is added to the ODA budget, even though they are only fictional appropriations and no actual money is transferred. It is therefore questionable whether they are really schemes which provide additional ODA.

If, however, the technique is applied transparently and is "additional", it could provide an important means by which ODA which helped achieve environmental objectives generally and biodiversity in particular could be increased. The appeal of this technique is also enhanced by the large amount of unsustainable debt which remains and the increasing pressure on donor countries to forgive this unsustainable debt. For example, paragraph 33.14 (e) of Agenda 21 calls upon donors "to provide debt relief for the poorest heavily indebted countries pursuing structural adjustment". Chapter 33 also provides that funding for Agenda 21 should use, as much as possible, existing sources of funding. Similar calls are evident in the recent Desertification Treaty.[20] Together these obligations not only imply a duty for donors to undertake debt relief, but also require that it should occur in a way which encourages, to the greatest possible extent, sustainable development. To some extent, this happens by merely allowing debt forgiveness, in that it reduces the unsustainable demands placed upon developing countries' resources by the need to service the external debt. It is questionable, however, whether this is using debt relief to its fullest extent to encourage sustainable development as required by Agenda 21. The need to maximise benefits of all sources could arguably require that debt relief be associated with an environmental programme suitable for the particular country, and require that the interest payments and/or the principal repayments on the remaining debt be used to finance environmental funds.

The requirement does, however, imply a type of conditionality and all the problems and criticisms which follow from this. For example, it is preferable that debt forgiveness be unconditional, thereby allowing the developing country itself to allocate the extra resources arising from the absence of debt servicing. To the extent that debt reductions will occur unconditionally, these arguments are persuasive. In the absence of total debt forgiveness, however, mechanisms which will deepen the levels of reduction provide potential benefits for both donors and recipients alike. Thus, if debt conversion for biodiversity conservation can be linked to deepening debt reductions, this might outweigh many of the problems arising from its implied conditionality.

Multilateral Aid

Multilateral disbursements of ODA accounts for some US$16.8 billion or 30% of all ODA and comes primarily from the multilateral development banks, such as the World Bank Group, the Asian Development Bank and the African Development Bank.[21] Most of these organisations belong to the Committee of International Development Institutions on the Environment which coordinates activities on the environment. The extent to which it develops policy and the degree to which multilateral development banks target biodiversity conservation has, however, so far been minimal.

---

[20] Article 20.2(d).
[21] US$6750 million or 38% of multilateral ODA in 1992. For a detailed outline see *op. cit.*, n. 3, at p. 188 (Table 24).

Generally, initiatives undertaken by many of these multilateral organisations to date indicate that efforts to control harmful projects and identify beneficial ones with regards to biodiversity are still in their formative stages. Some organisations have firm guidelines in place; others are still formalising such procedures. Some agencies prefer a less explicit approach, believing that concerns over the environment and biodiversity must become an integral part of the project cycle, instead of an extra component of the normal appraisal process. Although this situation may improve as these institutions implement Agenda 21, the aspirations of Agenda 21 are yet to be fully implemented.

The most important institution in this regard is the World Bank, not only because it disburses the largest amount of ODA,[22] but because many other institutions look to it for the lead in environmental matters. The World Bank's influence on biodiversity conservation issues will, however, be most directly felt through its major role in the operation of the Global Environmental Facility.[23] More generally, its overall environmental programme, although not explicitly focused on biodiversity conservation, does none the less have a significant effect on biodiversity and is typical of the type of measures now being undertaken by the other multilateral institutions.

The World Bank's general environmental programme has three main objectives.[24] First, it attempts to assist developing countries in setting priorities, building institutions and implementing programmes for sound environmental stewardship. It also attempts to help countries build on the synergies between poverty reduction, its principal goal, and economic efficiency and environmental protection. Finally, it attempts to ensure that potential adverse environmental impacts from World Bank financed projects are addressed. The World Bank's efforts to help countries improve their environmental management is centred around three elements: providing financial resources for environmental investments such as lending for natural resources and the rural environment; disseminating knowledge about environmentally sustainable development through its programme of policy, analysis and research; and providing support for the development of national and regional environmental action plans ("NEAPs"). Although the quality of the NEAPs has been quite varied, some of the latest plans have been more than the banal exhortatory documents which characterised the earliest NEAPs and some have even contained concrete and specific recommendations.[25] All of the NEAPs have recognised the importance of biodiversity conservation and to the extent that they have any effect upon the activities of international aid agencies and governmental policy, represent an important mechanism for biodiversity conservation.

The World Bank's efforts to build on the synergies between poverty reduction, economic efficiency and environmental protection are concentrated in the energy sector of their operations. Increasing energy efficiency and removing subsidies to natural resources which create wasteful distortions by encouraging full internalising of the environmental costs helps biodiversity indirectly by minimising the pressure on natural resources such as water and wood. Due to the attenuated nature of these

---

[22] 71.5% of ODA from all multilateral development banks or $4323 million in 1991: *ibid.*
[23] See Section 6.
[24] See generally, the World Bank, *The World Bank and the Environment: Fiscal 1993.*
[25] One of the best received NEAPs was the regional action plan developed for Central and Eastern Europe.

type of efforts and their affect on biodiversity, it is impossible to measure or mean-ingfully judge what impact, if any, they will have on biodiversity.

As with bilateral aid, perhaps the most important aspect of the World Bank's general programme is the attempts it makes to ensure that potentially adverse impacts on biodiversity from all World Bank financed projects are addressed by its EIA policy. The 1989 Environmental Assessment Operational Directive separates projects into four categories each of which receives varying degrees of assessment, with the extent of the EIA being proportional to the perceived degree of impact that a project may have on the environment. The EIA is the borrower's responsibil-ity. The Bank provides assistance in designing the terms of reference for the EIA and normally a field visit by Bank staff is suggested. The Bank usually recommends that borrowers hire experts not involved in the project to carry out the EIA. EIAs for large projects may take up to eighteen months to be completed with input from the ongoing EIA occurring at relevant points in the overall project cycle. The final report is submitted to the Bank for consideration with the project or loan applica-tion. Funding for EIAs may be accomplished by a Bank loan or grant and usually comes to 5–10% of the cost of project preparation. In the 1993 fiscal year, the Bank reported that of the projects at the pre-approval stage, 14% will require full EIAs, 42% limited EIAs and 33% will not require an EIA.

As noted above, EIAs as a means of biodiversity conservation suffer from a number of limitations. An internal Bank review of EIAs covering the period from October 1989 to June 1992[26] concluded that these problems also limit the effective-ness of EIAs conducted for World Bank projects. The most prevalent weaknesses involved analysis of alternatives, mitigation measures, monitoring and institutional arrangements. The review also found that, in many of the initial EIAs, public con-sultation was "disappointing" and consultation with the affected populations and local NGOs "had been limited at best". Although the review considered that progress had been made in incorporating the public, it concluded that more involvement was needed and recommended that extensive consultation with affected populations and local NGOs be routinely carried out in determining the scope and content of an EIA. It also found that although in theory people under-stood that EIAs were a preventative mechanisms as opposed to a reactive one, in practice this aspect of EIA was being ignored. For the most part the EIA was still being thought of as an additional extra to the project design, resulting in a lack of breadth in identifying relevant issues, limited attention to alternatives and weak mit-igation plans. The quality of EIA documents was found to be lacking. These short-comings are often due to borrowers' underdeveloped EIA capacity and to the tight timetables for the preparation of projects.

Similar problems have been found in the implementation of the Environmental Impact Assessment Directive within the European Union,[27] where funds and exper-tise are generally much greater. Therefore, despite the fact that the Bank is develop-ing a number of initiatives to improve the EIA process, it remains doubtful whether the full potential of the process will be realised in the near future.[28]

Apart from these general environmental programmes, the World Bank also has a number of projects targeted specifically at biodiversity conservation which are not

[26] *Op. cit.*, n. 24, at pp. 58–65.
[27] Directive 85/337 OJ L175/40, 5.7.85.
[28] *Op. cit.*, n. 15.

covered by the GEF. For instance, the World Bank has over a hundred current loans which involve biodiversity components.[29] It is also currently preparing regional biodiversity strategies for Asia and the Pacific, Latin America and the Caribbean and African regions. These strategies survey the rate of biodiversity loss; determine priorities for conservation, including establishing and maintaining protected areas; review the design of biodiversity components in project lending; identify future Bank and GEF biodiversity projects; and facilitate the mobilisation ad coordination of financial resources. The Bank also assisted in the development of the Global Marine Biodiversity Conservation Strategy. It is also providing technical assistance to many developing countries to produce their national biodiversity action plan as required by the CBD.

Another important project in which the Bank is involved is the Pilot Programme to Conserve the Brazilian Rain Forest in collaboration with the Brazilian government and the G7 Countries. This programme is funded by a group of donors comprised of Canada, France, Germany, Italy, Japan, the Netherlands, the UK, the USA and the EU. The finance is administered by the Bank through the Rain Forest Trust Fund. The programme supports an integrated set of projects designed to reduce the rate of deforestation in a manner consistent with the sustainable development of the region's natural and human resources. By June 1993, twelve projects were being prepared by Brazilian project teams and some US$280 million had been pledged in financial and technical assistance; US$58 million of this total had been pledged to the Rain Forest Trust Fund.[30]

Other multilateral institutions who disburse significant amounts of ODA are the Commission of the European Union through the Lomé IV Convention (US$4.156 billion in 1992) and the UN (primarily through the FAO, UNDP, UNEP, Unesco and the International Fund for Agricultural Development who disbursed US$5.886 billion in 1992).[31] All these institutions have begun to emphasis biodiversity issues in their policies and programme designs. Essentially, they all use techniques similar to those outlined above to attempt to ensure that biodiversity considerations are incorporated into their programmes. As a result, the gap between rhetoric and practice is, however, as prevalent for these organisations as it is for the multilateral development banks.

Other Sources of Multilateral Aid

An increasingly important source of multilateral ODA is funds disbursed under international environmental conventions. In the context of biodiversity conservation, two of the most important funds have been the International Oil Pollution Compensation Fund ("IOPC Fund") managed by the IMO and the World Heritage Fund managed by Unesco. The Ramsar Convention also established a new fund in 1991 to help its contracting parties preserve wetlands, which in the future should play an important role. Without doubt the most important example of this mechanism in years to come will be the fund for the CBD.

Details of the World Heritage Fund are well known and need not be covered in detail in this chapter. The annual budget of the WHF has been for the last few years

---

[29] Op. cit., n. 24, at p. 108.
[30] Ibid., p. 109.
[31] Op. cit., n. 3, at p. 188.

approaching US$3 million.[32] Despite the relatively small size of the WHF's resources, it has been an important factor in the wide membership which the Convention has enjoyed and consequently has been a major factor in the success of the Convention. The WHF also played an important role as a model for later conventions such as the 1985 Ozone Convention and its Montreal Protocol and the CBD and, in this way, has had an important indirect impact upon ODA and biodiversity conservation.

The IOPC Fund was established in 1971 pursuant to the International Convention on the Establishment of an International Fund for Compensation for Oil Pollution Damage and represents the earliest example of this type of funding mechanism. This Convention provides for a free standing fund that awards additional compensation to any person suffering oil pollution damage, to the extent that the protection offered by its companion treaty, the 1971 International Convention on Civil Liability for Oil Pollution Damage, is inadequate. The IOPC Fund is important in the context of this chapter because of the nature of its mandate, which is to provide additional compensation for damage not covered by a civil liability regime. Thus it has been called upon primarily to rectify damage to the unowned marine environment and its natural resources, usually the critical aspects of biodiversity damaged by such incidents.

The IOPC Fund is financed by initial and annual contributions.[33] Annual contributions are paid by any person who has received in the relevant calendar year more than 150,000 tonnes of crude oil in a Member State. Annual contributions are levied to meet the anticipated payments by the IOPC Fund and the administrative expenses of the Fund during the coming year. The levy of contributions is based on reports of oil receipts which are submitted by governments of Member States. The contributions are paid by the private parties who import the oil directly to the IOPC Fund. Governments have no responsibility for these payments.[34]

The IOPC Fund establishes two types of accounts or funds. The first is the general fund from which are paid the administrative expenses. The other type of funds are the major claims funds, which are established to meet any potential liability from major incidents, such as the sinking of an oil tanker. In October 1993, the Assembly levied annual contributions which amounted to £78 million; £8 million for the general fund, and £70 million for specific major claims funds. Payments made by the IOPC Fund vary considerably from year to year and consequently the level of contributions also fluctuate considerably.

The method of assessing contributions, although unusual, has a number of advantages over more conventionally structured funds such as the GEF or the WHF. More typically governments have only been willing to agree to some fixed scale of assessment (i.e. the Unesco scale). The variable scale it employs means that the IOPC Fund is more likely to have the necessary resources to fulfil its mandate than funds which are dependent upon contributions fixed upon some arbitrary scale. Another interesting feature of the IOPC Fund is that the money is levied directly against private persons. Normally, governments only agree to mechanisms which

---

[32] WHF Financial Statements available from Unesco or WCMC, Table 31.1, p. 492.
[33] Initial contributions are payable when a State becomes a Member of the IOPC Fund and are calculated on the basis of a fixed amount per tonne of oil received the year preceding the State's entry to the Convention.
[34] Article 13.

allow financial resources to be raised through the general taxation system and require special payment from the annual budget. Requiring payment from the private parties directly is a more efficient form of revenue raising in that it does away with unnecessary levels of bureaucracy. Less bureaucracy increases transparency and means that there is a greater degree of hypothecation between the purposes for which the revenue is raised and spent. Experience has generally shown that this increases the willingness of the potential contributors to make the necessary contributions. Finally, it increases the accountability of the IOPC Fund which in turn increases the likelihood of prudent and responsible management.

Although current levels of funding disbursed this way are relatively insignificant, the amounts which are being discussed in relation to the financial mechanism of the CBD mean that this mechanism will assume a much greater importance in the future.

The CBD and the GEF

Without question the two most important developments of recent years in the context of this chapter are the GEF and the CBD. The GEF has over the last three years allocated over US$700 million of ODA to "innovative" projects for the conservation of biodiversity and is set to allocate a similar amount in the coming three years. Furthermore, it has been appointed as the interim financial mechanism for the CBD. As a result, it has in a short time had a significant effect upon the way that ODA is being applied to biodiversity conservation. Serious doubts remain, however, regarding the appropriateness of the GEF as the funding mechanism for the CBD. This issue has dominated the work of the ICCBD and the Interim Secretariat.

The Global Environment Facility

The Pilot Phase

The origins of the GEF lie in a suggestion by France at the 1989 Annual IMF–World Bank Development Committee Meeting to create a global fund for promoting activities by developing countries that protect the environment and provide benefits to the global community. In November 1990, agreement was reached by twenty-five countries that the World Bank, UNDP and UNEP (designated as the Implementing Agencies) would co-operate in administering the GEF as a mechanism for distributing concessionary finance to help developing countries implement programmes that protect the global environment in four focal areas, one of which was the "protection of biodiversity".[35] By 31 March 1991, twenty-one countries had committed approximately US$1.4 billion to the fund to be applied over a three-year trial, which concluded in July 1994 and was known as the Pilot Phase.

Administration of the Pilot Phase emphasised co-operation between its Implementing Agencies, the participants and NGOs, in an attempt to marshal the best available skills without setting up new bureaucracies. Management of project design drew collaboratively on the experience and expertise of the three Implementing Agencies. UNDP was responsible for technical, operational and capacity-building activities and was charged with managing the Small Grants

---

[35] The others being global warming, international watercourses and ozone depletion.

Programme for NGOs. UNEP provided the secretariat for the Scientific and Technical Advisory Panel ("STAP"), offered environmental expertise for the Facility and supported research and information dissemination. The World Bank chaired and administered the pilot facility, managed the trust fund and was responsible for investment projects.

During the Pilot Phase, the GEF functioned under the collective policy guidance of its Participants, namely, States that contributed to the Facility or that expressed the intention to do so.[36] As of April 1994, there were ninety-four Participant States. The Participants held five meetings at which they reviewed and broadly endorsed successive "tranches" of a work programme, prepared jointly by the Implementing Agencies and submitted to the meeting by its Chairman. The Participants were supported in their functions of policy guidance and programme review by the STAP, composed of sixteen independent experts. The STAP drew up two sets of criteria to determine eligibility of projects for the GEF Biodiversity Programme. The STAP also reviewed individual projects submitted by the Implementing Agencies and advised on their conformity with its criteria and priorities.

By June 1994, some US$303.5 million, representing 42% of the GEF's total allocation (US$727.1 million), had been allocated to the Biodiversity Portfolio. Some additional funding could be considered as contributing to biodiversity conservation through cross-cutting projects that address the protection of international waters and biodiversity (six projects), and projects that address global warming and biodiversity (three projects). The Biodiversity Portfolio was made up of a total of fifty-four projects, involving activities in forty-three countries, and four projects of a global nature in support of the objectives of the CBD. The portfolio includes twenty-seven investment projects (World Bank), twenty-three technical assistance projects (UNDP), and four under the category of support to conventions and targeted research (UNEP).[37] The Pilot Phase, however, did not only provide "new and additional" ODA, but has also been the principal fora in which the meaning of "incremental" has been developed.

GEF II

The Pilot Phase of the GEF was the subject of strong criticism about almost every aspect of the programme. In April 1992, the participants agreed that its structure and modalities should be modified. This decision was supported by further calls for its restructuring in Agenda 21,[38] the FCCC[39] and the CBD.[40]

At UNCED, the donors wanted the GEF to be the conduit for all "new and additional" funding required by the new commitments with regard to the global environment such as Agenda 21 and the CBD. Many developing countries, however, argued that the GEF was unsuitable, not least because of its control by the World Bank, which they perceived to be part of the problem and most certainly not part of the solution. In the end, compromise was reached whereby the developing

---

[36] The Pilot Phase required a contribution before any State could become a member. The developing country contribution was set at a nominal $4000.

[37] See *Global Environment Facility: Scientific and Technical Advisory Panel: Summary report on the Review of the Biodiversity Portfolio: Tranches I–V*, April 1994.

[38] Para. 33.14(a)(iii).

[39] Article 11.1.

[40] Article 39.

countries would accept the GEF as one source of additional funding if, and only if, it was restructured so as to accommodate their concerns about the GEF and its trustee, the World Bank. These concerns were clearly articulated in paragraph 33.14(a)(iii) of Agenda 21 which stated:

> [the GEF] should be restructured so as to, *inter alia*:
>
> Encourage universal participation;
> Have sufficient flexibility to expand its scope and coverage to relevant programme areas of Agenda 21, with global environmental benefits, as agreed;
> Ensure a governance that is transparent and democratic in nature, including in terms of decision-making and operations, by guaranteeing a balanced and equitable representation of the interests of developing countries and giving due weight to the funding efforts of donor countries;
> Ensure new and additional financial resources on grant and concessional terms, in particular to developing countries;
> Ensure predictability in the flow of funds by contributions from developed countries, taking into account the importance of equitable burden-sharing;
> Ensure access to and disbursement of the funds under mutually agreed criteria without introducing new forms of conditionality.

The resulting negotiations for the restructuring and replenishment of GEF took place over the course of the following two years and concluded in Geneva, Switzerland, in March 1994.[41]

There the Instrument for the Establishment of the Restructured GEF was adopted and Contributing Participants also agreed to the first replenishment of the GEF Trust Fund totalling SDR 1.44 billion (US$2.02 billion).

Although the purpose and the methods of the programme design and implementation remain largely unchanged for the next phase of the GEF, there were a number of important changes to reflect the demands articulated above. One important change, as far as the Biodiversity Programme is concerned, was the broadening of the Facility's mandate to include "agreed incremental costs of activities concerning land degradation, primarily desertification and deforestation, as they relate to the four focal areas shall be eligible for funding" and the "agreed incremental costs of other relevant activities under Agenda 21 that may be agreed by the Council shall also be eligible for funding insofar as they achieve global environmental benefits by protecting the global environment in the four focal areas". The broadening of the mandate will have a direct impact upon the Biodiversity Programme and will presumably increase the amount of resources devoted by the GEF to biodiversity issues.

Another important development was the establishment of a complex administrative structure designed to ensure transparent and democratic governance of the GEF. Administration of the Facility is now made up of a Participants' Assembly consisting of all the Participants with responsibility for reviewing the general policies of the facility and which will meet every three years. The Implementing Agencies

---

[41] Meetings were held in Abidjan, Cote d'Ivoire, in December 1992, in Rome, Italy, in March 1993, in Beijing, China, in May 1993, in Washington, DC, USA, in September 1993, in Paris, France, in November 1993, in Cartagena, Colombia, in December 1993 and concluded in Geneva, Switzerland, in March 1994.

will provide similar functions as under the Pilot Phase. Finally, and perhaps most importantly, a Council was established which is to be "responsible for developing, adopting and evaluating the operational policies and programmes for GEF-financed activities, in conformity with the present Instrument and fully taking into account reviews carried out by the Assembly" and the directions it receives from the CBD and the FCCC. The Council consists of thirty-two members, representing constituency groupings formulated and distributed taking into "account the need for balanced and equitable representation of all Participants and giving due weight to the funding efforts of all donors". The Council is comprised of sixteen members from developing countries, fourteen members from developed countries and two members from the countries of Central and Eastern Europe and the former Soviet Union. The Council's first meeting was in July 1994, with its next meeting scheduled for October 1994 and plans to meet every six months thereafter, though it can meet as "frequently as necessary". The work of the Council and the administrative structure generally is to be supported a Secretariat and the STAP. The Council is to "act as the focal point for the purpose of relations with the CBD". Decisions of the Council will be made on the basis of a double weighted majority, representing both a 60% majority of the total number of Participants and a 60% majority of the total contributions.

Finally, the Instrument also explicitly acknowledged that "the GEF shall function under the guidance of, and be accountable to, the Conferences of the Parties which shall decide on policies, programme priorities and eligibility criteria for the purposes of the conventions".[42]

The CBD

From the outset of the negotiations for the CBD it was generally acknowledged that new and additional funding was required for developing countries if the aspirations of the Convention were to have any hope of being achieved. The financial provisions of the CBD were negotiated after the FCCC financial provisions had been settled. In the CBD negotiations, donors argued for identical provisions to the FCCC and developing countries argued for several material differences in relation to the institutional structure and the amounts to be available to the fund. Due to these tensions, the resulting provisions in Articles 20, 21 and 39, though largely similar to the provisions of the FCCC, slightly differ and contain some potentially problematic language.

Article 20.2 embodies the widely accepted obligation with regard to the amount of ODA, committing developed countries to "provide new and additional financial resources to enable developing country Parties to meet the agreed full incremental costs to them of implementing measures which fulfil the obligations of this Convention". The problems with this obligation have already been noted. Like the provisions of the FCCC, Article 20.2 also requires that this commitment must "take into account the need for adequacy, predictability and timely flow of funds".

With regard to the institutional mechanism for the CBD, donors had wanted the GEF to be appointed as the only mechanism, with the relationship between the COP and the GEF being identical to that used in the FCCC, which itself derived from the agreement of the GEF participants reached just weeks earlier. The

---

[42] Para. 6.

developing countries, on the other hand, who were dissatisfied with the language of the FCCC, sought to renegotiate the issue on the basis that: there was to be no mention of the GEF; financing should be by means of a fund administered by the COP; and to the extent that another financial mechanism was described, it should be subject to the authority of the COP and call for accountability, transparency and democratic governance.

One problem with the resulting compromise arises with regard to the possibility of the GEF being able to qualify as the permanent financial mechanism for the CBD. Whereas the FCCC designates the GEF as the interim financial mechanism with the exhortation that it "should be appropriately restructured"; the CBD in Article 39, as a result of the developing countries insistence, identified the GEF as the interim financial mechanism only if it "has been fully restructured in accordance with the requirements of Article 21". Article 21.1 requires the mechanism to "function under the authority and guidance of, and be accountable to, the Conference of the Parties" and that it "operate within a democratic and transparent system of governance". Whether the restructured GEF has these characteristics is considered by some parties to be questionable. Indeed, Pakistan stated at the last ICCBD that they believed that the restructured GEF does not qualify under the Convention, since it is not democratic or under the authority of the Convention.

More generally, the suitability of the GEF as the permanent or long term institution for the financial mechanisms was an issue which received an enormous amount of attention in the ICCBD process. Working Group II of the ICCBD was directed to develop an evaluation framework to determine the: "institution or institutions operating the financial mechanism; characteristics desired in the institution or institutions operating the financial mechanism under the convention; process for developing an evaluation framework to propose to the COP; process to examine the funding needs; and how to select an institution to operate the financial mechanism upon entry into force" in order to facilitate the work of the first COP.

Discussion of these issue at the ICCBDs was described as "highly sensitive and politically charged from the outset. With two strongly polarized views, divided along North/South lines, no texts could be negotiated and, as a result, no recommendations will be forwarded to the COP."[43] At the last ICCBD, in June 1994, donor countries again insisted on the recently restructured GEF becoming the permanent institutional structure to operate the Convention's financial mechanism. They argued that the GEF restructuring was negotiated by the countries present at the ICCBD. They also claimed that their governments have committed finances to the GEF for funding of biodiversity projects and that it is unlikely that any new money will be forthcoming. On the other hand, many developing countries did not want the GEF to become the permanent institutional structure until it has proved that the restructuring has responded to all the developing country concerns. Other developing countries refused to even consider the GEF on an interim basis. Pakistan, as noted above, said that the restructured GEF does not even qualify under the Convention. Kenya, Syria and Mauritius called for a different mechanism altogether. Many developing countries also pointed out the difficulty of the GEF living up to such expectations, as it is also answerable to other bodies. Debate over the restructuring of the GEF and its relationship to the COP overshadowed any

---

[43] Taken from the *Earth Negotiation Bulletin*, Vol. 9, No. 17, 7 July 1994.

discussion of the methodologies for estimating funding needs. In the end, consensus entirely eluded the Working Group. Furthermore, it is questionable how much progress can be expected in the near future given the divisiveness of the issue at the ICCBDs. The Interim Secretariat recognises the fact that the parties are still a long way apart on this matter and is hoping to hold further meetings on the issue prior to the first COP.

This intransigence is unfortunate and has diminished the ability of the CBD to make a significant contribution to the development of the Biodiversity Portfolio of the next phase of the GEF. In this regard the approach of the INC of the FCCC is illuminating. The INC referred the issue, in the context of its Article 11, which is almost identical to Article 21, to the United Nations Office of Legal Affairs and sought their legal opinion on the suitability of the restructured GEF as the financial mechanism for the Convention and the arrangements which might be entered into between the COP and the restructured GEF, if it was to be so appointed. The Office of Legal Affairs opined that generally the GEF satisfied the requirements of the FCCC. With respect to the requirement that the GEF function under the guidance of, and be accountable to, the COP which will decide on policies, programme priorities and eligibility criteria for the purposes of the Convention, it noted that this was "reaffirmed in paragraphs 15 and 26 of the Instrument" and that in "defining the eligibility criteria for the GEF funding, the Instrument makes it clear that GEF grants that are made available within the framework of the financial mechanism of the Convention shall be in conformity with the eligibility criteria decided by the COP". It also noted that, as far as the requirements for democratic and transparent governance are concerned, these were guaranteed in the preambular paragraph (c), and paragraphs 7 and 25 of the Instrument. In conclusion, they advised that "the restructured GEF is an entity which meets the requirements set forth in paragraph 1 of Article 11 of the Convention and, therefore, may be selected by the COP as an entity entrusted with the operation of the financial mechanism".

The Office of Legal Affairs did note, however, that there were problems associated with the appropriate arrangements which might be entered into between the Conference of the Parties and the restructured GEF as an operating entity. For example, it noted that with regard to the management of the portfolio, there is "reason to believe that under the Convention it is expected that the COP should play a slightly more active role in exercising control over the implementation of the policies, programme priorities and eligibility criteria established by the COP, than is envisaged for it in the GEF Instrument". It went on to note that this and other issues which required attention such as

> accountability, observance of compliance with the eligibility criteria for funding, procedures for the reconsideration of particular funding decisions and, last but not least, procedures for joint determination and periodical review of the aggregate GEF funding necessary and available for the implementation of the Convention, will have to be regulated in an agreement concluded for these purposes. In other words, in order to ensure the effective operation of the GEF as a source of funding of the activities under the Convention, the above-captioned issues should be spelled out in a legally binding treaty instrument.

Practical issues like these were obviously not addressed in the polemic which characterised the debate in the ICCBDs.

Ultimately, it must be recognised that there are problems with the "either, or" approach which has dominated the discussions so far. On the one hand, it must be recognised that, in the long term, the GEF will not be the principal mechanism for the transfer of financial resources required by the Convention. As noted in Chapter 3, the GEF is a public mechanism which, compared to private mechanisms, is neither an efficient nor transparent means of delivery because it is controlled by national and international, rather than local, structures. The GEF also retains an element of voluntariness, in the sense that payments made to it by donors are often thought of as aid and not as payment for services rendered. In the long term private mechanisms whereby the actual user pays the actual provider will be required to ensure that the aims of the CBD are achieved. On the other hand, it should be recognised that in the meantime ODA will be required to achieve the aims of the CBD and donors will only be willing to provide the necessary ODA if they believe that it will be used in a prudent and efficient manner, which in the current situation means relying upon an institutional structure similar to the GEF.

Finally, Article 20.3 reiterates the importance of other sources of ODA for the implementation of the CBD. This point should not be forgotten since, despite the importance of the GEF, the amounts which have so far passed through it when compared with overall ODA flows are insignificant. Given this situation, which will not change in the foreseeable future, it is probably more important to develop proper EIA measures for bilateral and multilateral aid agencies.

The attention devoted to the nature of the institution has meant that other important issues have largely been ignored. These include: the meaning of "agreed full incremental costs"; the burden sharing arrangements of the donor countries; the measures necessary to ensure that financial resources are available on a timely and predictable basis; access to the resources of the fund; and what role, if any, the GEF should have in implementation. These issues must, however, be settled before the financial mechanism of the CBD is operational.

Conclusions

Michael Bowman and Catherine Redgwell

The 1992 Convention on Biological Diversity establishes for the first time a unifying conceptual and practical framework for international efforts concerning the conservation and sustainable utilization of the living natural resources of the planet. Early examples of international co-operation in this field can be traced back to the latter part of the nineteenth century and it might well be argued that such an overarching instrument is long overdue, particularly given the widespread incidence of species extinctions and ecosystem degradation which has so troubled commentators in recent times.

Although it is easy to overestimate the practical significance and value of international legal measures in any context, and in the environmental field in particular, surely few would deny that international law must have some part to play if these worrying trends are to be retarded or reversed.

The collection of essays in this volume accordingly represents an exploration of many of the key issues relating to the conservation of biological diversity when viewed from a legal perspective.

Biodiversity and the Biodiversity Convention

In devising our initial programme of work with this project, an appropriate starting point seemed to be to define and analyse the biodiversity concept in international law. Without this philosophical foundation it would be difficult fully to appreciate the orientation of the many agreements referred to in subsequent chapters, including the Biodiversity Convention itself, not to mention the need to have a mental picture of just what it is that biodiversity encompasses. This task fell to Michael Bowman in the opening chapter of the volume.

He begins with the threefold approach to biodiversity reflected in Article 2 of the Convention, namely: (a) diversity of ecosystems; (b) diversity of species; and (c) genetic diversity within species. These are plainly not wholly distinct but mutually interdependent categories. Diversity at each of these levels is best conserved *in situ*, with entire ecosystems preserved along with the species they contain. Regard should also be had to genetic diversity within the conserved species to ensure its ability to withstand present and future threats; hence the need to protect each species throughout its range. *Ex situ* conservation has an important corollary role to play in this respect.

A number of existing wildlife treaties, while not employing the language of biological diversity, clearly contribute to its conservation. The recognition of biodiversity in international instruments is examined, including the extent to which the question of genetic diversity has been addressed.

This discussion sets the stage for the core of Bowman's treatment of the philosophical foundations of the biodiversity concept, starting with the beneficiaries of biodiversity. Is it meaningful to talk of human rights in this context? Collective rather than individual rights or interests in biodiversity seem more plausible, including possibly those of future generations, to whom reference is frequently made in the preambular portion of treaties, the Biodiversity Convention being no exception. It is difficult, however, to see how even the limited and controversial right to a clean environment may be readily translated into the realm of biological diversity. In any event the question may be asked whether such an approach might not be unduly anthropocentric. Drawing on writings primarily from the field of environmental ethics, Bowman points out that the elements of the natural world may be valued instrumentally, inherently, or intrinsically. Each of these forms of value gains some recognition in international environmental law generally, and, more particularly, in the preamble of the Biodiversity Convention itself. The latter, he concludes, appears to rank intrinsic value equally with various forms of instrumental and inherent value. Whilst the emphasis of the substantive provisions of the Convention is undoubtedly upon the instrumental value of nature, this recognition of intrinsic value is important because it is understood as the value which the entities have of themselves, for themselves, without the necessity of any external valuer. Hence Bowman's characterisation of this recognition as "bold" and a particularly striking feature of the Convention.

But where does that value reside – in biodiversity as a whole, or in the components thereof? One reading of the Convention is that, consistently with the views of many environmental ethicists, both instrumental and intrinsic value reside in individual plants and animals, and that it may also be possible to recognise species and ecosystems as the repositories of intrinsic value. In practical terms he views the conclusion that individual plants and species possess intrinsic value as the most problematic, particularly in respect of their exploitation. However, intrinsic value does not render them sacrosanct, but gives them a *prima facie* claim to our consideration. A plausible case may be made for assessing the strength of that claim by reference to the degree of complexity of the organism concerned.

Governments are notoriously unwilling to address these underlying philosophical issues, which lie at the heart of many of the problems of implementation of international conservation treaties. While some kind of uneasy compromise between opposing viewpoints can sometimes be found, this is no substitute for confronting these unresolved philosophical questions head on, without which Bowman doubts that the biodiversity concept will be able to shoulder the burden required of it in the realms of global conservation.

The text of the Biodiversity Convention itself is analysed in the chapter by Alan Boyle. He notes that it represents a significant step forward from the existing body of international agreements in the conservation field in that it constitutes, at least in principle, an attempt to internationalise the conservation and sustainable use of nature in a more comprehensive way, based upon the biodiversity concept itself. Despite the preambular recognition of the intrinsic value of biological diversity, the Convention is plainly not a preservationist treaty – it assumes human use and benefit as the primary purpose of conservation. Equally, although it declares that the conservation of biological diversity is a common concern of humankind, it strongly reaffirms the traditional sovereignty of states over their own biological resources, and their sovereign right to exploit these resources pursuant to their own environmental policies, though their freedom in that regard is now to be circumscribed by the obligations of conservation and sustainable use which the Convention establishes. A key feature of the text lies in the way it seeks to strike a balance between these latter obligations and the urgent demands of developing states for the fair and equitable sharing of the benefits of utilization of genetic resources, access to and transfer of technology, and financial assistance to enable such countries to meet the costs of implementation. Boyle notes, however, the distinct risk that expectations regarding the anticipated benefits to developing states in these areas may prove unrealistic. He also draws attention to a number of significant weaknesses in the Convention, including the diluted provisions regarding environmental impact assessment, the perfunctory treatment of transboundary issues, the vague and heavily-qualified nature of many obligations and the potential for contradiction between them, and finds force in many of the United States' criticisms of the ultimate text. He concludes that it "will not be clear for some time whether the Convention provides a viable framework for real progress or is merely an exercise in political symbolism".

One particular ambiguity in the text of the Convention is explored in more detail in Sam Johnston's chapter on sustainable use, where he examines the concept both from a legal and an economic viewpoint. As regards the former, Johnston concludes that the notion of sustainable use has been context-driven, with different facets emphasised according to the context. For example, in fisheries conventions the focus is on setting quotas, but this approach alone is clearly not sufficient for sustainable use. The Biodiversity Convention itself may be interpreted as calling for the establishment of quotas, but equally may be seen as requiring the adoption of other techniques including preservation, a holistic ecosystem approach to management, management on the basis of biological unity, rehabilitation of denuded aspects of biodiversity, and the precautionary approach. Johnston concludes that sustainable use is a concept employed inconsistently and enigmatically in the Biodiversity Convention, without clear normative content. Absent a clear articulation of sustainable use at customary law, Johnston sees limited scope to overcome these deficiencies. At best "sustainable use as a legal principle can be considered to be [no] more than a guiding philosophy".

From an economic perspective, Johnston favours the creation of some form of internationally recognised property right in the components of biodiversity, internalising benefits and motivating proper management of biodiversity. For this reason he views the attempt legally to define the term "sustainable use" with reference to its economic meaning as a positive feature of the Convention. This may be a reflection of the fact that the concept adopted in the Convention owes its origins to the first World Conservation Strategy (1980) which described sustainable use as "analogous to spending the interest while keeping the capital".

Existing International Agreements

As many of the chapters in this book amply demonstrate, the conservation of biological diversity was the subject of a great number of international agreements prior to the conclusion of the Convention on Biological Diversity in 1992. The existence of these agreements is explicitly acknowledged both in the preamble, and in Article 22(1) of the Convention. The former refers to the desire of the Parties to "enhance and complement existing international arrangements for the conservation of biological diversity", while the latter provides that the Convention "shall not affect the rights and obligations of any Contracting Party deriving from any existing international agreement, except where the exercise of those rights and obligations would cause a serious damage or threat to biological diversity".

Three essays in particular have focused on existing international agreements, which the Biodiversity Convention is not generally intended to supersede and which will indeed continue to bear the brunt of international efforts at conservation. Existing agreements for the conservation of terrestrial species and ecosystems are reviewed in the chapter by Robin Churchill. As to the former, he notes the prevalence in these agreements of certain key conservation techniques, such as the protection of those species found within designated nature reserves or similar areas, and the protection of "listed" species wherever found. These are commonly backed by supporting measures, including the prohibition of the use of indiscriminate means of capture or the introduction of alien species. As regards ecosystem protection, the principal techniques involve the establishment of reserves or other protected areas and the adoption of conservation measures regarding particular habitat types. Supporting mechanisms here include the creation of buffer zones, the conduct of environmental impact assessments prior to the commencement of development projects and the control of pollution and other forms of environmental degradation. He observes that, while the establishment of protected areas is in principle an effective device, a great deal is commonly left to the discretion of individual parties. More generally, he detects a number of underlying weaknesses in this network of treaty arrangements, including a lack of effective machinery for monitoring and ensuring compliance and a sometimes disappointingly low level of acceptance, occasionally insufficient even to procure the treaty's entry into force. Furthermore, much effort has hitherto been devoted towards the protection of charismatic species or spectacular vistas, rather than the conservation of biological diversity as such. It is perhaps significant that the agreements he identifies as arguably the most successful in practice – those concerning polar bears and vicuna – involve protection for a single species of large mammal, and are therefore unlikely to serve as models for future efforts for the conservation of biological diversity. Indeed, he concludes that existing agreements neither address this issue squarely nor tackle the prime causes of biodiversity loss. The sectoral and/or regional coverage of many of these instruments also means that many biodiversity-rich areas are not currently subject to any form of protection at all. There is a need for these weaknesses to be confronted, and above all for the transfer of resources to developing countries to assist them in the fulfilment of their conservation obligations.

Qualitatively different from the maintenance of biodiversity in terrestrial systems is the maintenance of biodiversity in the oceans. The Convention clearly includes marine biodiversity within its ambit, and its jurisdictional scope covers the high seas as well as national territory. None the less, David Freestone points out that the Convention, with its emphasis on, *inter alia*, finance and biotechnology aspects, bypasses some of the key issues of marine biodiversity conservation. He views the "guiding agenda" of the biodiversity debate as being the issue of ownership and exploitation of biotechnology rather than conservation. In this regard, Chapter 17 of Agenda 21 on the Protection of Oceans provides more guidance on the protection of marine biodiversity than Chapter 15 on the Conservation of Biological Diversity.

The extent to which existing international law has addressed marine biodiversity conservation is revealed through Freestone's survey of existing measures taken to conserve its various components. He identifies regulation or prohibition of the taking of designated species and the protection of habitat by designation of protected areas as the two main techniques for the conservation of marine species, which in more modern treaties are increasingly used in combination (e.g. the 1990 Agreement on the Conservation of Seals in the Wadden Sea). A third, regulation of trade in endangered species, was added in 1973 under CITES to address the growth of the aquarium and aquatic curio trade. All three techniques demonstrate the influence of terrestrial approaches to the conservation of ecosystem components; less common are examples of marine ecosystem management, the notable exception being the 1980 Convention on the Regulation of Antarctic Marine Living Resources. However, Freestone observes that the practice under some of the species and habitat conservation conventions has been to adopt a *de facto* ecosystem approach. For example, the imaginative designation of corals, mangroves and sea grasses as "protected species" under the Caribbean SPAW Protocol should contribute to the protection and recovery of fragile and vulnerable ecosystems.

At the level of customary international law, Part XII of the 1982 Law of the Sea Convention (on Protection and Preservation of the Marine Environment) will assist in the crystallisation of the general obligations of States to protect the marine environment. The general obligation on States to ensure that activities within their jurisdiction or control do not cause damage to the environment of other states or of areas beyond the limits of national jurisdiction is of clear application to the marine environment within and beyond national jurisdiction, or even straddling the two. This is consistent with a more holistic view of the protection of the marine environment, evidenced in Chapter 17 of Agenda 21, for example. Further elaboration of these and other principles, such as the precautionary principle, will ensure preventive action to protect marine biological diversity is taken even in the face of scientific uncertainty.

Conventional arrangements for the protection of the Antarctic environment are considered in the chapter by Catherine Redgwell. Quite apart from the geographical isolation of the southern continent and the many unique features of its ecosystems, treaties concerning Antarctica are particularly deserving of separate treatment on the grounds that several of them exhibit the unusual characteristic of having been adopted *in advance* of the onset of the various environmental pressures they seek to regulate. Indeed, in the case of the Convention on the Regulation of Antarctic Mineral Resources it was not only concluded but effectively abandoned before the commencement of commercial mining activities. Of particular interest in the context of the present study is the 1980 Convention on the Conservation of Antarctic Marine Living Resources (CCAMLR), which was widely commended upon its adoption for pioneering an "ecosystem" approach in relation to the harvesting of marine living resources. Yet Redgwell draws attention to the very considerable practical difficulties in obtaining sufficiently reliable scientific data to enable such an approach to be effectively implemented, as well as the political and procedural problems involved in securing consensus on specific conservation measures. She also explores the relationship between the Biodiversity Convention and the treaties concerning Antarctica, based upon the terms of Article 22 of the former, identifying a number of uncertainties concerning, for example, the ambiguities inherent in the expression "law of the sea" in Article 22(2) and the application of such concepts as "national jurisdiction" in the light of the unresolved nature of sovereignty claims in Antarctica. The possibility of action, or inaction, under CCAMLR amounting to a violation of obligations under the Biodiversity

Convention is also canvassed. Her overall conclusion, however, is that there is scope for the Antarctic Treaty System to make significant contributions to the conservation of biological diversity.

Most of these existing conservation agreements address *in situ* techniques, which are recognised in the preamble to the Biodiversity Convention as a "fundamental requirement for the conservation of biological diversity". Also granted "an important role to play", and complementary to *in situ* approaches, are *ex situ* conservation measures. These form the focus of the chapter by Lynda Warren. Despite the Convention's incorporation of a definition of the term "*ex situ* conservation" she detects a number of ambiguities surrounding the concept, stemming in part from a failure to identify and distinguish the diverse motives inspiring the adoption of *ex situ* techniques, which may include conservation for its own sake, education and research and commercial exploitation. She also notes that the provisions of Article 9 involve some confusion between scientific needs for *ex situ* conservation and the question of exploitation rights regarding the use of genetic material, and that the expression of preference for *ex situ* measures to be adopted in the country of origin may prove problematic. She further traces references to the use of such measures in existing treaties, and concludes that the vital issue is to achieve the optimum blend of *in situ* and *ex situ* measures, cautioning against over-reliance on the latter. Attempts at the reintroduction of declining species require in particular to be accompanied by careful consideration of the reasons for the original decline. In general, *ex situ* conservation should be seen as part of a wider concept of restoration ecology; otherwise it may simply become a source of false hopes.

The protection, conservation and development of plant genetic resources is a topic which embraces both *in situ* and *ex situ* approaches and extends its reach into the areas of wildlife conservation, agricultural development and intellectual property protection. It has hitherto been one of the "Cinderella" subjects of international environmental law, but seems certain to be brought into much sharper relief by the Biodiversity Convention itself. In his chapter on plant genetic resources (PGRs) Greg Rose compares the general approaches manifest in such instruments as the Convention and Agenda 21 with the more narrowly focused activities of the FAO and its Commission on Plant Genetic Resources. He also considers both national and international arrangements for the establishment and maintenance of gene banks and botanical gardens, and the significance of property rights in plant genetic resources. Throughout these areas, controversy is evident regarding questions of ownership, access and control. Rose's assessment is that the Biodiversity Convention has already reinvigorated progress towards a legally binding regime for the conservation of PGRs which reflects an appropriate balance between control and access. FAO is serving as the main vehicle for this progress, particularly through its development of the notion of Farmers' Rights and the establishment of an International Fund to enhance the capacity of developing states in this area. He further suggests that the 1983 International Undertaking on Plant Genetic Resources may even be reconstituted as a Protocol to the Convention. Yet plainly certain tensions still remain, and in particular it is unclear whether the need for conservation of these vital resources has yet been fully perceived by all the many protagonists in the debate.

One theme touched on is the chapter by Rose, and explored more fully in Ian Walden's chapter, is the relevance of intellectual property rights to biodiversity conservation. The Biodiversity Convention affirms the sovereignty of States over their natural resources in Articles 3 and 15, whilst the latter further recognises the sovereign right of States to determine access to genetic resources subject to national legislation. Given that significant biodiversity is located within national borders, and largely within developing States, the Convention also addresses, *inter alia*, incentive measures for implementation. In keeping with the current vogue for market mechanisms for environmental regulation, Article 11 provides that: "Each Contracting Party shall, as far as possible and as appropriate, adopt economically and socially sound measures that act as incentives for the conservation and sustainable use of components of biological diversity." Granting States, groups or individuals property rights, particularly intellectual property rights, over natural genetic material is one method for providing an economic incentive for policies aimed at preserving biodiversity which is explored by Walden.

Intellectual property rights already play an important role in the growing biotechnology industry. Walden draws a critical distinction between property rights which protect the access to the information derived from genetic material, and property rights which may be used to regulate physical access to the material itself. Examples of the former include patents, plant breeding rights and trade secret laws, all of which are used in the biotechnology industry as a means of protecting the commercial exploitation of genetic material.

Intellectual property rights are of more limited application to the general conservation of biological diversity since they provide an economic incentive only to preserve collections of species for research and development purposes, rather than for wider conservation policies. Moreover, historically, intellectual property law has distinguished between human creations and creations of nature, with only the former entitled to protection. As a consequence, a patent will not be available to protect "plant or animal varieties or essentially biological processes for the production of plants or animals"; some degree of interference by humans with such processes will be necessary for intellectual property rights capable of protection to arise. Thus a substance freely available in nature is not patentable unless it can be isolated, characterised, and shown to be 'new'. Walden therefore considers the creation of a *sui generis* "intellectual property-style" right in *the discovery of wild (unmodified) genetic material* to surmount these difficulties, and to provide a real economic incentive for the institution of policies preserving biodiversity. Whilst clearly recognising that restrictions on access, whether through tangible or intellectual property laws, enable economic returns to be achieved, Walden cautions that the opportunity costs for the international community will need to be weighed in the balance.

Implementation and Financial Issues

A useful case study on legislative approaches to implementing the Biodiversity Convention is afforded by Kristina Gjerde's essay on biodiversity conservation in the United States. As the home of the first national park, the United States may be considered in some respects to be in the forefront of nature conservation and environmental legislation. This legislative record was one of the reasons that the avowed policy of the United States in negotiating the Convention was to ensure that it did not go beyond existing United States law. The Convention was to be one which the United States could implement through existing federal legislation, a point reinforced in the message from the President upon transmission of the Convention to Senate on 20 November 1993.

Gjerde evaluates the four major areas of federal US law which impact most upon biodiversity conservation, namely: species protection; habitat preservation; environmental impact assessment; and pollution prevention and control. From this she concludes that the unwillingness to create new legislation ignores the *lacunae* in existing US law. In particular, there is no legislative mandate for the protection of ecosystems nor to prevent species from becoming endangered in the first place; the Endangered Species Act cannot accomplish the former since it does not extend to habitat. Nor does the present pattern of protected areas, terrestrial and marine, pay sufficient regard to biodiversity goals either in the criteria for initial selection or in the context of ongoing management. The insufficient size of existing protected areas to sustain species, and the lack of representation of all important ecosystems, are particularly cogent criticisms of existing legislation on habitat protection. A further concern is the lack of inter-agency coordination on federal lands which, in Gjerde's view, could afford significant ecosystem protection. Finally, the anthropocentric focus of pollution laws, though changing, has some way to go before ecological criteria are considered to be as important as human use of resources or human health risks.

Similar concerns regarding the suitability of using existing legislation to implement the Convention may be expressed in connection with the European Community's efforts at the conservation of biological diversity, discussed in the essay by Patricia Birnie. A party to the Biodiversity Convention in its own right, the EC has a clear legislative mandate for environmental protection which was first implied from and then explicitly added to the EEC Treaty in 1986 with the Single European Act. The environmental policy of the EC has evolved since the early 1970s through framework Environment Action Programmes. The Fifth, titled "Towards Sustainability", was issued on the eve of UNCED, and includes for the first time direct reference to "biological diversity".

Despite expressing regret at Rio that the Biodiversity Convention did not contain measures of the same rigour as its own directives, the EC in fact has only two directives which relate directly to *in situ* and *ex situ* conservation, the 1979 Wild Birds Directive and the 1992 Habitats Directive. The latter was introduced to ensure effective implementation of the regional Berne Convention on the Conservation of European Wildlife and Natural Habitats and establishes a "Natura 2000 network". This is a European ecological network of special areas of conservation, the main aim of which is to promote the maintenance of biodiversity, but qualified by the permissibility of "taking account of economic, social, cultural and regional requirements". Member States have until June 1995 to indicate which species and natural habitats are suitable for designation, from which the EC Commission will draw up a list of sites of Community importance. Species and sites in danger of disappearance are the main priorities. Once a site of Community importance has been identified by the Commission, the Member State in which that site is located must designate it a special area of conservation with priorities for restoration and maintenance thereof. Effective implementation and enforcement are, Birnie notes, the key to achieving the conservation objectives of this innovatory instrument. Parallel UK practice is not encouraging in this regard, given the "minimalist approach towards implementation" of the directive. However, against this may be weighed the advantages of a Community legal system with independent powers of enforcement through the EC Commission and the European Court of Justice. There is not, however, any independent environmental inspectorate in the EC, so much reliance must therefore be placed on NGOs to monitor and report violations of Community law. In this regard other related legislation in the Community will impact on biodiversity directly or indirectly through pollution control, environmental impact assessment, and access to environmental information.

The potential for the EC to finance implementation of biodiversity conservation through its Lomé IV Agreement with ACP States is also present, as is the means of ensuring that development aid is subject to prior environmental impact assessment.

Of critical importance to any attempt to establish global measures of biodiversity conservation will be the level of the commitment displayed by developing countries. Naturally, it is not to be supposed that all developing countries will exhibit a unanimity of approach to such questions, and the chapter by Jayakumar Nayar and David Ong suggests that even within such nations there may be a considerable divergence of view between their official stance as seen through the pronouncements of governments, which tend to emphasize the importance of traditional, Western-style concepts of development, and grassroots perspectives, which may embrace a quite difference model based upon "human-centred" approaches to conservation. The chapter traces the considerable influence exerted by developing countries in the negotiations leading to the adoption of the Convention, reflected in such features as the linkage between access to genetic material and the transfer of biotechnology, research and development, but questions whether this approach pays sufficient attention to the needs of indigenous communities, which depend for their very survival on the variety of plant and animal life around them. The link between natural and cultural diversity is strongly underlined in this context. It is pointed out that the harnessing of conservation and sustainable use which the Convention envisages could be realised through a variety of paradigms, ranging from resource exploitation for traditional economic growth to human use of resources for sustainable livelihood within a broader context of human development. The fear is expressed that inadequate attention has been paid both to the interests of local communities and to the threats posed to biodiversity by traditional patterns of development, the construction of dams and other prestige projects and the perpetuation of monocultures in commercial, export-oriented agriculture. These deficiencies are in turn traced to the perspective that views the relationship between humans and the natural environment "not as an intimate subsistence-defined symbiosis, but rather as one of domination whereby biodiversity is a commodity to be exploited for profit".

The relationship between biodiversity and indigenous peoples is further explored in the chapter by John Woodliffe, who indicates how much the developed world may have to learn from such communities. Having identified the defining characteristics of such peoples, he reaffirms both their traditional familiarity with sustainable utilization and their particular vulnerability to biodiversity loss. He notes the general absence of references to indigenous peoples in conservation treaties and, in respect of those agreements where such references do appear, raises the issue of their compatibility with the Biodiversity Convention itself, pointing out that the provisions of Article 8(j) are not sufficiently precisely drafted to dispel the uncertainties in that regard. He also identifies a general policy vacuum regarding wildland biodiversity prospecting and in that context discusses both the Merck–INBio Agreement and certain proposed model terms for contracts governing the exploitation of genetic resources, which incorporate such notions as environmental impact assessment and the conservation of biodiversity, informed consent and consultation with local communities. He observes, however, that indigenous peoples are not themselves envisaged as parties to such agreements. He concludes that while the rights of indigenous populations unquestionably command a place on the international agenda, there are powerful countervailing forces in favour of Western, industrialized models of development and traditional patterns of overexploitation of natural habitat. The position is further complicated by the fact that it may be judged unrealistic or unacceptable to cast indigenous peoples in the permanent role of exemplars of the simple life in harmony with nature, and that arguably nothing can prevent their ultimate absorption into a wider, inclusive international society.

Agenda 21 calculates the cost of implementing measures necessary to conserve biodiversity at US$3.5 billion per annum, US$1.75 billion of which should be derived "from the international community on grant or concessional terms". Multilateral and bilateral aid is the focus of the final chapter by Sam Johnston. He examines both the earmarking of funds for biodiversity conservation – a difficult sum to quantify in terms of overseas development assistance since biodiversity conservation is not a separate head in accounting for expenditure – and the effect which introducing an awareness of biodiversity conservation into funding other activities, e.g. in the agricultural or transport sectors, may have because of cross-sectoral impact on biodiversity. The latter is most usefully achieved through environmental impact assessment of ODA-funded projects. Johnston graphically illustrates his point regarding cross-sectoral impacts through reference to the German Government's aid agency, BMZ, which in 1991 earmarked DM3.5 million for biodiversity while expending DM325 million in the forestry sector. If a large portion of that forestry expenditure is devoted to supporting sustainable exploitation, the contribution which such budgets may make to biodiversity conservation is potentially vast compared with the relatively small sums directly allocated.

One of the defects which Johnston identifies in the Biodiversity Convention is ambiguity in the commitment of developed States to "provide new and additional financial resources to enable developing country Parties to meet the agreed full incremental costs to them of implementing [the Convention]" (Article 20(2)). The phrase "agreed full incremental costs" lacks clarity. Similar wording in the Global Environmental Facility has been interpreted and applied with difficulty and has resulted in a project-by-project approach to allocation which does little to develop the normative content of the commitment.

Johnston examines the difficulties in restructuring the GEF to take on the role of the financial mechanism under the Convention. He concludes that the focus on the nature of the institution itself has been at the expense of both enhancement of the role that the Convention has played in developing the GEF Biodiversity Portfolio and of further articulation of "agreed full incremental costs" without which the financial mechanism of the Convention cannot become fully operational. At any rate, in the long term, private financial mechanisms will need to be developed whereby the user pays the provider directly in order to ensure that the aims of the Convention are achieved.

Possible Future Action

The 1992 Rio Earth Summit was widely described, at least in the rhetoric of politicians, as finding the international community on the threshold of a new era of international co-operation and as marking the beginnings of a fresh approach to sustainable development. Like Janus, the ancient Italian guardian of thresholds and god of all beginnings, the Biodiversity Convention, opened for signature at that conference, faces in two directions at once. Looking back upon the existing network of international conservation arrangements, it seeks to enhance and complement them by means of its own more comprehensive approach. Essentially these arrangements are left intact, except where the exercise of rights and obligations thereunder "would cause a serious damage or threat to biological diversity" within the meaning of Article 22(1). Our contributors are generally agreed that the majority of existing conservation treaties are unlikely to fall foul of this provision, although Alan Boyle points out that parties to the Biodiversity Convention may find themselves forced to reconsider the operation of fisheries or similar agreements under which excessive exploitation occurs. There may also be scope for considering whether existing treaties might be brought into still closer harmony with the Convention, for example by developing their capacity to protect and conserve diversity as such, and in particular genetic diversity within species, which has been something of a neglected area to date. It is also worthy of mention that there may well be agreements from *outside* the field of conservation – trade or development treaties, for example – which *do* pose a significant threat to biodiversity, and these may well require reconsideration in the light of Article 22(1).

At the same time the Biodiversity Convention looks to the future in envisaging the elaboration of protocols for the more detailed implementation of its objectives. Article 19(3) specifically enjoins the parties to consider the need for a protocol concerning the safe transfer, handling and use of any living modified organism resulting from biotechnology, and other possibilities are referred to in this work. David Freestone and Greg Rose, for example, discuss respectively the need for a protocol concerning the conservation of marine biodiversity and the possible reconstitution of the 1983 International Undertaking on Plant Genetic Resources as a protocol to the Convention. Some thought will plainly also need to be given to the co-ordination of efforts under the Biodiversity and Climate Change Conventions, given the interlinkage of their subject matters.

Yet it is not only at the international level that further action may be required. Much will obviously depend upon measures taken to implement the Convention at the national level. In that regard, our contributors suggest that certain parties may be adopting an unduly optimistic view if they believe that existing domestic legislation will be sufficient to enable them fully to comply with their obligations under the Convention. Furthermore, to the extent that the Biodiversity Convention is seen as a vehicle for economic development, major questions may have to be resolved regarding the nature and form that such development should take, since a pursuit of Western, industrialized models may simply constitute a perpetuation of current problems rather than the means to a solution.

Final Thoughts

It is plausible to argue that in the realms of international relations, and particularly where environmental issues are concerned, the decade is the smallest meaningful chronological unit for the measurement of progress. On that view, it would plainly be premature to come to any definite conclusions regarding the Biodiversity Convention at this stage. Nevertheless, a few provisional observations may be tentatively offered.

From an optimistic perspective, the Biodiversity Convention may be seen as something of a triumph of vision, balance and comprehensiveness. For the first time a treaty of intended global scope, which has already attracted the participation of a significant proportion of the international community, has sought to accord protection to the full range of nature's variety at each of several key levels of biological organisation. Furthermore, some recognition is accorded to all the forms of value – instrumental, inherent and intrinsic – which such diversity might be judged to possess. In addition, a balance has been sought between a range of potentially competing interests – conservation and utilization, sovereign rights and common concern, access to genetic resources and appropriate compensation for granting the same, intellectual property protection and the transfer of technology. Few would deny that these are amongst the ingredients of any viable conservation regime.

Yet at the same time there is room for an altogether more pessimistic assessment. A significant factor here is that the precise detail of many of these arrangements has still to be determined. When seen in that light, the carefully crafted compromises which the Convention establishes may rather begin to take on the appearance of the battle-lines for future wars of attrition. In that regard, it is by no means encouraging to see press reports[1] of the first meeting of the Conference of the Parties, held at Nassau in November/December 1994, entitled "Political paralysis stalls biodiversity talks" and containing vivid descriptions of the Parties' inability to agree on virtually anything. No doubt progress will be achieved in time, but this needs to be seen in the context that, as the report itself laconically concludes, "since the Rio summit, the world has lost an estimated 40,000 to 80,000 species".

Concerned observers of this process might well reflect upon how poorly the natural world has been served by its own somewhat random, territorial fragmentation into discrete political units and by the consequent bedevilment of the conservation debate by short-sighted striving for national political advantage. Perhaps the participants in this struggle would benefit from a more careful consideration of the subject matter of their discourse. This might serve to remind them that, just as nation states are at present the fundamental units of the international community, so the basic building block of the natural world is Richard Dawkins' "immortal replicator", the gene.[2] The gene, as Dawkins himself has so eloquently explained, is, above all, selfish and yet genes, he also observes, display a propensity to "gang up" into cells, which themselves "gang together".[3] The reason, as Colin Tudge has recently reminded us,[4] is that selfishness generally favours co-operation: "selection favours individual cells that join forces with others in multicellular organisms, and individuals who team up to form societies. In all cases the rogues may flourish up to a point ... but roguery is generally self-defeating".

Regrettably, there is little sign to date that political leaders have grasped this essential truth. Indeed, they all too frequently appear rather to resemble the six blind men who, in J.G. Saxe's poem,[5] went to see an elephant. Each, it will be remembered, grasped hold of or blundered into one particular part of the beast's anatomy – its trunk, tusk, knee, ear, tail or side – and thus respectively concluded that it most resembled a snake, spear, tree, fan, rope or wall. Armed with these vivid, but wholly selective and misleading, impressions, the six

> Disputed loud and long,
> Each in his own opinion
> Exceeding stiff and strong
> Though each was partly in the right
> And all were in the wrong.

---

[1] F. Pearce, "Political paralysis stalls biodiversity talks", *New Scientist*, 17 December 1994, at p. 5. See also "Fist Meeting of the Confeence of the Parties to the Rio Convention" (1995) 25/1–2 EPL 38.

[2] R. Dawkins, *The Selfish Gene* (1976; new edn. 1989).

[3] *Ibid.*, Chapter 13. See also by the same author, *The Extended Phenotype* (1982).

[4] C. Tudge, "In the beginning was the gene", a review of J. Maynard Smith and E. Szathmary, *The Major Transitions in Evolution* (1994), *New Scientist* 18 March 1995, at p. 39.

[5] J.G. Saxe, "The Blind Men and the Elephant", in D. Hall (ed.), *Oxford Book of Children's Verse in America* (1985).

Until such time as governments prove capable of grasping the true nature of the beast in question, which would seem to lie in a full and impartial acceptance of *all* the elements of the biodiversity equation, rather than those which reflect merely national advantage, effective measures for the conservation and sustainable utilization of nature will remain a pious aspiration, with the Biodiversity Convention itself, at best, an "exercise in political symbolism" and, at worst, an unintended monument to human folly.

1992 UNITED NATIONS FRAMEWORK CONVENTION ON BIOLOGICAL DIVERSITY

The Contracting Parties

Conscious of the intrinsic value of biological diversity and of the ecological, genetic, social, economic, scientific, educational, cultural, recreational and aesthetic values of biological diversity and its components,

Conscious also of the importance of biological diversity for evolution and for maintaining life sustaining systems of the biosphere,

Affirming that the conservation of biological diversity is a common concern of humankind,

Reaffirming that States have sovereign rights over their own biological resources,

Reaffirming also that States are responsible for conserving their biological diversity and for using their biological resources in a sustainable manner,

Concerned that biological diversity is being significantly reduced by certain human activities,

Aware of the general lack of information and knowledge regarding biological diversity and of the urgent need to develop scientific, technical and institutional capacities to provide the basic understanding upon which to plan and implement appropriate measures,

Noting that it is vital to anticipate, prevent and attack the causes of significant reduction or loss of biological diversity at source,

Noting also that where there is a threat of significant reduction or loss of biological diversity, lack of full scientific certainty should not be used as a reason for postponing measures to avoid or minimize such a threat,

Noting further that the fundamental requirement for the conservation of biological diversity is the *in-situ* conservation of ecosystems and natural habitats and the maintenance and recovery of viable populations of species in their natural surroundings,

Noting further that *ex-situ* measures, preferably in the country of origin, also have an important role to play,

Recognizing the close and traditional dependence of many indigenous and local communities embodying traditional lifestyles on biological resources, and the desirability of sharing equitably benefits arising from the use of traditional knowledge, innovations and practices relevant to the conservation of biological diversity and the sustainable use of its components,

Recognizing also the vital role that women play in the conservation and sustainable use of biological diversity and affirming the need for the full participation of women at all levels of policy-making and implementation for biological diversity conservation,

Stressing the importance of, and the need to promote, international, regional and global cooperation among States and intergovernmental organizations and the non-governmental sector for the conservation of biological diversity and the sustainable use of its components,

Acknowledging that the provision of new and additional financial resources and appropriate access to relevant technologies can be expected to make a substantial difference in the world's ability to address the loss of biological diversity,

Acknowledging further that special provision is required to meet the needs of developing countries, including the provision of new and additional financial resources, and appropriate access to relevant technologies,

Noting in this regard the special conditions of the least developed countries and small island States,

Acknowledging that substantial investments are required to conserve biological diversity and that there is the expectation of a broad range of environmental, economic and social benefits from those investments,

Recognizing that economic and social development and poverty eradication are the first and overriding priorities of developing countries.

Aware that conservation and sustainable use of biological diversity is of critical importance for meeting the food, health and other needs of the growing world population, for which purpose access to and sharing of both genetic resources and technologies are essential,

Noting that, ultimately, the conservation and sustainable use of biological diversity will strengthen friendly relations among States and contribute to peace for humankind,

Desiring to enhance and complement existing international arrangements for the conservation of biological diversity and sustainable use of its components and

Determined to conserve and sustainably use biological diversity for the benefit of present and future generations,

Have agreed as follows:

Article 1

OBJECTIVES

The objectives of this Convention, to be pursued in accordance with its relevant provisions, are the conservation of biological diversity, the sustainable use of its components and the fair and equitable sharing of the benefits arising out of the utilization of genetic resources, including by appropriate access to genetic resources and by appropriate transfer of relevant technologies, taking into account all rights over those resources and to technologies, and by appropriate funding.

Article 2

USE OF TERMS

For the purposes of this Convention:

"Biological diversity" means the variability among living organisms from all sources including, *inter alia,* terrestrial, marine and other aquatic ecosystems and the ecological complexes

of which they are part: this includes diversity within species, between species and of ecosystems.

"Biological resources" includes genetic resources, organisms of parts thereof, populations, or any other biotic component of ecosystems with actual or potential use or value for humanity.

Biotechnology" means any technological application that uses biological systems, living organisms, or derivatives thereof, to make or modify products or processes for specific use.

"Country of origin of genetic resources" means the country which possesses those genetic resources in *in-situ* conditions.

"County providing genetic resources" means the country supplying genetic resources collected from *in-situ* resources, including populations of both wild and domesticated species, or taken from *ex-situ* sources, which may or may not have originated in that country.

"Domesticated or cultivated species" means species in which the evolutionary process has been influenced by humans to meet their needs.

"Ecosystem" means a dynamic complex of plant, animal and micro-organism communities and their non-living environment interacting as a functional unit.

"Ex-situ conservation" means the conservation of components of biological diversity outside their natural habitats.

"Genetic material" means any material of plant, animal, microbial or other origin containing functional units of heredity.

"Genetic resources" means genetic material of actual or potential value.

"Habitat" means the place or type of site where an organism or population naturally occurs.

"In-situ conditions" means conditions where genetic resources exist within ecosystems and natural habitats, and, in the case of domesticated or cultivated species, in the surroundings where they have developed their distinctive properties.

"In-situ conservation" means the conservation of ecosystems and natural habitats and the maintenance and recovery of viable populations of species in their natural surroundings and, in the case of domesticated or cultivated species, in the surroundings where they have developed their distinctive properties.

"Protected area" means a geographically defined area which is designated or regulated and managed to achieve specific conservation objectives.

"Regional economic integration organization" means an organization constituted by sovereign States of a given region, to which its member States have transferred competence in respect of matters governed by this Convention and which has been duly authorized , in accordance with its internal procedures, to sign, ratify, accept, approve or accede to it.

"Sustainable use" means the use of components of biological diversity in a way and at a rate that does not lead to the long-term decline of biological diversity, thereby maintaining its potential to meet the needs and aspirations of present and future generations.

"Technology" includes biotechnology.

Article 3

PRINCIPLE

States have, in accordance with the Charter of the United Nations and the principles of international law, the sovereign right to exploit their own resources pursuant to their own environmental policies, and the responsibility to ensure that activities within their jurisdiction or control do not cause damage to the environment of other States or of areas beyond the limits of national jurisdiction.

Article 4

JURISDICTIONAL SCOPE

Subject to the rights of other States, and except as otherwise expressly provided in this Convention, the provisions of this Convention apply, in relation to each Contracting Party:

(a) In the case of components of biological diversity, in areas within the limits of its national jurisdiction; and
(b) In the case of processes and activities, regardless of where their effects occur, carried out under its jurisdiction or control, within the area of its national jurisdiction or beyond the limits of national jurisdiction.

Article 5

COOPERATION

Each Contracting Party shall, as far as possible and as appropriate, cooperate with other Contracting Parties, directly or, where appropriate, through competent international organizations, in respect of areas beyond national jurisdiction and on other matters of mutual interest, for the conservation and sustainable use of biological diversity.

Article 6

GENERAL MEASURES FOR CONSERVATION AND SUSTAINABLE USE

Each Contracting Party shall, in accordance with its particular conditions and capabilities:

(a) Develop national strategies, plans or programmes for the conservation and sustainable use of biological diversity or adapt for this purpose existing strategies, plans or programmes which shall reflect, *inter alia*, the measures set out in this Convention relevant to the Contracting Party concerned; and
(b) Integrate, as far as possible and as appropriate, the conservation and sustainable use of biological diversity into relevant sectoral or cross-sectoral plans, programmes and policies.

Article 7

IDENTIFICATION AND MONITORING

Each Contracting Party shall, as far as possible and as appropriate, in particular for the purposes of Articles 8 to 10:

(a) Identify components of biological diversity important for its conservation and sustainable use having regard to the indicative list of categories set down in Annex I;
(b) Monitor, through sampling and other techniques, the components of biological diversity identified pursuant to subparagraph (a) above, paying particular attention to those requiring urgent conservation measures and those which offer the greatest potential for sustainable use;
(c) Identify processes and categories of activities which have or are likely to have significant adverse impacts on the conservation and sustainable use of biological diversity, and monitor their effects through sampling and other techniques; and
(d) Maintain and organize, by any mechanism data, derived from identification and monitoring activities pursuant to subparagraphs (a), (b) and (c) above.

Article 8

IN-SITU CONSERVATION

Each Contracting Party shall, as far as possible and as appropriate:

(a) Establish a system of protected areas or areas where special measures need to be taken to conserve biological diversity;
(b) Develop, where necessary, guidelines for the selection, establishment and management of protected areas or areas where special measures need to be taken to conserve biological diversity;
(c) Regulate or manage biological resources important for the conservation of biological diversity whether within or outside protected areas, with a view to ensuring their conservation and sustainable use;
(d) Promote the protection of ecosystems, natural habitats and the maintenance of viable populations in natural surroundings;
(e) Promote environmentally sound and sustainable development in areas adjacent to protected areas with a view to furthering protection of these areas;
(f) Rehabilitate and restore degraded ecosystems and promote the recovery of threatened species *inter alia* through the development and implementation of plans or other management strategies;
(g) Establish or maintain means to regulate, manage or control the risks associated with the use and release of living modified organisms resulting from biotechnology which are likely to have adverse environmental impacts that could affect the conservation and sustainable use of biological diversity, taking also into account the risks to human health;
(h) Prevent the introduction of, control or eradicate those alien species which threaten ecosystems, habitats or species;
(i) Endeavour to provide the conditions needed for compatibility between present uses and the conservation of biological diversity and the sustainable use of its components;
(j) Subject to its national legislation, respect, preserve and maintain knowledge, innovations and practices of indigenous and local communities embodying traditional lifestyles relevant for the conservation and sustainable use of biological diversity and promote their wider application with the approval and involvement of the holders of such knowledge, innovations and practices and encourage the equitable sharing of the benefits arising from the utilization of such knowledge, innovations and practices;
(k) Develop or maintain necessary legislation and/or other regulatory provisions for the protection of threatened species and populations;
(l) Where a significant adverse effect on biological diversity has been determined pursuant to Article 7, regulate or manage the relevant processes and categories of activities; and
(m) Cooperate in providing financial and other support for *in-situ* conservation outlined in subparagraphs (a) to (l) above, particularly to developing countries.

Article 9

EX-SITU CONSERVATION

Each Contracting Party shall, as far as possible and as appropriate, and predominantly for the purpose of complementing *in-situ* measures:

(a) Adopt measures for the *ex-situ* conservation of components of biological diversity, preferably in the country of origin of such components;

(b) Establish and maintain facilities for *ex-situ* conservation of and research on plants, animals and micro-organisms, preferably in the country of origin of genetic resources;

(c) Adopt measures for the recovery and rehabilitation of threatened species and for their reintroduction into their natural habitats under appropriate conditions;

(d) Regulate and manage collection of biological resources from natural habitats for *ex-situ* conservation purposes so as not to threaten ecosystems and *in-situ* populations of species, except where special temporary *ex-situ* measures are required under subparagraph (c) above; and

(e) Cooperate in providing financial and other support for *ex-situ* conservation outlined in subparagraphs (a) to (d) above and in the establishment and maintenance of *ex-situ* conservation facilities in developing countries.

Article 10

SUSTAINABLE USE OF COMPONENTS OF BIOLOGICAL DIVERSITY

Each Contracting Party shall, as far as possible and as appropriate:

(a) Integrate consideration of the conservation and sustainable use of biological resources into national decision-making;

(b) Adopt measures relating to the use of biological resources to avoid or minimize adverse impacts on biological diversity;

(c) Protect and encourage customary use of biological resources in accordance with traditional cultural practices that are compatible with conservation or sustainable use requirements;

(d) Support local populations to develop and implement remedial action in degraded areas where biological biodiversity has been reduced; and

(e) Encourage cooperation between its governmental authorities and its private sector in developing methods for sustainable use of biological resources.

Article 11

INCENTIVE MEASURES

Each Contracting Party shall, as far as possible and as appropriate, adopt economically and socially sound measures that act as incentives for the conservation and sustainable use of components of biological diversity.

Article 12

RESEARCH AND TRAINING

The Contracting Parties, taking into account the special needs of developing countries, shall:

(a) Establish and maintain programmes for scientific and technical education and training in measures for the identification, conservation and sustainable use of biological diversity and its components and provide support for such education and training for the specific needs of developing countries;
(b) Promote and encourage research which contributes to the conservation and sustainable use of biological diversity, particularly in developing countries, *inter alia*, in accordance with decisions of the Conference of the Parties taken in consequence of recommendations of the Subsidiary Body on Scientific, Technical and Technological Advice; and
(c) In keeping with the provisions of Articles 16, 18 and 20, promote and cooperate in the use of scientific advances in biological diversity research in developing methods for conservation and sustainable use of biological resources.

Article 13

PUBLIC EDUCATION AND AWARENESS

Each Contracting Party shall:

(a) Promote and encourage understanding of the importance of, and the measures required for, the conservation of biological diversity, as well as its propagation through media and the inclusion of these topics in educational programmes; and
(b) Cooperate, as appropriate, with other States and international organizations in developing educational and public awareness programmes, with respect to conservation and sustainable use of biological diversity.

Article 14

IMPACT ASSESSMENT AND MINIMIZING ADVERSE IMPACT

1   Each Contracting Party, as far as possible and as appropriate, shall:

(a) Introduce appropriate procedures requiring environmental impact assessment of its proposed projects that are likely to have significant adverse effects on biological diversity with a view to avoiding or minimizing such effects and, where appropriate, allow for public participation in such procedures;
(b) Introduce appropriate arrangements to ensure that the environmental consequences of its programmes and policies that are likely to have significant adverse impacts on biological diversity are duly taken into account;
(c) Promote, on the basis of reciprocity, notification, exchange of information and consultation on activities under their jurisdiction or control which are likely to significantly affect adversely the biological diversity of other States or areas beyond the limits of national jurisdiction by encouraging the conclusion of bilateral or multilateral arrangements, as appropriate;
(d) In the case of imminent or grave danger or damage, originating under its jurisdiction or control, to biological diversity within the area under jurisdiction of other States or in areas beyond the limits of national jurisdiction, notify immediately the potentially affected States of such danger or damage, as well as initiate action to prevent or minimize such danger or damage; and
(e) Promote national arrangements for emergency responses to activities or events, whether caused naturally or otherwise, which present a grave and imminent danger to biological diversity and encourage international cooperation to supplement such national efforts and, where appropriate and agreed by the States or regional economic integration organizations concerned, to establish joint contingency plans.

2   The Conference of the Parties shall examine, on the basis of studies to be carried out, the issue of liability and redress, including restoration and compensation, for damage to biological diversity, except where such liability is a purely internal matter.

Article 15

ACCESS TO GENETIC RESOURCES

1   Recognizing the sovereign rights of States over their natural resources, the authority to determine access to genetic resources rests with the national governments and is subject to national legislation.

2   Each Contracting Party shall endeavour to create conditions to facilitate access to genetic resources for environmentally sound uses by other Contracting Parties and not to impose restrictions that run counter to the objectives of this Convention.

3   For the purpose of this Convention, the genetic resources being provided by a Contracting Party, as referred to in this Article and Articles 16 and 19, are only those that are provided by Contracting Parties that are countries of origin of such resources or by the Parties that have acquired the genetic resources in accordance with this Convention.

4   Access, where granted, shall be on mutually agreed terms, and subject to the provisions of this Article.

5   Access to genetic resources shall be subject to prior informed consent of the Contracting Party providing such resources unless otherwise determined by that Party.

6   Each Contracting Party shall endeavour to develop and carry out scientific research based on genetic resources provided by other Contracting Parties with the full participation of, and where possible in, such Contracting Parties.

7   Each Contracting Party shall take legislative, administrative or policy measures, as appropriate, and in accordance with Article 16 and 19 and, where necessary, through the financial mechanism established by Articles 20 and 21 with the aim of sharing in a fair and equitable way the results of research and development and the benefits arising from the commercial and other utilization of genetic resources with the Contracting Party providing such resources. Such sharing shall be upon mutually agreed terms.

Article 16

ACCESS TO AND TRANSFER OF TECHNOLOGY

1   Each Contracting Party, recognizing that technology includes biotechnology, and that both access to and transfer of technology among Contracting Parties are essential elements for the attainment of the objectives of this Convention, undertakes subject to the provisions of this Article to provide and/or facilitate access for and transfer to other Contracting Parties of technologies that are relevant to the conservation and sustainable use of biological diversity or make use of genetic resources and do not cause significant damage to the environment.

2   Access to and transfer of technology referred to in paragraph 1 above to developing countries shall be provided and/or facilitated under fair and most favourable terms, including on concessional and preferential terms where mutually agreed and, where necessary, in accordance with the financial mechanism established by Articles 20 and 21. In the case of technology subject to patents and other intellectual property rights, such access and transfer shall be provided on terms which recognize and are consistent with the adequate and effective protection of intellectual property rights. The application of this paragraph shall be consistent with paragraphs 3, 4 and 5 below.

3   Each Contracting Party shall take legislative, administrative or policy measures, as appropriate, with the aim that Contracting Parties, in particular those that are developing countries, which provide genetic resources are provided access to and transfer of technology which makes use of those resources, on mutually agreed terms, including technology protected by patents and other intellectual property rights, where necessary through the provisions of Articles 20 and 21 and in accordance with international law and consistent with paragraphs 4 and 5 below.

4   Each Contracting Party shall take legislative, administrative or policy measures, as appropriate, with the aim that the private sector facilitates access to, joint development and transfer of technology referred to in paragraph 1 above for the benefit of both governmental institutions and the private sector of developing countries and in this regard shall abide by the obligations included in paragraphs 1, 2 and 3 above.

5   The Contracting Parties, recognizing that patents and other intellectual property rights may have an influence on the implementation of this convention, shall cooperate in this regard subject to national legislation and international law in order to ensure that such rights are supportive of and do not run counter to its objectives.

Article 17

EXCHANGE OF INFORMATION

1   The Contracting Parties shall facilitate the exchange of information, from all publicly available sources, relevant to the conservation and sustainable use of biological diversity, taking into account the special needs of developing countries.

2   Such exchange of information shall include exchange of results of technical, scientific and socio-economic research, as well as information on training and surveying programmes, specialized knowledge, indigenous and traditional knowledge as such and in combination with the technologies referred to in Article 16, paragraph 1. It shall also, where feasible include repatriation of information.

Article 18

TECHNICAL AND SCIENTIFIC COOPERATION

1   The Contracting Parties shall promote international technical and scientific cooperation in the field of conservation and sustainable use of biological diversity where necessary, through the appropriate international and national institutions.

2   Each Contracting Party shall promote technical and scientific cooperation with other Contracting Parties, in particular developing countries, in implementing this Convention *inter alia* through the development and implementation of national policies. In promoting such cooperation, special attention should be given to the development and strengthening of national capabilities, by means of human resources development and institution building.

3   The Conference of the Parties, at its first meeting, shall determine how to establish a clearing-house mechanism to promote and facilitate technical and scientific cooperation.

4   The Contracting Parties shall, in accordance with national legislation and policies, encourage and develop methods of cooperation for the development and use of technologies, including indigenous and traditional technologies, in pursuance of the objectives of this Convention. For this purpose, the Contracting Parties shall also promote cooperation in the training of personnel and exchange of experts.

5   The Contracting Parties shall, subject to mutual agreement, promote the establishment of joint research programmes and joint ventures for the development of technologies relevant to the objectives of this Convention.

Article 19

HANDLING OF BIOTECHNOLOGY AND DISTRIBUTION OF ITS BENEFITS

1 Each Contracting Party shall take legislative, administrative or policy measures, as appropriate, to provide for the effective participation in biotechnological research activities by those Contracting Parties, especially developing countries, which provide the genetic resources for such research, and where feasible in such Contracting Parties.

2 Each Contracting Party shall take all practicable measures to promote and advance priority access on a fair and equitable basis by Contracting Parties, especially developing countries, to the results and benefits arising from biotechnologies based upon genetic resources provided by those Contracting Parties. Such access shall be on mutually agreed terms.

3 The Parties shall consider the need for and modalities of a protocol setting out appropriate procedures, including, in particular, advance informed agreement, in the field of the safe transfer, handling and use of any living modified organism resulting from biotechnology that may have adverse effect on the conservation and sustainable use of biological diversity.

4 Each Contracting Party shall, directly or by requiring any natural or legal person under its jurisdiction providing the organisms referred to in paragraph 3 above, provide any available information about the use and safety regulations required by that Contracting Party in handling such organisms, as well as any available information on the potential adverse impact of the specific organisms concerned to the Contracting Party into which those organisms are to be introduced.

Article 20

FINANCIAL RESOURCES

1 Each Contracting Party undertakes to provide, in accordance with it capabilities, financial support and incentives in respect of those national activities which are intended to achieve the objectives of this Convention, in accordance with its national plans, priorities and programmes.

2 The developed country Parties shall provide new and additional financial resources to enable developing country Parties to meet the agreed full incremental costs to them of implementing measures which fulfil the obligations of this Convention and to benefit from its provisions and which costs are agreed between a developing country Party and the institutional structure referred to in Article 21, in accordance with policy, strategy, programme priorities and eligibility criteria and an indicative list of incremental costs established by the Conference of the Parties. Other Parties, including countries undergoing the process of transition to a market economy, may voluntarily assume the obligations of the developed country Parties. For the purpose of this Article, the Conference of the Parties, shall at its first meeting establish a list of developed country Parties and other Parties which voluntarily assume the obligations of the developed country Parties. The Conference of the Parties shall periodically review and if necessary amend the list. Contributions from other countries and sources on a voluntary basis would also be encouraged. The implementation of these commitments shall take into account the need for adequacy, predictability and timely flow of funds and the importance of burden-sharing among the contributing Parties included in the list.

3 The developed country Parties may also provide, and developing country Parties avail themselves of, financial resources related to the implementation of this Convention through bilateral, regional and other multilateral channels.

4 The extent to which developing country Parties will effectively implement their commitments under this Convention will depend on the effective implementation by developed country Parties of their commitments under this Convention related to financial resources

and transfer of technology and will take fully into account the fact that economic and social development and eradication of poverty are the first and overriding priorities of the developing country Parties.

5   The Parties shall take full account of the specific needs and special situation of least developed countries in their actions with regard to funding and transfer of technology.

6   The Contracting Parties shall also take into consideration the special conditions resulting from the dependence on, distribution and location, biological diversity within developing country Parties, in particular small island States.

7   Consideration shall also be given to the special situation of developing countries, including those that are most environmentally vulnerable, such as those with arid and semi-arid zones, coastal and mountainous areas.

Article 21

FINANCIAL MECHANISM

1   There shall be a mechanism for the provision of financial resources to developing country Parties for purposes of this Convention on a grant or concessional basis the essential elements of which are described in this Article. The mechanism shall function under the authority and guidance of, and be accountable to, the Conference of the Parties for purposes of this Convention. The operations of the mechanism shall be carried out by such institutional structure as may be decided upon by the Conference of the Parties at its first meeting. For purposes of this Convention, the Conference of the Parties shall determine the policy, strategy, programme priorities and eligibility criteria relating to the access to and utilization of such resources. The contributions shall be such as to take into account the need for predictability, adequacy and timely flow of funds referred to in Article 20 in accordance with the amount of resources needed to be decided periodically by the Conference of the Parties and the importance of burden-sharing among the contributing Parties included in the list referred to in Article 20, paragraph 2. Voluntary contributions may also be made by the developed country Parties and by other countries and sources. The mechanism shall operate within a democratic and transparent system of governance.

2   Pursuant to the objectives of this Convention, the Conference of the Parties shall at its first meeting determine the policy, strategy and programme priorities, as well as detailed criteria and guidelines for eligibility for access to and utilization of the financial resources including monitoring and evaluation on a regular basis of such utilization. The Conference of the Parties shall decide on the arrangements to give effect to paragraph 1 above after consultation with the institutional structure entrusted with the operation of the financial mechanism.

3   The Conference of the Parties shall review the effectiveness of the mechanism established under this Article, including the criteria and guidelines referred to in paragraph 2 above, not less than two years after the entry into force of this Convention and thereafter on a regular basis. Based on such review, it shall take appropriate action to improve the effectiveness of the mechanism if necessary.

4   The Contracting Parties shall consider strengthening existing financial institutions to provide financial resources for the conservation and sustainable use of biological diversity.

Article 22

RELATIONSHIP WITH OTHER INTERNATIONAL CONVENTIONS

1   The provisions of this Convention shall not affect the rights and obligations of any Contracting Party deriving from any existing international agreement except where the exercise of those rights and obligations would cause a serious damage or threat to biological diversity.

2   Contracting Parties shall implement this Convention with respect to the marine environment consistently with the rights and obligations of States under the law of the sea.

Article 23

CONFERENCE OF THE PARTIES

1   A Conference of the Parties is hereby established. The first meeting of the Conference of the Parties shall be convened by the Executive Director of the United Nations Environment Programme not later than one year after the entry into force of this Convention. Thereafter, ordinary meetings of the Conference of the Parties shall be held at regular intervals to be determined by the Conference at its first meeting.

2   Extraordinary meetings of the Conference of the Parties shall be held at such other times as may be deemed necessary by the Conference, or at the written request of any Party, provided that, within six months of the request being communicated to them by the Secretariat, it is supported by at least one third of the Parties.

3   The Conference of the Parties shall by consensus agree upon and adopt rules of procedure for itself and for any subsidiary body it may establish, as well as financial rules governing the funding of the Secretariat. At each ordinary meeting, it shall adopt a budget for the financial period until the next ordinary meeting.

4   The Conference of the Parties shall keep under review the implementation of this Convention, and, for this purpose, shall:

(a) Establish the form and the intervals for transmitting the information be submitted in accordance with Article 26 and consider such information as well as reports submitted by any subsidiary body;

(b) Review scientific, technical and technological advice on biological diversity provided in accordance with Article 25;

(c) Consider and adopt, as required, protocols in accordance with Article 28;

(d) Consider and adopt, as required, in accordance with Articles 29 and 30, amendments to this Convention and its annexes;

(e) Consider amendments to any protocol, as well as to any annexes thereto, and, if so decided, recommend their adoption to the parties to the protocol concerned;

(f) Consider and adopt, as required, in accordance with Article 30, additional annexes to this Convention;

(g) Establish such subsidiary bodies, particularly to provide scientific and technical advice, as are deemed necessary for the implementation of this Convention;

(h) Contract, through the Secretariat, the executive bodies of conventions dealing with matters covered by this Convention with a view to establishing appropriate forms of cooperation with them; and

(i) Consider and undertake any additional action that may be required for the achievement of the purposes of this Convention in the light of experience gained in its operation.

5   The United Nations, its specialized agencies and the International Atomic Energy Agency, as well as any State not party to this Convention, may be represented as observers at meetings of the Conference of the Parties. Any other body or agency, whether governmental or non-governmental, qualified in fields relating to conservation and sustainable use of biological diversity, which has informed the Secretariat of its wish to be represented as an observer at a meeting of the Conference of the Parties, may be admitted unless at least one third of the Parties present object. The admission and participation of observers shall be subject to the rules of procedure adopted by the Conference of the Parties.

Article 24

SECRETARIAT

1   A Secretariat is hereby established. Its functions shall be:

(a) To arrange for and service meetings of the Conference of the Parties provided for in Article 23;
(b) To perform the functions assigned to it by any protocol;
(c) To prepare reports on the execution of its functions under this Convention and present them to the Conference of the Parties;
(d) To coordinate with other relevant international bodies and, in particular to enter into such administrative and contractual arrangements as may be required for the effective discharge of its functions; and
(e) To perform such other functions as may be determined by the Conference of the Parties.

2   At its first ordinary meeting, the Conference of the Parties shall designate the secretariat from amongst those existing competent international organizations which have signified their willingness to carry out the secretariat functions under this Convention.

Article 25

SUBSIDIARY BODY ON SCIENTIFIC, TECHNICAL AND TECHNOLOGICAL ADVICE

1   A subsidiary body for the provision of scientific, technical and technological advice is hereby established to provide the Conference of the Parties and, as appropriate, its other subsidiary bodies with timely advice relating to the implementation of this Convention. This body shall be open to participation by all Parties and shall be multidisciplinary. It shall comprise government representatives competent in the relevant field of expertise. It shall report regularly to the Conference of the Parties on all aspects of its work.

2   Under the authority of and in accordance with guidelines laid down by the Conference of the Parties, and upon its request, this body shall:

(a) Provide scientific and technical assessments of the status of biological diversity;
(b) Prepare scientific and technical assessments of the effects of types of measures taken in accordance with the provisions of this Convention;
(c) Identify innovative, efficient and state-of-the-art technologies and know-how relating to the conservation and sustainable use of biological diversity and advise on the ways and means of promoting development and/or transferring such technologies;
(d) Provide advice on scientific programmes and international cooperation in research and development related to conservation and sustainable use of biological diversity; and
(e) Respond to scientific, technical, technological and methodological questions that the Conference of the Parties and its subsidiary bodies may put to the body.

3   The functions, terms of reference, organization and operation of this body may be further elaborated by the Conference of the Parties.

Article 26

REPORTS

Each Contracting Party shall, at intervals to be determined by the Conference of the Parties, present to the Conference of the Parties, reports on measures which it has taken for the implementation of the provisions of this Convention and their effectiveness in meeting the objectives of this Convention.

Article 27

SETTLEMENT OF DISPUTES

1   In the event of a dispute between Contracting Parties concerning the interpretation or application of this Convention, the parties concerned shall seek solution by negotiation.
2   If the parties concerned cannot reach agreement by negotiation, they may jointly seek the good offices of, or request mediation by, a third party.
3   When ratifying, accepting, approving or acceding to this Convention, or at any time thereafter, a State or regional economic integration organization may declare in writing to the Depositary that for a dispute not resolved in accordance with paragraph 1 or paragraph 2 above, it accepts one or both of the following means of dispute settlement as compulsory:

(a)   Arbitration in accordance with the procedure laid down in Part 1 of Annex II;
(b)   Submission of the dispute to the International Court of Justice.

4   If the parties to the dispute have not, in accordance with paragraph 3 above, accepted the same or any procedure, the dispute shall be submitted to conciliation in accordance with Part 2 of Annex II unless the parties otherwise agree.
5   The provisions of this Article shall apply with respect to any protocol except as otherwise provided in the protocol concerned.

Article 28

ADOPTION OF PROTOCOLS

1   The Contracting Parties shall cooperate in the formulation and adoption of protocols to this Convention.
2   Protocols shall be adopted at a meeting of the Conference of the Parties.
3   The text of any proposed protocol shall be communicated to the Contracting Parties by the Secretariat at least six months before such a meeting.

Article 29

AMENDMENT OF THE CONVENTION OR PROTOCOLS

1   Amendments to this Convention may be proposed by any Contracting Party. Amendments to any protocol may be proposed by any Party to that protocol.
2   Amendments to this Convention shall be adopted at a meeting of the Conference of the Parties. Amendments to any protocol shall be adopted at a meeting of the Parties to the Protocol in question. The text of any proposed amendment to this Convention or to any protocol, except as may otherwise be provided in such protocol, shall be communicated to the Parties to the instrument in question by the secretariat at least six months before the meeting at which it is proposed for adoption. The secretariat shall also communicate proposed amendments to the signatories to this Convention for information.
3   The Parties shall make every effort to reach agreement on any proposed amendment to this Convention or to any protocol by consensus. If all efforts at consensus have been exhausted, and no agreement reached, the amendment shall as a last resort be adopted by a two-third majority vote of the Parties to the instrument in question present and voting at the meeting, and shall be submitted by the Depositary to all Parties for ratification, acceptance or approval.

4   Ratification, acceptance or approval of amendments shall be notified to the Depositary in writing. Amendments adopted in accordance with paragraph 3 above shall enter into force among Parties having accepted them on the ninetieth day after the deposit of instruments of ratification, acceptance or approval by at least two thirds of the Contracting Parties to this Convention or of the Parties to the protocol concerned, except as may otherwise be provided in such protocol. Thereafter the amendments shall enter into force for any other Party on the ninetieth day after that Party deposits its instrument of ratification, acceptance or approval of the amendments.

5   For the purposes of this Article "Parties present and voting" means Parties present and casting an affirmative or negative vote.

Article 30

ADOPTION AND AMENDMENT OF ANNEXES

1   The annexes to this Convention or to any protocol shall form an integral part of the Convention or of such protocol, as the case may be, and, unless expressly provided otherwise, a reference to this Convention or its protocols constitutes at the same time a reference to any annexes thereto. Such annexes shall be restricted to procedural, scientific, technical and administrative matters.

2   Except as may be otherwise provided in any protocol with respect to its annexes, the following procedure shall apply to the proposal, adoption and entry into force of additional annexes to this Convention or of annexes to any protocol:

(a) Annexes to this Convention or to any protocol shall be proposed and adopted according to the procedure laid down in Article 29;

(b) Any Party that is unable to approve an additional annex to this Convention or an annex to any protocol to which it is Party shall so notify the Depositary, in writing, within one year from the date of the communication of the adoption by the Depositary. The Depositary shall without delay notify all Parties of any such notification received. A Party may at any time withdraw a previous declaration of objection and the annexes shall thereupon enter into force for that Party subject to subparagraph (c) below;

(c) On the expiry of one year from the date of the communication of the adoption by the Depositary, the annex shall enter into force for all Parties to this Convention or to any protocol concerned which have not submitted a notification in accordance with the provisions of subparagraph (b) above.

3   The proposal, adoption and entry into force of amendments to annexes to this Convention or to any protocol shall be subject to the same procedure as for the proposal, adoption and entry into force of annexes to the Convention or annexes to any protocol.

4   If an additional annex or an amendment to an annex is related to an amendment to this Convention or to any protocol, the additional annex or amendment shall not enter into force until such time as the amendment to the Convention or to the protocol concerned enters into force.

Article 31

RIGHT TO VOTE

1   Except as provided for in paragraph 2 below, each Contracting Party to this Convention or to any protocol shall have one vote.

2   Regional economic integration organizations, in matters within their competence, shall exercise their right to vote with a number of votes equal to the number of their member States which are Contracting Parties to this Convention or the relevant protocol. Such

organizations shall not exercise their right to vote if their member States exercise theirs, and vice versa.

Article 32

RELATIONSHIP BETWEEN THIS CONVENTION AND ITS PROTOCOLS

1   A State or a regional economic integration organization may not become a Party to a protocol unless it is, or becomes at the same time, a Contracting Party to this Convention.

2   Decisions under any protocol shall be taken only by the Parties to the protocol concerned. Any Contracting Party that has not ratified, accepted or approved a protocol may participate as an observer in any meeting of the parties to that protocol.

Article 33

SIGNATURE

This Convention shall be open for signature at Rio de Janeiro by all States and any regional economic integration organization from 5 June 1992 until 14 June 1992, and at the United Nations Headquarters in New York from 15 June 1992 to 4 June 1993.

Article 34

RATIFICATION, ACCEPTANCE OR APPROVAL

1   This Convention and any protocol shall be subject to ratification, acceptance or approval by States and by regional economic integration organizations. Instruments of ratification, acceptance or approval shall be deposited with the Depositary.

2   Any organization referred to in paragraph 1 above which becomes a Contracting Party to this convention or any protocol without any of its member States being a Contracting Party shall be bound by all the obligations under the Convention or the protocol, as the case may be. In the case of such organizations, one or more of whose member States is a Contracting Party to this Convention or relevant protocol, the organization and its member States shall decide on their respective responsibilities for the performance of their obligations under the Convention or protocol, as the case may be. In such cases, the organization and the member States shall not be entitled to exercise rights under the Convention or relevant protocol concurrently.

3   In their instruments of ratification, acceptance or approval, the organizations referred to in paragraph 1 above shall declare the extent of their competence with respect to the matters governed by the Convention or the relevant protocol. These organizations shall also inform the Depositary of any relevant modification in the extent of their competence.

Article 35

ACCESSION

1   This Convention and any protocol shall be open for accession by States and by regional economic integration organizations from the date on which the Convention or the protocol concerned is closed for signature. The instruments of accession shall be deposited with the Depositary.

2   In their instruments of accession, the organizations referred to in paragraph 1 above shall declare the extent of their competence with respect to the matters governed by the Convention or the relevant protocol. These organizations shall also inform the Depositary of any relevant modification in the extent of their competence.

3   The provisions of Article 34, paragraph 2, shall apply to regional economic integration organizations which accede to this Convention or any protocol.

Article 36

ENTRY INTO FORCE

1   This Convention shall enter into force on the ninetieth day after the date of deposit of the thirtieth instrument of ratification, acceptance, approval or accession.
2   Any protocol shall enter into force on the ninetieth day after the date of deposit of the number of instruments of ratification, acceptance, approval or accession, specified in that protocol, has been deposited.
3   For each Contracting Party, which ratifies, accepts or approves this Convention or accedes thereto after the deposit of the thirtieth instrument of ratification, acceptance, approval or accession, it shall enter into force on the ninetieth day after the date of deposit by such Contracting Party of its instrument of ratification, acceptance, approval or accession.
4   Any protocol, except as otherwise provided in such protocol, shall enter into force for a Contracting Party that ratifies, accepts or approves that protocol or accedes thereto after its entry into force pursuant to paragraph 2 above, on the ninetieth day after the date on which that Contracting Party deposits its instrument of ratification, acceptance, approval or accession, or on the date on which this Convention enters into force for that Contracting Party, whichever shall be the later.
5   For the purposes of paragraphs 1 and 2 above, any instrument deposited by a regional economic integration organization shall not be counted as additional to those deposited by member States of such organization.

Article 37

RESERVATIONS

No reservations may be made to this Convention.

Article 38

WITHDRAWALS

1   At any time after two years from the date on which this Convention has entered into force for a Contracting Party, that Contracting Party may withdraw from the Convention by giving written notification to the Depositary.
2   Any such withdrawal shall take place upon expiry of one year after the date of its receipt by the Depositary, or on such later date as may be specified in the notification of the withdrawal.
3   Any contracting Party which withdraws from this Convention shall be considered as also having withdrawn from any protocol to which it is party.

Article 39

FINANCIAL INTERIM ARRANGEMENTS

Provided that it has been fully restructured in accordance with the requirements of Article 21, the Global Environment Facility of the United Nations Development Programme, the United Nations Environment Programme and the International Bank for Reconstruction and Development shall be the institutional structure referred to in Article 21 on an interim basis, for the period between the entry into force of this Convention and the first meeting of the Conference of the Parties or until the Conference of the Parties decides which institutional structure will be designated in accordance with Article 21.

Article 40

SECRETARIAT INTERIM ARRANGEMENTS

The Secretariat to be provided by the Executive Director of the United Nations Environment Programme shall be the secretariat referred to in Article 24, paragraph 2, on an interim basis for the period between the entry into force of this Convention and the first meeting of the Conference of the Parties.

Article 41

DEPOSITARY

The Secretary-General of the United Nations shall assume the functions of Depositary of this Convention and any protocols.

Article 42

AUTHENTIC TEXTS

The original of this Convention, of which the Arabic, Chinese, English, French, Russian and Spanish texts are equally authentic, shall be deposited with the Secretary-General of the United Nations.

IN WITNESS WHEREOF the undersigned, being duly authorized to that effect, have signed this Convention.

Done at Rio de Janeiro on this fifth day of June, one thousand nine hundred and ninety-two.

ANNEX I

IDENTIFICATION AND MONITORING

1   Ecosystems and habitats: containing high diversity, large numbers of endemic or threatened species, or wilderness; required by migratory species of social, economic, cultural or scientific importance; or, which are representative, unique or associated with key evolutionary or other biological processes;
2   Species and communities which are: threatened; wild relatives of domesticated or cultivated species; of medicinal, agricultural or other economic value; or social, scientific or cultural importance; or importance for research into the conservation and sustainable use of biological diversity, such as indicator species; and
3   Described genomes and genes of social, scientific or economic importance.

ANNEX II

Part 1
ARBITRATION

Article 1

The claimant party shall notify the secretariat that the parties are referring a dispute to arbitration pursuant to Article 27. The notification shall state the subject matter of arbitration and include, in particular, the articles of the Convention or the protocol, the interpretation or application of which are at issue. If the parties do not agree on the subject matter of the dispute before the President of the tribunal is designated, the arbitral tribunal shall determine the subject matter. The secretariat shall forward the information thus received to all Contracting Parties to this Convention or to the protocol concerned.

Article 2

1   In disputes between two parties, the arbitral tribunal shall consist of three members. Each of the parties to the dispute shall appoint an arbitrator and the two arbitrators so appointed shall designate by common agreement the third arbitrator who shall be the President of the tribunal. The latter shall not be a national of one of the parties to the dispute, nor have his or her usual place of residence in the territory of one of these parties, nor be employed by any of them, nor have dealt with the case in any other capacity.
2   In disputes between more than two parties, parties in the same interest shall appoint one arbitrator jointly by agreement.
3   Any vacancy shall be filled in the manner prescribed for the initial appointment.

Article 3

1   If the President of the arbitral tribunal has not been designated within two months of the appointment of the second arbitrator, the Secretary-General of the United Nations shall, at the request of a party, designate the President within a further two-month period.
2   If one of the parties to the dispute does not appoint an arbitrator within two months of receipt of the request, the other party may inform the Secretary-General who shall make the designation within a further two-month period.

Article 4

The arbitral tribunal shall render its decision in accordance with the provisions of this Convention, any protocols concerned, and international law.

Article 5

Unless the parties to the dispute otherwise agree, the arbitral tribunal shall determine its own rules of procedure.

Article 6

The arbitral tribunal may, at the request of one of the parties, recommend essential interim measures of protection.

Article 7

The parties to the dispute shall facilitate the work of the arbitral tribunal and, in particular, using all means at their disposal, shall:

(a) Provide it with all relevant documents, information and facilities; and
(b) Enable it, when necessary, to call witnesses or experts and receive their evidence.

Article 8

The parties and the arbitrators are under an obligation to protect the confidentiality of any information they receive in confidence during the proceedings of the arbitral tribunal.

Article 9

Unless the arbitral tribunal determines otherwise because of the particular circumstances of the case, the costs of the tribunal shall be borne by the parties to the dispute in equal shares. The tribunal shall keep a record of all its costs, and shall furnish a final statement thereof to the parties.

Article 10

Any Contracting Party that has an interest of a legal nature in the subject matter of the dispute which may be affected by the decision in the case, may intervene in the proceedings with the consent of the tribunal.

Article 11

The tribunal may hear and determine counterclaims arising directly out of the subject matter of the dispute.

Article 12

Decisions both on procedure and substance of the arbitral tribunal shall be taken by a majority vote of its members.

Article 13

If one of the parties to the dispute does not appear before the arbitral tribunal or fails to defend its case, the other party may request the tribunal to continue the proceedings and to make its award. Absence of a party or a failure of a party to defend its case shall not constitute a bar to the proceedings. Before rendering its final decision, the arbitral tribunal must satisfy itself that the claim is well founded in fact and law.

Article 14

The tribunal shall render its final decision within five months of the date on which it is fully constituted unless it finds it necessary to extend the time-limit for a period which should not exceed five more months.

Article 15

The final decision of the arbitral tribunal shall be confined to the subject matter of the dispute and shall state the reasons on which it is based. It shall contain the names of the members who have participated and the date of the final decision. Any member of the tribunal may attach a separate or dissenting opinion to the final decision.

Article 16

The award shall be binding on the parties to the dispute. It shall be without appeal unless the parties to the dispute have agreed in advance to an appellate procedure.

Article 17

Any controversy which may arise between the parties to the dispute as regards the interpretation or manner of implementation of the final decision may be submitted by either party for decision to the arbitral tribunal which rendered it.

Part 2
CONCILIATION

Article 1

A conciliation commission shall be created upon the request of one of the parties to the dispute. The commission shall, unless the parties otherwise agree, be composed of five members, two appointed by each Party concerned and a President chosen jointly by those members.

Article 2

In disputes between more than two parties, parties in the same interest shall appoint their members of the commission jointly by agreement. Where two or more parties have separate interests or there is a disagreement as to whether they are of the same interest, they shall appoint their members separately.

Article 3

If any appointments by the parties are not made within two months of the date of the request to create a conciliation commission, the Secretary-General of the United Nations shall, if asked to do so by the party that made the request, make those appointments within a further two-month period.

Article 4

If a President of the conciliation commission has not been chosen within two months of the last of the members of the commission being appointed, the Secretary-General of the United Nations shall, if asked to do so by a party, designate a President within a further two-month period.

Article 5

The conciliation commission shall take its decisions by majority vote of its members. It shall, unless the parties to the dispute otherwise agree, determine its own procedure. It shall render a proposal for resolution of the dispute, which the parties shall consider in good faith.

Article 6

A disagreement as to whether the conciliation commission has competence shall be decided by the commission.

Select Bibliography

Books, Monographs, Reports

Alvares, C. and R. Billorey, *Damming the Narmada: India's Greatest Planned Environmental Disaster*, Penang, Third World Network/APPEN, 1988.

Attfield, R., *The Ethics of Environmental Concern*, Athens, GA/London, University of Georgia Press, 2nd edn., 1991.

Attfield, R. and A. Belsey (eds.), *Philosophy and the Natural Environment*, Cambridge, CUP, 1994.

Ball, S. and S. Bell, *Environmental Law*, London, Blackstone Press, 2nd edn., 1991.

Bilderbeek, S. (ed.), *Biodiversity and International Law*, Amsterdam, IOS Press, 1992.

Birnie, P.W., *International Regulation of Whaling*, 2 vols., New York, Oceana, 1985.

Birnie, P.W. and A.E. Boyle, *International Law and the Environment*, Oxford, Clarendon Press, 1992.

Boardman, R., *International Organisation and the Conservation of Nature*, London, Macmillan, 1981.

Burhenne, W.E. and W.A. Irwin, *The World Charter for Nature*, Berlin, Erich Schmidt Verlag, 2nd rev. edn., 1986.

Burke, W.T., *The New International Law of Fisheries*, Oxford, OUP, 1994.

Byrne, N., *Commentary on the Substantive Law of the 1991 UPOV Convention for the Protection of Plant Varieties*, London, Centre for Commercial Law Studies, 1991.

Campiglio, L. *et al.*, *The Environment After Rio*, London/Dordrecht/Boston, Graham and Trotman/Martinus Nijhoff, 1994.

Caufield, C., *In the Rainforest*, London, Pan Books (Picador edn.), 1985.

Cavalieri, P. and P. Singer (eds.), *The Great Ape Project*, London, Fourth Estate, 1993.

CGIAR, *Geneflow: A Publication About the Earth's Plant Genetic Resources*, Washington, CGIAR, 1992.

Churchill, R.R. and D.A.C. Freestone (eds.), *International Law and Global Climate Change*, London, Graham and Trotman, 1991.

Churchill, R.R. and A.V. Lowe, *The Law of The Sea*, Manchester University Press, 2nd edn., 1989.

Clark, S.R.L., *The Moral Status of Animals*, Oxford, Clarendon, 1977.

Clean Water Network, *Briefing Papers on the Clean Water Act Reauthorization*, March 1993.

Colchester, M. and L. Lohmann, *Tropical Forestry Action Plan: What Progress?*, Penang, World Rainforest Movement and *The Ecologist*, 1990.

Cooper, I.P., *Biotechnology Law*, Clark Boardman Callaghan, 1992 revision.

Dawkins, R., *The Selfish Gene*, Oxford/New York, OUP, 1976 and new 1989 edn.

Dawkins, R., *The Extended Phenotype*, Oxford/New York, OUP, 1982.

Dawkins, R., *The Blind Watchmaker*, Harlow, Longman, 1986.

de Klemm, C., *Wild Plant Conservation and the Law*, IUCN Environmental Policy and Law Paper No. 24, IUCN, 1990.

de Klemm, C. and C. Shine, *Biological Diversity Conservation and the Law*, IUCN Environmental Policy and Law Paper No. 29, IUCN, 1993.

de Silva, L., *Development Aid: A Guide to Facts and Issues*, 1982.

Devall, W. and G. Sessions, *Deep Ecology: Living as if Nature Mattered*, Salt Lake City, Gibbs M. Smith, 1985.

Dugan, P.J. (ed.), *Wetland Conservation: A Review of Current Issues and Required Action*, Gland, IUCN, 1990.

Fowler, C. and P.R. Mooney, *The Threatened Gene: Food, Politics and the Loss of Genetic Diversity*, Cambridge, Lutterworth Press, 1990.

Francioni, F. and T. Scovazzi (eds.), *International Law for Antarctica*, Milan, Giuffre, 1987.

Freestone, D., *The Requirements of Proof for Conservation in High Seas Fisheries*, UN FAO Legal Office (forthcoming).

Gjerde, K. and D. Freestone, *Particularly Sensitive Areas: An Important Environmental Concept At a Turning Point*, London, Graham and Trotman/Martinus Nijhoff, 1994 (a special issue of the *International Journal of Marine and Coastal Law*).

Glowka, L. *et al.*, *The Convention on Biological Diversity: An Explanatory Guide*, IUCN Environmental Policy and Law Paper No. 30, IUCN, 1994.

Goldsmith, E. and N. Hildyard, *The Social and Environmental Effects of Large Dams*, Vol. 1: *Overview*, Wadebridge, Wadebridge Ecological Centre, 1984.

Goodwin, B., *How the Leopard Changed its Spots: the Evolution of Complexity*, London, Weidenfeld and Nicolson, 1994.

Groves, M., M. Read and B.A. Thomas (eds.), *Species Endangered by Trade: A Role for Horticulture?* London, FFPS, 1993.

Haigh, N., *EEC Environmental Policy and Britain*, Harlow, Longman, 1990.

Handmer, J. and M. Wilder (eds.), *Towards a Conservation Strategy for the Australian Antarctic Territory*, Canberra Centre for Resource and Environmental Studies, ANU, 1993.

Hargrove, E.C. (ed.), *The Animal Rights/Environmental Ethics Debate*, Albany, State University of New York Press, 1992.

Heap, J., *Handbook of the Antarctic Treaty System*, Cambridge, Polar Publications, 7th edn., 1990.

Hecht, S. and A. Cockburn, *The Fate of the Forest: Developers, Destroyers and Defenders of the Amazon*, London/New York, Verso, 1989.

IUCN, *A Strategy for Antarctic Conservation*, Cambridge/Gland, IUCN, 1991.

IUCN, *Implementation of the Berne Convention*, Gland, IUCN, 1986.

IUCN/UNEP/WWF, *World Conservation Strategy: Living Resource Conservation for Sustainable Development*, IUCN/UNEP/WWF, 1980.

IUCN/UNEP/WWF, *Caring for the Earth: A Strategy for Sustainable Living*, Gland, IUCN/UNEP/WWF, 1991.

Johnson, S.P., *The Earth Summit: The United Nations Conference on Environment and Development (UNCED)*, Dordrecht/London, Martinus Nijhoff and Graham and Trotman, 1993.

Johnson, S. and G. Corcelle, *The Environmental Policies of the European Communities*, London, Graham and Trotman, 1989.

Jorgensen-Dahl, A. and W. Ostreng (eds.), *The Antarctic Treaty System in World Politics*, London, Macmillan, 1991.

Joyner, C.C., *Antarctica and the Law of the Sea*, Dordrecht, Martinus Nijhoff, 1992.

Joyner, C.C. and S.K. Chopra (eds.), *The Antarctic Legal Regime*, London, Martinus Nijhoff, 1988.

Juma, C. *The Gene Hunters*, London, Zed Books, 1989.

Kaufman, L. and K. Mallory (eds.), *The Last Extinction*, 1986.

Kerry, K.R. and G. Hempsel (eds.), *Antarctic Ecosystems: Ecological Change and Conservation*, Springer-Verlag, 1990.

Kiss, A. and D. Shelton, *International Environmental Law*, New York, Transnational Publishers, 1991.

Kiss, A. and D. Shelton, *Manual of European Environmental Law*, Cambridge, Grotius, 1993.

Kloppenburg, J.R., *First the Seed*, Cambridge, CUP, 1988.

Krämer, L., *Focus on European Environmental Law*, London, Sweet and Maxwell, 1992.

Krattiger, A.F. (ed.), *Widening Perspectives on Biodiversity*, Gland, IUCN, 1994.

Leopold, A., *A Sand County Almanac*, New York, OUP, 1949.

Leopold, A., *A Sand County Almanac with Other Essays on Conservation*, New York, OUP (2nd edn. of the above), 1966.

Lovelock, J.E., *Gaia: A New Look at Life on Earth*, Oxford, OUP, 1979.

Lowe, A.V., *Development Co-operation: Aid in Transition*, 1993 Report of the OECD's Development Assistance Committee.

Lyster, S., *International Wildlife Law*, Cambridge, Grotius, 1985.

Markham, A., N. Dudley and S. Stolton, *Some Like it Hot: Climate Change, Biodiversity and the Survival of Species*, Gland, WWF, 1993.

Matthews, F., *The Ecological Self*, London, Routledge, 1991.

McNeeley, J.A. (ed.), *Parks for Life*, Report of the IVth World Congress on National Parks and Protected Areas, Gland, IUCN, 1993.

McNeeley, J.A. *et al.*, *Conserving the World's Biological Diversity*, Gland/Washington, IUCN/WRI/CI/WWF-US/World Bank. 1990.

Munro, J.D. and J.G. Lammers (eds.), *Environmental Protection and Sustainable Development*, London/Dordrecht, Graham and Trotman/Martinus Nijhoff, 1986.

Norse, E.A. (ed.), *Global Marine Biological Diversity*, Washington, Island Press, 1993.

OECD, *Economic Instruments for Environmental Protection*, Paris, OECD, 1989.

OECD, *The Environmental Effects of Trade*, Paris, OECD, 1994.

Paterson, D. and M. Palmer (eds.), *The Status of Animals: Ethics, Education and Welfare*, Oxford, CAB International, 1989.

Pearce, D. *et al.*, *Blueprint for a Green Economy*, London, Earthscan Publications, 1989.

Pearce, D. (ed.), *Blueprint 2: Greening the World Economy*, London, Earthscan Publications, 1991.

Pearce, D. *et al.*, *Blueprint 3: Measuring Sustainable Development*, London, Earthscan Publications, 1993.

Pearce, D. and D. Moran, *The Economic Value of Biodiversity*, London, Earthscan Publications/IUCN, 1994.

Perry, A.R. and R.G. Ellis, *The Common Ground of Wild and Cultivated Plants*, Cardiff, National Museum of Wales, 1994.

Phillips, J. and A. Firth, *Introduction to Intellectual Property Law*, London, Butterworths, 1990.

Pritchard, D., *Implementation of the EC Directive on the Conservation of Natural Habitats and of Wild Fauna and Flora*, London, RSPB/Birdlife International, 1993.

Regan, T., *The Case for Animal Rights*, London/New York, Routledge, 1984.

Reid, W.V. *et al.*, *Biodiversity Prospecting: Using Genetic Resources for Sustainable Development*, Baltimore, World Resources Institute, 1993.

Rodd, R., *Biology, Ethics and Animals*, Oxford, Clarendon, 1990.

Rural Advancement Foundation International, *Conserving Indigenous Knowledge: Integrating Two Systems of Innovation*, New York, UNDP, 1994.

Rural Advancement Foundation International, *A Report of Germplasm Embargoes*, RAFI Communiqués, September 1988.

Sand, P.H., *The Effectiveness of International Environmental Agreements*, Cambridge, Grotius, 1992.

Sandlund, O.T., K. Hindar and H.D. Brown (eds.), *Conservation of Biodiversity for Sustainable Development*, Oslo, Scandinavian University Press, 1992.

Sands, P. (ed.), *Greening International Law*, London, Earthscan Publications, 1993.

Sherman, K., L. Alexander and B. Gold (eds.), *Large Marine Ecosystems: patterns, processes and yields*, Washington, AAAS, 1990.

Shiva, V., *Biodiversity: A Third World Perspective*, Penang, Third World Network.

Shiva, V. *et al.*, *Biodiversity: Social and Ecological Perspectives*, Penang, World Rainforest Movement, 1991.

South Centre, *Environment and Development: Towards a Common Strategy of the South in the UNCED Negotiations and Beyond*, Geneva, 1991.

Swanson, T.M., *The International Regulation of Extinction*, London, Macmillan, 1994.

Swanson, T.M. and E. Barbier (eds.), *Economics for the Wilds: Wildlife and Wildlands, Diversity and Development*, London, Earthscan Books, 1992.

Switzerland, Federal Office of the Environment, *Results of the Questionnaire regarding Seven Environmental Conventions: Participation and Implementation*, 1993.

Thorne-Miller, B. and J. Catena, *The Living Ocean: Understanding and Protecting Marine Biodiversity*, Washington, Island Press, 1991.

Tudge, C., *Last Animals at the Zoo*, London/Sydney/Auckland/Johannesburg, Hutchinson Radius, 1991.

UNCTAD, *The Effects of the Internalization of External Costs on Sustainable Development*, 1994.

UN Economic Commission for Europe, *Code of Practice for the Conservation of Threatened Animals and Plants*, UN Doc.ECE/ENVWA/25, New York, United Nations, 1992.

United Kingdom, Department of the Environment, *Biodiversity: The UK Action Plan*, Cm 2428, London, HMSO, 1994.

United States, Office of Technology Assessment, *Technologies to Maintain Biological Diversity*, US Government Printing Office, 1987.

Verhoeven, J., P. Sands and M. Bruce (eds.), *The Antarctic Environment and International Law*, London/Dordrecht/Boston, Graham and Trotman/Martinus Nijhoff, 1992.

Vogel, J.H., *Genes for Sale: Privatisation as a Conservation Policy*, New York, OUP, 1994.

Watts, Sir A., *International Law and the Antarctic Treaty System*, Cambridge, Grotius Publications, 1992.

Weiss, E. Brown, *In Fairness to Future Generations: International Law, Common Patrimony and Intergenerational Equity*, New York, Transnational, 1989.

Western, D. and M.C. Pearl, *Conservation for the Twenty-First Century*, New York/Oxford, OUP, 1989.

Wilson, E.O., *The Diversity of Life*, London/New York, Allen Lane/Penguin, 1992.

Wilson, E.O. (ed.), *Biodiversity*, Washington, DC, National Academy Press, 1988.

Wilson, E.O., *Biophilia*, Cambridge, MA, Harvard University Press, 1984.

Wood, W. and C.E. Jones, *Monitoring Environmental Assessment and Planning*, London, HMSO, 1991.

World Bank, *World Debt Tables 1993-94: External Finance for Developing Countries*, Washington DC, 1994.

WCED, *Our Common Future*, Oxford, OUP, 1987.

World Conservation Monitoring Centre (B. Groombridge, ed.), *Global Biodiversity: Status of the Earth's Living Resources*, London, Chapman and Hall, 1992.

World Rainforest Movement, *Rainforest Destruction: Causes, Effects and False Solutions*, Penang, World Rainforest Movement, 1990.

WRI/IUCN/UNEP, *Global Biodiversity Strategy*, WRI/IUCN/UNEP, 1992.

Articles

Alexander, D., "Some Themes in Intellectual Property and Environment" (1993) 2 *Review of European Community and International Environmental* 113.

Angel, M.V., "Biodiversity in the Pelagic Ocean" (1993) 7/4 *Conservation Biology* 760.

Anon., "First Meeting of the Conference of the Parties to the Rio Convention" (1995) 25/1–2 *Environmental Policy and Law* 38.

Belsky, M.H., "Management of Large Marine Ecosystems: Developing a New Rule of Customary International Law" (1985) 22 *San Diego Law Review* 733.

Bowman, M.J., "The Ramsar Convention Comes of Age" (1995) 42 *Netherlands International Law Review* (forthcoming).

Bragdon, S.H., "National Sovereignty and Global Environmental Responsibility: Can the Tension be Reconciled for the Conservation of Biological Diversity?" (1992) 33 *Harvard International Law Journal* 381.

Brownlie, I.M., "Legal Status of Natural Resources in International Law" (1979/I) 162 *Recueil des Cours* 267.

Burhenne-Guilmin, F. and S. Casey-Lefkowitz, "The New Law of Biodiversity" (1992) 3 *Yearbook of International Environmental Law* 43.

Byrne, N.J., "Plant Breeding and UPOV" (1993) 2 *Review of European Community and International Environmental Law* 136.

Canal-Forgues, E., "Code of Conduct for Plant Germplasm Collecting and Transfer" (1993) 2 *Review of European Community and International Environmental Law* 171.

Chandler, M., "The Biodiversity Convention: Selected Issues of Interest to the International Lawyer" (1993) 4 *Colorado Journal of International Environmental Law and Policy* 141.

Charney, J.I., "The Marine Environment and the 1982 United Nations Convention on the Law of the Sea" (1994) *International Lawyer* 879.

Cole, C.A. "Species Conservation in the United States: The Ultimate Failure of The Endangered Species Act and Other Land Use Laws" (1992) 72 *Boston University Law Review* 343.

Cooper, D., "The International Undertaking on Plant Genetic Resources" (1993) 2 *Review of European Community and International Environmental Law* 158.

Correa, C., "Biological Resources and Intellectual Property Rights" (1992) 4(5) *European Intellectual Property Review* 154.

Coughlin, M.D., "Using the Merck–INBio Agreement to Clarify the Convention on Biological Diversity" (1993) *Columbia Journal of Transnational Law* 337.

D'Amato, A. and S.K. Chopra, "Whales: Their Emerging Right to Life" (1991) 85 *American Journal of International Law* 21.

Doremus, H., "Patching the Ark: Improving Legal Protection of Biological Diversity" (1991) 18 *Ecology Law Quarterly* 265.

Esquinas-Alcazar, J., "The Global System on Plant Genetic Resources" (1993) 2 *Review of European Community and International Environmental Law* 153.

Fikkan, A., "Polar Bears: Hot Topic in a Cold Climate" (1990) 10 *International Challenges* 32.

Fischmann, R.L., "Biological Diversity and Environmental Protection: Authorities to Reduce Risk" (1992) 22 *Environmental Law* 435.

Freestone, D., "The Road from Rio: International Environmental Law after the Earth Summit" (1994) 6 *Journal of Environmental Law* 193.

Freestone, D., "The Effective Conservation and Management of High Seas Living Resources: Towards a New Regime?" (1995) *University of Canterbury Law Journal* (forthcoming).

Gjerde, K., "The Law of the Sea" in J. Broadus and R. Vartanov (eds.), *The Oceans and Environmental Security: Shared US/USSR Perspectives*, Washington, Island Press, 1994.

Goodpaster, K., "On Being Morally Considerable" (1978) 75 *Journal of Philosophy* 308.

Handl, G., "Environmental Security and Global Change: The Challenge to International Law" (1990) 1 *Yearbook of International Environmental Law* 32.

Hardin, G., "The Tragedy of the Commons" (1969) *Science* 1243.

Hartshorn, G., "Key Environmental Issues for Developing Countries" (1991) 44 *Journal of International Affairs* 393.

Holst, J.D., "The Unforseeability Factor: Federal Lands, Managing for Uncertainty, and the Preservation of Biological Diversity" (1992) 13 *Public Land Law Review* 113.

Houck, O.A., "The Endangered Species Act and its Implementation by the U.S. Departments of Interior and Commerce" (1993) 64 *University of Colorado Law Review* 277.

Irvin, W.R., "The Endangered Species Act: Keeping Every Cog and Wheel" (1993) *National Resources and the Environment* 36.

Johnston, S., "Conservation Role of Botanic Gardens and Genebanks" (1993) 2 *Review of European Community and International Environmental Law* 175.

Kamara, B., "The Role of Indigenous Knowledge in Biological Diversity Conservation", in V. Sanchez and C. Juma (eds.), *Biodiplomacy*, 1993.

Keiter, R.B., "NEPA and the Emerging Concept of Ecosystem Management on Public Lands" (1990) 25 *Land and Water Law Review* 43.

Kilbourne, J.C. "The Endangered Species Act under the Microscope: A Closeup Look from a Litigator's Perspective" (1991) 21 *Environmental Law* 499.

Kock, G., "Fishing and Conservation in Southern Waters" (1994) 30 (172) *Polar Record* 3.

Maebius, S.B., "Novel DNA Sequences and the Utility Requirement" (1992) 74 *Journal of the Patent and Trademark Society Office* 651.

Maffei, M.C., "Evolving Trends in the International Protection of Species" (1993) 36 *German Yearbook of International Law* 131.

McNeeley, J.A., "Common Property Resource Management or Government Ownership: Improving the Conservation of Biological Resources" (1991) *International Relations* 211.

Naess, A., "The Shallow and the Deep, Long-Range Ecology Movement" (1973) 16 *Inquiry* 95.

Nijar, G.S. and Chee Yoke Ling, "Intellectual Property Rights: The Threat to Farmers and Biodiversity", *Third World Resurgence*, Issue No. 39, November 1993, p. 35.

Noonan, W., "Ownership of Biological Tissues" (1990) 72 *Journal of the Patent and Trademark Society Office* 110.

Olwell, P., "Restoring Diversity: Strategies for Rare Plant Reintroductions" (1993) 7(3) *Plant Conservation* 1.

Payne, R.W.J., "The Emergence of Trade Secret Protection in Biotechnology" (1988) 6 *Bio/Technology* 130.

Pearce, F., "Political Paralysis Stalls Biodiversity Talks", *New Scientist*, 17 December 1994.

Peterson, K., "Recent Intellectual Property Trends in Developing Countries" (1992) 33 *Harvard International Law Journal* 277.

Redgwell, C., "Environmental Protection in Antarctica: The 1991 Protocol" (1994) 43 *International and Comparative Law Quarterly* 599.

Sax, J.L., "Nature and Habitat Conservation and Protection in the United States" (1993) 20 *Ecology Law Quarterly* 47.

Scully, T., "The Protection of the Marine Environment and the UN Conference on Environment and Development", in *The Law of the Sea: New Worlds, New Discoveries*, Proceedings of the 26th Annual Conference of the Law of the Sea Institute, Genoa, 1992.

Sedjo, R., "Property Rights, Genetic Resources and Biotechnological Change" (1992) 35 *Journal of Law and Economics* 199.

Shine, C., and P. Kohona, "The Convention on Biological Diversity: Bridging the Gap between Conservation and Development" (1992) 1 *Review of European Community and International Environmental Law* 278.

Smith, E.M., "The Endangered Species Act and Biological Conservation" (1984) 57 *Southern California Law Review* 361.

Sohn, L.B., "The Stockholm Declaration and the Human Environment" (1973) 15 *Harvard International Law Journal* 423.

Soulé, M.E. "Conservation: Tactics for a Constant Crisis" (1991) *Science* 253.

Stuyt, A., "The UN Year of Indigenous Peoples 1993: Some Latin American Perspectives" (1993) 40 *Netherlands International Law Review* 449.

Swanson, T.M., "Economics of a Biodiversity Convention" (1992) 21: 3 *Ambio* 250.

Swepston, L. and R. Plant, "International Standards and the Protection of the Land Rights of Indigenous and Tribal Populations" (1985) 124 *International Labour Review* 91.

Tarlock, A.D., "Local Government Protection of Biodiversity: What is its Niche?" (1993) 60 *University of Chicago Law Review* 555.

Teschemacher, R., "Patentability of Microorganisms per se" (1982) 13:1 *International Review of Industrial Property and Copyright Law* 27.

Thornton, R.D., "Searching for Consensus and Predictability: Habitat Conservation Planning under The Endangered Species Act of 1973" (1991) 21 *Environmental Law* 605.

Tydemans, C. and L. Warren, "Legal Mechanisms within the European Community for the Protection of Wildlife and Habitats" (1995) *Journal of Wildlife Management* (forthcoming).

Vatikiotis, M., "Priming for Rio: Malaysia sets tone for Earth Summit Agenda", *Far Eastern Economic Review*, 14 May 1992, p. 22.

Walden, I., "Intellectual Property in Genetic Sequences" (1993) 2 *Review of European Community and International Environmental Law* 126.

Winter, G., "Patent Law Policy in Biotechnology" (1992) 2 *Journal of Environmental Law* 167.

Worthy, J., "Intellectual Property Protection after GATT" (1994) 5 *European Intellectual Property Review* 195.

Yagerman, K.S., "Protecting Critical Habitat Under the Federal Endangered Species Act" (1990) 20 *Environmental Law* 811.

Yamin, F. and D. Posey, "Indigenous Peoples, Biotechnology and Intellectual Property Rights" (1993) 2 *Review of European Community and International Environmental Law* 141.

Index